T0231309

Also in the Variorum Collected Studies Series:

MARTIN J.S. RUDWICK
The New Science of Geology
Studies in the Earth Sciences in the Age of Revolution

EKMELEDDIN IHSANOGLU
Science, Technology and Learning in the Ottoman Empire
Western Influence, Local Institutions, and the Transfer of Knowledge

DONALD CARDWELL (Ed. Richard L. Hills)
The Development of Science and Technology in Nineteenth-Century Britain
The Importance of Manchester

HUGH TORRENS
The Practice of British Geology, 1750–1850

RICHARD YEO
Science in the Public Sphere
Natural Knowledge in British Culture 1800–1860

DAVID ELLISTON ALLEN
Naturalists and Society
The Culture of Natural History in Britain, 1700–1900

ROY M. MACLEOD
The 'Creed of Science' in Victorian England

DAVID OLDROYD
Sciences of the Earth
Studies in the History of Mineralogy and Geology

IAN INKSTER
Technology and Industrialisation
Historical Case Studies and International Perspectives

DAVID J. JEREMY
Artisans, Entrepreneurs and Machines
Essays on the Early Anglo-American Textile Industries, 1770–1840s

DAVID M. KNIGHT
Science in the Romantic Era

IAN INKSTER
Scientific Culture and Urbanisation in Industrialising Britain

VARIORUM COLLECTED STUDIES SERIES

Lyell and Darwin, Geologists

Professor Martin J.S. Rudwick

(Courtesy of the Scripps Institution of Oceanography
Archives, University of California, San Diego)

Martin J.S. Rudwick

Lyell and Darwin, Geologists

Studies in the Earth Sciences in the Age of Reform

Routledge
Taylor & Francis Group

LONDON AND NEW YORK

First published 2005 by Ashgate Publishing

2 Park Square, Milton Park, Abingdon, Oxfordshire OX14 4RN
711 Third Avenue, New York, NY 10017

Routledge is an imprint of the Taylor & Francis Group, an informa business

First issued in paperback 2018

ISBN 978-0-86078-959-8 (hbk)
ISBN 978-1-138-37566-6 (pbk)

British Library Cataloguing in Publication Data
Rudwick, M. J. S.
    Lyell and Darwin, geologists : studies in the earth sciences in the
    age of reform. – (Variorum collected studies series)
    1. Lyell, Charles, Sir, 1797–1875 2. Darwin, Charles,
    1809–1882 – Knowledge – Geology 3. Geology – History – 19th
    century
    I. Title
    551'.0922

Library of Congress Cataloging-in-Publication Data
Rudwick, M.J.S.
    Lyell and Darwin, geologists : studies in the earth sciences in the
    age of reform / Martin J.S. Rudwick.
        p. cm. – (Variorum collected studies ; CS818)
    Includes bibliographical references.
    ISBN 0 86078 959 4
    1. Lyell, Charles, Sir, 1797–1875 – Influence. 2. Darwin, Charles,
    1809–1882 – Influence. 3. Geologists–Great Britain–History. 4. Geology –
    History. I. Title. II. Collected studies ; CS818.

    QE11.R83 2005
    551'.092'241–dc22   2004062693

VARIORUM COLLECTED STUDIES SERIES CS818

CONTENTS

Introduction   vii–xi

Notes on the articles   xiii–xiv

Bibliography   xv–xvi

Acknowledgements   xvii

LYELL'S CONCEPT OF UNIFORMITY

I   Uniformity and progression: reflections on the structure
        of geological theory in the age of Lyell   209–227
        Perspectives in the History of Science and Technology, ed.
        Duane H.D. Roller. Norman, OK: University of Oklahoma
        Press, 1971

II   Lyell and the *Principles of Geology*   1–24
        Lyell: The Past is the Key to the Present, eds. D.J. Blundell
        and A.C. Scott (Special Publications 143). London: The
        Geological Society, 1998, pp. 3–15

THE MAKING OF THE *PRINCIPLES*

III   Poulett Scrope on the volcanoes of the Auvergne:
        Lyellian time and political economy   205–242
        The British Journal for the History of Science 7. London, 1974

IV   Lyell on Etna, and the antiquity of the earth   288–304
        Toward a History of Geology, ed. Cecil J. Schneer. Cambridge,
        MA: The M.I.T. Press, 1969

V   Historical analogies in the geological work of Charles Lyell 89–107
        Janus 64. Hooykaas and the History of Science, Symposium
        held at Utrecht on 3–4 March 1997. Utrecht, The Netherlands:
        Utrecht State University Press, 1977

CONTENTS

VI   Charles Lyell's dream of a statistical palaeontology   225–244
      Palaeontology 21. London, 1978

THE RECEPTION OF THE *PRINCIPLES*

VII   Caricature as a source for the history of science: De la
      Beche's anti-Lyellian sketches of 1831   534–560
      Isis 66. Washington, D.C., 1975

VIII Charles Lyell, F.R.S. (1797–1875) and his London lectures
      on geology, 1832–33   231–263
      Notes and Records of the Royal Society of London 29. London,
      1975

DARWIN AS A GEOLOGIST

IX   Charles Darwin in London: the integration of public
      and private science   186–206
      Isis 73. Washington, D.C., 1982

X   Darwin and Glen Roy: a "great failure" in scientific
      method?   97–185
      Studies in the History and Philosophy of Science 5.
      Cambridge, 1974

Index   1–6

This volume contains xviii + 316 pages

INTRODUCTION

This volume, which is a sequel to *The New Science of Geology*, reproduces some more of my published articles on the history of the sciences. In terms of period, the articles in this volume are all focussed on the second quarter of the nineteenth century – or, roughly, on what political historians often call "the age of reform" – which was also, coincidentally or not, an exceptionally fruitful period in the history of all the natural sciences. Second, they deal with what was clearly recognised by that time as an exciting new science, "geology", newly defined by its ambition not only to describe the earth and to try to understand terrestrial processes and their causes, but also to reconstruct the earth's own history (or what geologists and other earth scientists now call *geohistory*). And third, these particular articles seem to have stood the test of time, in that they continue to be cited by historians and geologists and to be used in teaching, and I continue to be asked for offprints of them. This brief introduction summarises their contents and explains the relations between them; it is followed by brief notes on the circumstances in which they were written and published, and a bibliography giving full details of all the books and articles mentioned. Most of the articles, which are identified by Roman numerals, are reproduced here exactly as they were published, with their original pagination, to facilitate citation etc.; only one (Article II) has had to be reset to ensure legibility, but apart from new pagination it is unchanged.

The historical problem at the centre of my research, ever since I turned myself in mid-career from a geologist into a historian, has been to try to understand how this new kind of science, with a sense of nature's own history at its core, was first constructed: initially in a quite tentative way, but eventually on such firm foundations that earth scientists now take it completely for granted. In the introduction to *The New Science of Geology*, and in some of the articles reprinted in it, I criticised the traditional stereotype, according to which the early history of this science had been above all a struggle between geology and Genesis, or more generally between Science and Religion. I suggested instead that it involved a much more subtle – and much more interesting – reconstitution of earlier scientific practices into a new science with its own distinctive methods and genres. Above all, those who came to be called "geologists" hoped to reconstruct geohistory, to gain reliable knowledge of what happened on earth during the "deep time" before the first appearance of human beings.

This ambition demanded a close analysis of the relation between the observable present world of nature and the distant past worlds that could only be known through material relics such as rocks, fossils and so on. It was clear to all geologists that the present was the best key to the deep past, in the sense that the processes observably at work on earth at the present day ("actual causes") were obviously the most reliable guide for interpreting the surviving traces of past ages; this was (and is) the almost self-evident principle of "*actualism*". But there was profound disagreement about how far these present processes on their own were *adequate* to account for everything in the past, and about the methods of scientific reasoning that should be applied in such cases. The more the sheer magnitude of deep time was appreciated – as it was by all geologists in the early nineteenth century, if not yet by the general public – the more questionable it was to assume that the tiny slice of time comprised by recorded human history was an adequate or representative sample of those literally unimaginable spans of geohistory. It was conceivable that some geological processes or "causes" might have been confined to earlier phases of geohistory; others might operate only rarely, and perhaps – by chance – never within the short span of human history; but in neither case would there be any necessary violation of the presumably perennial "laws of nature".

During the "age of reform" these debates were brought into sharp focus through the work of Charles Lyell (1797–1875), and particularly through his most famous publication the *Principles of Geology*. Lyell's work was far from being the first truly scientific study of the earth; nor did it split geologists into two opposing camps, though for polemical purposes he sometimes implied that it did. In fact a kind of synthesis between his approach and that of his critics emerged over the years, and it is this that has come to be taken for granted by modern geologists.

The first two articles in this collection set the scene for this debate. "Uniformity and progression" (Article I) analyses the several distinct meanings that Lyell and his critics attached to the general concept of "the uniformity of nature", when it was applied to the specific problems raised by geology as the first *historical* science of nature. Lyell's formulation of a "*uniformitarian*" or steady-state interpretation of geohistory challenged an already well articulated alternative, which was certainly no less scientific in character. I define the latter as "*directionalist*"; it included, but was much broader than, the "progressionist" interpretation of the fossil record that later provided support for an evolutionary interpretation of the history of life. "Lyell and the *Principles of Geology*" (II) outlines the origin, content and development of his greatest work, which during the rest of his life – nearly half a century – he repeatedly revised and ultimately transformed. In both articles I emphasise the contrast between Lyell's forceful case for the adequacy of "actual causes", and his quite separate claim that the earth had been broadly in a steady or cyclic state as far back as the evidence could be traced: on the first point he gained increasing support from other geologists, but the second they found deeply implausible and Lyell himself eventually abandoned it.

The next four articles focus on different aspects of the making of the *Principles*. Of course, Lyell's approach did not spring fully fledged from nowhere. "Poulett Scrope on the volcanos of Auvergne" (III) describes the work of one of his most important early associates, who was later more prominent as an economist. Following first-hand experience of active volcanoes such as Vesuvius, Scrope's exploration of the extinct volcanoes of central France – already famous among Continental geologists though not to the insular British – convinced him of the immensity of deep time during which present processes ("actual causes" such as vulcanism and erosion) had operated, and hence of their explanatory power. This was a perspective that Lyell adopted with great enthusiasm when composing the *Principles*. I also suggest how an apparently casual metaphor drawn from the world of economics reveals the structure of Scrope's – and Lyell's – concept of time, as an explanatory resource that was in effect unlimited. "Lyell on Etna" (IV) analyses the decisive role of Lyell's fieldwork on and around Europe's greatest active volcano; this was the culmination of the geological Grand Tour (1828–29) that immediately preceded the writing of the *Principles* and was, as it were, his *Beagle* voyage. It convinced him that the whole of human history comprised a minuscule postscript to even the most recent portion of geohistory, so that very slow geological processes might be able to account for huge cumulative effects, without the need to postulate exceptional "catastrophes" in the deep past. "Historical analogies in the early geological work of Charles Lyell" (V) explores how, when constructing the argument of the *Principles*, he made creative use of metaphorical resources from right outside the natural sciences – in contemporary work on historiography, philology, demography and political economy – to develop his concept of geology as a historical science. "Charles Lyell's dream of a statistical palaeontology" (VI) reconstructs the research that lay at the heart of the third and culminating volume of the *Principles* (1833), namely the uniformitarian model of steady faunal change – though not evolution – that justified the terms ("Eocene" etc.) which he bequeathed to modern stratigraphical geologists.

The next two articles are concerned with reactions to the publication of the *Principles*, and with Lyell's own efforts to propagate his ideas to wider audiences. "Caricature as a source for the history of science" (VII) deals briefly with the critical reception of Lyell's work among his fellow geologists. It reproduces and analyses a sequence of jocular sketches by his contemporary Henry De la Beche, in which he sought to ridicule Lyell's steady-state or cyclic model. They culminated in a famous and widely circulated caricature of an imaginary "Professor Ichthyosaurus" lecturing to an audience of Jurassic reptiles during a future – and *post-human* – cycle of a directionless geohistory. Even before the final volume of the *Principles* was published, Lyell himself became a real professor at the newly founded King's College in London, and also lectured at the Royal Institution. "Charles Lyell and his London lectures on geology" (VIII) reconstructs, from his extant manuscript notes, how at this point he expounded his ideas to general audiences, which, after some hesitation, were allowed to include women. He only held his academic position for two years, but I conclude that his resignation had nothing to do with supposed religious opposition to his views, and everything to do with his own prudent calculations about time and money.

The last two articles reprinted in this volume are concerned with Lyell's most loyal and most famous follower. "Charles Darwin in London" (IX) analyses Darwin's work after he returned from the *Beagle* voyage, during the brief but highly creative period when he was living in London and making his reputation in the scientific world: not primarily as a biologist but as a Lyellian geologist. I propose two interpretative concepts, and offer two corresponding diagrams, to make sense of Darwin's work (they could equally well be applied to other scientific figures at other periods, provided the available sources are sufficiently rich and varied). The geological community was structured at this time around what I call a tacit "gradient of competence", on which Darwin's trajectory can be traced from amateur status into the acknowledged elite of the science. The detailed modern research of Darwin scholars also makes it possible to analyse his work in terms of what I call a "scale of relative privacy", ranging from the most private of theoretical notebooks through discussion and correspondence to fully published books and scientific articles. This allows his "public" geological research to be related to his parallel but strictly "private" projects on the origin of species and on "Man and mind", without the sharp dichotomy that the terms public and private usually imply.

Finally, "Darwin and Glen Roy" (X) analyses in detail his only substantial geological fieldwork in Britain, his first major scientific paper, and his much later and reluctant conclusion that it had been a "great failure", or even a "gigantic blunder", because he had failed to allow for the possible effects of Ice Age glaciers. The purpose of this historical analysis is not to topple Darwin from his well-earned pedestal but to gain a deeper understanding of his scientific thinking from a case in which he admitted he had failed. His flawed interpretation of the famous terraces in the Scottish Highlands was intimately linked to his adoption of Lyell's concept of the role of slow crustal mobility in geohistory. The latter made him (and Lyell) reluctant to accept Louis Agassiz's initially rather extravagant notion of a "catastrophic" Ice Age in the geologically recent past. I also argue that the structure of Darwin's argument about Glen Roy was closely analogous to that of his argument for the natural origin of new species, which he was constructing at just the same time, but which remained strictly private for another twenty years. This final article – which is by far the longest in the collection – is also, I believe, an another example of the value of engaging in geological *fieldwork* for historical purposes, treating what the historical actors observed in the field as a primary source that can be as important and revealing as any manuscript or printed text.

MARTIN J.S. RUDWICK

Cambridge
May 2004

PUBLISHER'S NOTE

The articles in this volume, as in all others in the Variorum Collected Studies Series, have not been given a new, continuous pagination. In order to avoid confusion, and to facilitate their use where these same studies have been referred to elsewhere, the original pagination has been maintained wherever possible.

Each article has been given a Roman number in order of appearance, as listed in the Contents. This number is repeated on each page and is quoted in the index entries.

NOTES ON THE ARTICLES

"Uniformity and progression" (I) is based on a paper given at a conference on the history of science and technology held at the University of Oklahoma in 1969; it was published in Duane Roller's *Perspectives in the History of Science and Technology* (1971). "Lyell and the *Principles of Geology*" (II) was my contribution to the Geological Society's meeting in 1997 to mark the bicentenary of Lyell's birth; it was published in *Lyell: The Past is the Key to the Present* (1998). (A more detailed analysis of Lyell's original argument can be found in my introduction to the facsimile reprint [1990–91] of the first edition of 1830–33.)

"Poulett Scrope on the volcanos of Auvergne" (III) was based on one of my first attempts (in 1967) to use geological fieldwork as a serious historical tool; it was published by the British Society for the History of Science (1974). "Lyell on Etna, and the antiquity of the earth" (IV), based on an even earlier field trip of the same kind, was written for the first New Hampshire conference on the history of geology, held in 1967 in memorably congenial seashore surroundings; it was published in Cecil Schneer's *Toward a History of Geology* (1969), a volume that, perhaps more than any other, put this research field on the modern scholarly map. "Historical analogies in the early geological work of Charles Lyell" (V) was written for a symposium at Utrecht in 1976 in honour of the late R. Hooykaas, my predecessor in the chair at the Vrije Universiteit in Amsterdam; it was published in *Janus* (1977). (A rather different version was published in 1979, in *Images of the earth*, edited by Ludmilla Jordanova and the late Roy Porter.) "Charles Lyell's dream of a statistical palaeontology" (VI) was given in 1977 as the Palaeontological Association's 20th Annual Address, and was published in its journal *Palaeontology* (1978).

"Caricature as a source for the history of science" (VII) was prompted by sketches that I discovered by chance in the back of one of De la Beche's field notebooks, while I was collecting material for *Great Devonian Controversy* (1985); it was published in *Isis* (1975), and the climactic caricature, which has since become famous, had the distinction of being chosen – though not by me – as the very first pictorial cover design for that venerable periodical. (At the last moment, while the article was in proof,

I altered its title to date the caricature as 1831, having long hesitated between that and 1830, but I have since concluded that the earlier date is more
likely; as explained in the article, the year affects the context but not the
interpretation of the caricature itself.) "Charles Lyell and his London
lectures on geology" (VIII) was based on manuscripts, now conserved in
Edinburgh, that had not previously been exploited by historians; it was
published in the Royal Society's *Notes and Records* (1975). (Longer
samples of Lyell's lecturing text are transcribed in "Lyell speaks in the
lecture theatre" [1976], one of my contributions to the London conference
in 1975 marking the centenary of his death; I used them to construct a
performing script for a memorable costumed re-enactment by the late John
Thackray in the theatre at the Royal Institution.)

"Charles Darwin in London" (IX) was a product of a graduate
seminar on Darwin that I taught at Princeton in 1981, jointly with the late
Gerald Geison; it was published in *Isis* (1982). Finally, "Darwin and Glen
Roy" (X), my only substantial contribution to the flourishing historical
"Darwin industry", was a modest attempt to emulate the example of the late
Gerd Buchdahl, who more than anyone else had encouraged me to turn
myself into a historian; it was published (1974) in *Studies in the History
and Philosophy of Science*, the journal that he founded to promote the
integration that its title proclaimed and that he himself embodied. (Darwin
scholars should note that this paper includes a full transcription of Adam
Sedgwick's important report to the Royal Society on Darwin's paper on
Glen Roy, which, unlike his reports to the Geological Society on Darwin's
other early papers, was not printed in the relevant volume [1986] of the
great modern edition of the *Correspondence of Charles Darwin*.)

BIBLIOGRAPHY

Blundell, D.J., and A.C. Scott (eds.). 1998. *Lyell: The Past is the Key to the Present*. London: The Geological Society (Special Publication 143).

Burkhardt, Frederick, and Sydney Smith (eds.). 1986. *The Correspondence of Charles Darwin: vol. 2, 1837–1843*. Cambridge: Cambridge University Press.

Jordanova, L.J., and R.S. Porter (eds.). 1979. *Images of the Earth: Essays in the History of the Environmental Sciences*. Chalfont St Giles: British Society for the History of Science (Monograph 1). [Revised ed., 1997].

Roller, Duane H.D. (ed.). *Perspectives in the History of Science and Technology*. Norman, Oklahoma: University of Oklahoma Press.

Rudwick, Martin J.S. 1969. Lyell on Etna, and the antiquity of the earth. *In* Schneer, *Toward a History of Geology*, 288–304. [ARTICLE IV]

———. 1971. Uniformity and progression: reflections on the structure of geological theory in the age of Lyell. *In* Roller, *Perspectives in the History of Science and Technology*, 209–27. [ARTICLE I]

———. 1974. Poulett Scrope on the volcanos of Auvergne: Lyellian time and political economy. *British Journal for the History of Science* 7: 205–42. [ARTICLE III]

———. 1974. Darwin and Glen Roy: a "great failure" in scientific method? *Studies in the History and Philosophy of Science* 5: 97–185. [ARTICLE X]

———. 1975. Caricature as a source for the history of science: De la Beche's anti-Lyellian sketches of 1831. *Isis* 66: 534–60. [ARTICLE VII]

———. 1975. Charles Lyell F.R.S. (1797–1875) and his London lectures on geology, 1832–33. *Notes and Records of the Royal Society of London* 29: 231–63. [ARTICLE VIII]

———. 1976. Charles Lyell speaks in the lecture theatre. *British Journal for the History of Science* 9: 147–55.

———. 1977. Historical analogies in the early geological work of Charles Lyell. *Janus* 64: 89–107. [ARTICLE V]

———. 1978. Charles Lyell's dream of a statistical palaeontology. *Palaeontology* 21: 225–44. [ARTICLE VI]

———. 1979. Transposed concepts from the human sciences in the early work of Charles Lyell. *In* Jordanova and Porter, *Images of the Earth*, 67–83 [revised ed. (1997), 77–91].

———. 1982. Charles Darwin in London: the integration of public and private science. *Isis* 63: 186–206. [ARTICLE IX]

———. 1985. *The Great Devonian Controversy: The Shaping of Scientific Knowledge among Gentlemanly Specialists*. Chicago and London: University of Chicago Press.

———. 1990–91. Introduction (1: vii–lviii) to facsimile reprint of first ed. (1830–33) of Lyell, *Principles of Geology*. Chicago: University of Chicago Press.

———. 1998. Lyell and the *Principles of Geology*. *In* Blundell and Scott, *Lyell: The Past is the Key to the Present*, 3–15. [ARTICLE II]

———. 2004. *The New Science of Geology: Studies in the Earth Sciences in the Age of Revolution*. Aldershot: Ashgate.

Schneer, Cecil J. (ed.). 1969. *Toward a History of Geology*. Cambridge, Mass.: M.I.T. Press.

ACKNOWLEDGEMENTS

Grateful acknowledgement is made to the following persons, institutions and publishers for their kind permission to reproduce the papers included in this volume: The University of Oklahoma Press, Norman, Oklahoma (article I); The Geological Society, London (article II); The British Society for the History of Science, London (article III); Massachusetts Institute of Technology Press, Cambridge, Massachusetts (article IV); The Palaeontological Association, London (article VI); The University of Chicago Press, Chicago, Illinois (articles VII, IX); The Royal Society, London (article VIII); *Studies in the History and Philosophy of Science*, Cambridge (article X).

I

Uniformity and Progression: Reflections on the Structure of Geological Theory in the Age of Lyell

"I ALWAYS FEEL," wrote Darwin in 1844, "as if my books came half out of Lyell's brain." No comment summarizes more succinctly the immense influence of Lyell's writing on the scientific thought of his age. Few men of science working in natural history—a term not yet become derogatory—were unaffected by the persuasive impact, imaginative power and sheer scope of Lyell's work. But the very magnitude of Lyell's achievement raises acute problems of interpretation.

How revolutionary was Lyell's geology? Did the *Principles of Geology* reform that science—or even create it as a scientific discipline—in the manner that Lyell intended? Was its argument persuasive to all whose minds had not been made impervious by earlier modes of thought? How much did Lyell's approach to geology differ from that of his contemporaries? In asking such questions I am not trying to play the fruitless game of devaluing a great man's achievements by pointing to his "forerunners." I want instead to ask what I think is a legitimate and important historical question: what is the contemporary scientific context of Lyell's geology?

I will start by quoting an opinion which deserves respect, for it comes from a scholar who is making a comprehensive study of Lyell.

> The year 1830 [writes Professor Leonard Wilson] marked a great divide in the history of geology. Before that year there had been many great accomplishments, but there had been no critical assessment of the meaning of geological phenomena . . . the interpretation of the past history of the earth . . . remained fanciful and speculative. After 1830 geology became a science.

Here we have a clear statement of a radical claim on behalf of the *Principles of Geology*: namely, that the year of publication of the first volume marks a genuine historical watershed. To use the cur-

I

rently fashionable jargon, it is a claim that Lyell's work did not merely set up a new paradigm for research—as Professor Kuhn himself has suggested—but also that it formed the *first* paradigm in geology, and that it was preceded only by work which was in some sense not "scientific." This claim, if valid, would be historically of the greatest importance, and therefore merits careful consideration.

Certainly there is much in Lyell's own writing to suggest such an interpretation. Lyell frequently claimed that any departure from his conception of the uniformity of nature was "unphilosophical," or at least that it had an effect that was the reverse of heuristic. But it was not for nothing that Lyell's early training was as a lawyer, or that his contemporaries saw in the *Principles* all too much of the "language of an advocate." Lyell's polarization between himself and his opponents—a polarization he took so far as to express it in the Pauline imagery of a cosmic battle between the forces of good and evil—certainly tells us much of importance about the complex character of Lyell himself. But there is a danger here that as historians we may allow ourselves to be swayed—perhaps even brainwashed—by Lyell's persuasive oratory. We need surely to try to get behind the simplicities of the courtroom drama to the more complex, less clear-cut situation that generally exists before the case comes to court.

Unfortunately we are not aided in this task by the categories we have inherited from the period. That Victorian polymath William Whewell can hardly be blamed for the fate of the scientific terms he coined. But it is increasingly clear that the labels with which he characterized the Lyellian debate have now become a pair of polysyllabic millstones around the neck of the history of geology.

"Uniformitarianism" and "Catastrophism" were useful terms to catch some of the striking features of the debate; but they left unclear the meaning of "uniformity" and the importance of "catastrophe," and more seriously they accepted too uncritically the battle-lines that Lyell had drawn. Moreover, since Whewell's time there has been a tendency for the terms to degenerate into mere labels of—respectively —praise and abuse. In our present age of enlightened historiography, historians of science may be unwilling to admit openly to indulging in such "Whiggish" activity; but I think it is clear that if we are to get out of this stalemate in the history of geology we must first agree to abandon the use of Whewell's terms, except within the strict meaning and historical context in which he first applied them.

That negative prescription must surely be coupled with a positive one: we need to make a semantic analysis—however simple-minded it may be in the eyes of philosophers of science—of the concept of "uniformity" in geology, and of the concepts to which it may stand opposed. No one who attempts such an analysis can do so without feeling a great debt to the pioneer work of Professor Hooykaas in this field; and my own indebtedness, in what follows, will be clear enough to all who are familiar with his work. Following the sincerest form of flattery, however, I want to suggest that his analysis can be taken further, and that this will help to clarify the historical questions I started with.

The problem of uniformity arises in geology from its membership of what Whewell usefully classified as "palaetiological" sciences: it is concerned with reasoning about the nature and causation of events in the past. That Lyell clearly recognized the implications of this classification is shown by his frequent use of illustrative analogies drawn from the historical and archaeological studies of his own day. He also recognized that in geology, as in other palaetiological studies, there must be a basic commitment to a belief in the uniformity of the fundamental "laws of nature" (whatever those are!), since without this recognition all research would be impossible. But for his own polemical purposes he often chose to imply that just this basic commitment was lacking in his opponents; whereas in fact they adhered as strongly as he did to "uniformity" in this sense.

The debate only becomes philosophically interesting at a higher and more complex level of causation, that is, in considering the uniformity of geological processes or agencies. Such "causes," as they were termed—for example marine erosion or vulcanism—were agreed to be conformable to, and indeed based upon, the physico-chemical "laws of nature"; but they were evidently causes of far greater complexity. Only about these geological causes can a semantic analysis usefully be made.

I think we can identify *four* logically distinct types of uniformity on this level. All four, and their opposites, were discussed in the Lyellian debate, but Lyell himself either confused them, or believed that each was logically entailed by the others, or—most probably—he was convinced that the reforms he was advocating could only be secured in practice by holding firmly to "uniformity" all along the line.

First, there is the *theological status* of a past geological "cause," in relation to the creative activity of God. It might be *naturalistic*, achieved by "secondary" or "intermediate" means and therefore potentially within the realm of positive knowledge, even if the nature of the means remained obscure. This was not of course incompatible with a belief that its action was divinely sustained and providentially ordered. Or it might be *supranaturalistic*, not attributable to any secondary means, and therefore referable to a "primary" act of "creative" power.

Second, there is the *methodological status* of a past geological "cause." It might be *actualistic*, corresponding in some way to "actual" "causes now in operation," so that observation of those causes could provide a reliable guide to its interpretation. Or it might be *non-actualistic*, in that it failed to have any valid analogy with the present day, being either a "former" or "ancient" cause that had ceased to act, or an intermittent cause that happened not to have operated during the short span of human history. Then interpretation in terms of actual causes would simply be misleading.

But it is important to note that actualism and its opposite can both be applied with varying degrees of rigour: a cause might be unobservable at the present day in the *degree* to which it formerly acted, and yet be perfectly actualistic in the sense that in *kind* it continues to act, perhaps on a much smaller scale. Obviously there is no sharp boundary in practice between degree and kind, since it is merely a matter of definition at what point a past cause becomes so different in degree from its present analogue that it deserves to be called different in kind.

Thirdly, there is the *rate* at which a past geological "cause" may have acted. It might be *gradualistic*, being perhaps insensibly slow within a man's lifetime, or even within the span of recorded human history. Or it might be *saltatory*, or in more familiar terms "paroxysmal" or "catastrophic," acting relatively suddenly and on a large scale. Here again there is clearly no dividing line between these terms, which are merely end-members of a continuum of varying rates or intensities.

Fourth and last, there is the overall *pattern* of a past geological "cause," when its action is traced over the whole known time-span of earth-history. It might be *steady-state*, exhibiting relatively minor fluctuations around a constant mean, at least when the overall state of the whole globe is taken into account. This is clearly the original meaning of "uniformitarian." Or the pattern might be *directional*, or in more familiar terms "developmental" or "progressive," in that, underlying the local fluctuations in its activity, a general overall trend can be detected on a global scale.

Here there are four meanings of geological "uniformity," and their opposites. It is important to see that they *are* logically distinct. A coherent geological synthesis could be constructed from *any* combination of these four pairs of categories. Moreover, there would be no logical absurdity, or empirical improbability, in assigning different causes to different categories in this scheme.

Using this analysis, I want to suggest for discussion that Lyell's *Principles of Geology* was not launched upon a pre-paradigm world of random observations whose meaning was not clearly understood, or upon a geology that had no articulated methods. I want to suggest instead that the scientific context of Lyell's work is *another* geological synthesis, of equally great scope, sophistication and explanatory power; a synthesis that shared the essentials of Lyell's method but differed significantly in its conclusions.

If this is true, why has this alternative paradigm not been more fully recognized? I think there are several reasons for its neglect by historians.

First, as I have already suggested, Lyell's eloquent but distorted account of his predecessors is highly persuasive. His readers are all too readily swayed into believing that Lyell alone, apart from a few commendable forerunners, was bringing light into a dark world of obscurantism and prejudice.

Second, Lyell's account has some slight plausibility if historical attention is confined to geological work in the English language. But geology was self-consciously an international activity, and England was not the center of the scientific world in the 1820's.

But third, the synthesis that Lyell's work opposed lacks the most desirable element for historical identification: a great name. It was a coherent research tradition, but it is expressed more adequately in the scores of papers in the French, German, and English scientific periodicals, than in any single book or in the work of any single geologist.

If this paradigm lacks a great personal name to identify it, what are we to call it? To term it "catastrophist" would be to give too great prominence to its saltatory elements, besides introducing extraneous overtones of calamity and disaster. To term it "progressionist" is equally misleading, because the overtones of improvement in that term are only applicable to certain aspects of the theory. I shall use instead the more neutral term *directionalist*. This will emphasize at once what I believe to be the most significant element in it, namely the directional pattern that it traced within the whole time span of earth-history.

The foundation of this directionalism was an essentially geophysical theory. Buffon's *Epoques de la Nature* had long before popularized the idea of a gradually cooling earth. But such theories gained great scientific prestige in the 1820's from two related directions.

First there was the increasing empirical evidence for the reality of a "central heat" in the earth. It had long been known that the temperature in mines increased steadily with depth, but there were many objections to the interpretation of this as evidence for a central heat. In the 1820's, however, Louis Cordier's rigorous analysis of the evidence proved clearly that the geothermal gradient was not only real but universal. This suggested very strongly that its source must be very deep-seated and global in extent, although local variations in the gradient were fairly clearly related to volcanic activity.

This empirical work seemed to establish the reality of the central heat; but its interpretation as a *residual* heat was powerfully supported by the scientific authority of Fourier's physics. Applying his heat theory to the problem of a cooling earth, Fourier showed that it could account for the observed geothermal gradient. At the same time it could explain why there had been no perceptible change in the earth's diurnal rotation, and therefore in the earth's radius, since the time of the Ancients. At earlier geological epochs the rate of cooling of the earth, and therefore the rate of contraction, would have been relatively rapid; but these rates would have diminished progressively. Thus the temperature at the earth's surface, plotted against time, would have fallen along an exponential curve; and in more recent epochs the heat-loss from the earth's interior would have been very small compared to the effects of solar radiation, so that conditions would have become fairly stable and uniform.

It is only in the light of this theory of a residual central heat that we can understand the directionalist interpretation given to other phenomena.

For example, the earlier controversies on basalt had been settled effectively in favor of its volcanic origin. But this only served to emphasize the wide extent and extremely large scale of volcanic activity at earlier epochs: there was no modern vulcanism to match the scale of the great basalt flows of Iceland and the northwest of the British Isles.

This alone would have suggested that vulcanism might have diminished in the course of time; but such a conclusion was powerfully strengthened by the belief that only the earth's central heat offered an adequate source for volcanic heat. If the central heat was residual, what was more natural than to conclude that its surface manifestation in vulcanism would have decreased correspondingly? This line of reasoning underlies all the directionalist volcanic theories of the period, including for example those of Lyell's friend George Poulett Scrope and Buckland's colleague Charles Daubeny.

The theory of a cooling earth implied that conditions on the surface and in the earth's crust would have been very different in the earliest epochs from what they later became. It should be unnecessary to point out that this did not imply any suspension of the laws of physics and chemistry. It did imply, however, that in remote epochs some processes might have occurred on a large scale, which now could only be imitated with difficulty—if at all—in the laboratory.

This was particularly relevant to the problem of the origin of granite and gneiss. Granite, for example, was known to be sometimes intrusive into earlier rocks, as Hutton had showed long before. But such occurrences were notably commoner among the oldest rocks, and became progressively rarer among the more recent. Since it was agreed that granite was the product of hot conditions in the depths of the earth, it seemed natural to conclude that its formation was a process that had acted on a large scale in the earliest epochs, and had become progressively more restricted as time went on.

But not all granites could be interpreted so simply: many seemed to grade insensibly into banded gneisses, and thence into schists and even more clearly sedimentary rocks. These complex masses of unfossiliferous rocks generally underlay even the Transition strata, so it was reasonable to attribute them to conditions peculiar to very ancient periods of the earth's history. Under the hot conditions then prevalent, it was argued, such rocks could well have been formed by crystallization from a melt, or by precipitation from hot solutions. With further cooling, these rocks of chemical origin would have been replaced gradually by more normal sedimentary rocks of mechanical origin. In such rocks of the Transition series the earliest traces of organisms were found.

The mention of fossils brings us to one of the most important features of the directionalist synthesis. It is well known that a "progressionist" history of life was one of Lyell's main targets for attack. But this biological interpretation was extremely closely integrated with the view of the history of the earth which I have just outlined.

Directionalism in the organic world did not merely run parallel to directionalism in the inorganic world; there was a kind of causal relation between them. In effect, the Cuvierian view of the close integration of the organism to its environment was applied to the concept of a directionally changing environment, and so became a means of explaining the directionally changing nature of successive faunas.

The gradually cooling globe had presented a succession of different environments, at first too hot for life of any kind, then (as we would say) hypertropical, then tropical in all parts of the earth, and finally

approximating to its present condition. In this succession of environments, appropriately adapted organisms had successively come into
existence, had flourished for a time, and then, as the environment
continued to change, they had become extinct and had been replaced
by others better adapted to the new conditions. Moreover, it was
commonly believed that this succession of environments had been
such as to allow progressively higher and more elaborate forms of
life to exist.

In this way, a directionally changing environment could explain
the empirical phenomena of characteristic fossils, on which the
spectacularly successful development of stratigraphy was increasingly
based. But it could also explain the phenomena of the progressive
development of life.

It was not surprising, on this view, that Man had appeared very
recently on Earth, that mammals were confined to the Tertiary strata,
that reptiles extended much further back into the Secondary strata, or
that the Transition strata seemed to contain few if any vertebrates at
all. This view of the adaptation of the forms of life to directionally
changing conditions was so firmly held that there was much skepticism when an anomaly was reported.

Some small mammalian jaws had been found in the English
Secondary strata. The discovery was authenticated both by the detailed stratigraphy of William Smith's home area, and by the authority
of Cuvier's vertebrate anatomy. Yet it was so unexpected that both
the stratigraphy and the anatomy were challenged. This example is
significant, for the resultant controversy was conducted entirely
within the directionalist paradigm; and the anomaly was finally seen
as an unexpected confirmation of the progressive development of life.
The fossils were of *marsupial* type, and not placental mammals; so it
did not seem surprising that this "lower" form of mammalian organization had appeared on earth before the true mammals of the Tertiary
epoch.

Thus the increasing elaboration and diversity of animal life, when
traced through the history of the earth, was seen as a correlate of the
directionally changing environment. This view of the history of life
was confirmed unexpectedly in the 1820's by Adolphe Brongniart's
pioneer work on fossil plants; for this proved that the earth's floras,
like its faunas, had shown increasing elaboration and diversity in the
course of time. Moreover, Brongniart's work provided further confirmation of the theory of a cooling earth.

The early floras of the Coal period were analogous in general
character to some of the tropical floras of the present day, yet their

fossil remains were found even in cool-temperate latitudes. This seemed good evidence not only for a higher temperature at that period in the past, but also for a more uniform temperature all over the earth's surface. This was precisely as expected on Fourier's heat theory, for the present climatic zonation of the earth was seen as a consequence of the present regime of solar radiation, whereas at earlier epochs this would have been masked by the greater heat flow from the earth's interior.

A "progressive" succession of adapted forms of life was thus well-established in the 1820's. What remained obscure, of course, were the means by which these forms had appeared on earth and later disappeared. Extinctions could still be explained by invoking Cuvierian *révolutions* (to which I shall turn shortly), but the problem of origins could no longer be evaded simply by invoking the mechanism of faunal and floral migrations.

Generally, of course, naturalists were content—as Lyell was too—to leave the problem unexplained, and to use the deliberately non-committal language of species "coming into existence" at successive epochs. Yet it is important to notice that when this empiricist agreement not to speculate on the problem was defied, as it was by Étienne Geoffroy St. Hilaire, his mechanism of transmutation by the influence of the environment was fully integrated into the directionalist synthesis. Geoffroy had no need to invoke the primary Lamarckian "tendency to improvement," for unlike Lamarck he accepted the idea of a directionally changing environment. Geoffroy's transformism was geared to the evolution of the environment: successive forms of organization had developed in response to the successive changes in the environment.

So far I have emphasized the primarily directional nature of this synthesis of geology and biology in the 1820's. But it was as "catastrophism" that it later became known, so we must next consider the place of saltatory elements within it. How essential to the synthesis was the postulate of occasional sudden events of great magnitude?

I have already commented that the word "catastrophe" carries overtones of disaster which are largely irrelevant. But of course Whewell's choice of that word, when "paroxysm" and "revolution" were also available, does reflect the importance given to the most recent event of the kind, which had been widely believed to be the catastrophic Flood recorded in all the ancient literatures known at the time (not only the biblical). In fact, however, the identification of the geological Deluge with the Flood had generally been abandoned by the time that Whewell coined the term "catastrophism." It is probable, however, that he was conscious of other overtones in the word. He must have been well aware that in Greek drama the καταστροφή is the final dénouement which brings the action to its conclusion. Given that a dramatic physical event seemed to have occurred on earth in the geologically recent past, it would be natural for Whewell to view this as the καταστροφή of a long directional development of the earth and its inhabitants.

Certainly there *was* strong evidence for some highly peculiar episode in the geologically recent past. In assessing the diluvial theory of the 1820's, we must never forget that nature played a harsh trick on the geologists of the period. With the benefit of scientific hindsight we can see that there had indeed been an extremely unusual event. A glacial period is a rarity in the history of the earth; and since it came to an end (temporarily?!) a mere ten thousand years ago, its effects were bound to seem all the more dramatic.

The north German plain, for example, was strewn with boulders of distinctive rocks that could be traced right across the Baltic to their sources in Scandinavia. On a smaller but still impressive scale, such erratic blocks could also be traced across Britain, often perched on hill-tops and showing no relation to the present drainage system. Such phenomena clearly witnessed to a discontinuity between the present and the geologically recent past; they implied a period of conditions very different from those of the present day.

What agencies could have been responsible for these spectacular effects? It was no use appealing to the magnitude of geological time, for no amount of time could shift boulders the size of houses by the agency of the processes now at work in the areas concerned. Moreover, the organic world seemed to have been affected as radically as the inorganic: as Cuvier had shown, the "diluvial" gravels yielded a whole fauna of extinct mammalia. The only plausible explanation was that a violent rush of water had swept suddenly over the relevant land-areas, causing both the transport of erratic blocks and the annihilation of the indigenous fauna.

It is unfortunate that the provincialism of some English-speaking historians has made William Buckland seem to be the archetypal catastrophist. Buckland was a brilliant lecturer, an entertaining showman and publicist, but a somewhat mediocre geologist. His reconstruction of the diluvial cave fauna was generally admired; but his straining of the evidence, in order to prove the uniqueness, suddenness and universality of the diluvial episode, was equally generally criticized. During the 1820's, it gradually became clearer that not all the diluvium was of the same date, that several diluvial episodes must be postulated, and that none of them had been universal. But this did little to lessen the puzzle of the diluvium.

With such a dramatic event or events in the recent past, it was natural to infer that similar but earlier paroxysmal episodes might have been responsible for some other problematical phenomena. For example Leopold von Buch, one of the best geologists of the period, put forward his influential theory of *Erhebungscratere* (craters of elevation) to account for the puzzling features of volcanic calderas; only episodes of sudden elevation seemed adequate to explain them.

Similarly, when strata of sedimentary rocks were found in mountain regions torn and crumpled like putty, it was difficult to believe that any ordinary processes had been responsible, no matter how long the time that had been available. Such spectacular effects seemed to bear witness to occasional sudden episodes of mountain elevation. This was developed and given greater scope in one of the most important theories of the 1820's, Léonce Élie de Beaumont's theory of epochs of elevation. Élie de Beaumont used the latest refinements in stratigraphy to date the successive periods of orogeny (as we would say) with greater precision than ever before. This proved conclusively that mountain building had occurred at intervals throughout earth-history.

But Élie de Beaumont went further, and proposed a causal hypothesis for these successive paroxysmal episodes, integrating them into the directionalist synthesis. With a gradually cooling and shrinking earth, he believed that compressional strains would slowly build up in the earth's crust. When these strains exceeded the strength of the crust, it would suddenly shear and buckle along lines of weakness, and linear mountain chains would thus be elevated. The strains would thereby be relieved, but in due course they would build up once more, only to be relieved by another episode of mountain formation along some different set of directions. This hypothesis was extended —characteristically—to explain biological problems too. Sudden episodes of mountain elevation would inevitably cause huge tsunamis ("tidal waves") to sweep across the continents. This would naturally cause dramatic changes in the ecology of the animals and plants, and could be responsible for the apparently sudden episodes of mass extinction between successive series of strata. The diluvial episode became merely the last of many such occasional episodes, and was tentatively linked to the recent elevation of the Andes.

It was not by any quirk of jealousy or pique that Adam Sedgwick acclaimed Élie de Beaumont's theory while criticizing Lyell's: to Sedgwick as to most other good geologists of the time, Élie de Beaumont's theory seemed an exciting synthesis of hitherto unrelated problems, even if they had reservations about how far it would stand up to detailed testing.

In Élie de Beaumont's theory, then, we can see how occasional saltatory events could be generated from an essentially gradualistic background (i.e., the gradually cooling earth), without any suspension of the ordinary "laws of nature." In the intervals between these saltatory events, Élie de Beaumont believed there had been immensely long periods of tranquil conditions, during which the ordinary strata had been deposited. This should serve to emphasize how saltatory events were only invoked when and where they seemed necessary to explain the observed phenomena: no "catastrophist" believed the past history of the earth had been an unrelieved succession of violent events. To use the phrase of the time, they were no more "prodigal of violence" than the phenomena seemed to require.

Using the same epigram, it should be clear that they were not "parsimonious of time" either. Of course they did not use the vastness of geological time as a device for eliminating the suddenness of saltatory events, as Lyell was to do; but it by no means follows that they did not believe in the enormous time-scale of earth-history.

This is a difficult point to prove by quotation, because all geologists were justifiably reluctant to commit themselves to statements about the magnitude of geological time, when they had nothing but their geological intuition to go on. But I think it is clear, nevertheless, from frequent comments on the slow tranquil deposition of ordinary strata, that they envisaged an extremely lengthy time scale. Paroxysmal events were postulated, not in order to accommodate large effects within a constricted time-span, but because the nature of the effects themselves seemed to call for more than just the long-continued action of slow processes.

If there remains any residual suspicion that "catastrophism" was necessarily linked to a belief in sudden episodes of species-creation, that too should be dispelled by noting how completely Geoffroy St. Hilaire assimilated his naturalistic transformism to the "catastrophism" of his geological contemporaries. Having geared organic change to environmental change, he was able to use occasional episodes of rapid environmental change as an explanation for corresponding episodes of rapid organic change. The embarrassing difficulties of transforming one major type of organization into another, without unadapted intermediates, were thus at least alleviated.

One final point about the saltatory element of the directionalist synthesis. Although Élie de Beaumont and others believed that paroxysmal episodes had punctuated the whole of the earth's history, including the recent past, other directionalists stuck more closely to the implications of the theory of central heat, and believed that paroxysmal events must have become less and less frequent in the course of time, as the earth cooled towards its present virtually stable condition. Thus Poulett Scrope, for example, could be highly critical of some of the more sweeping assertions of the diluvialists, and yet feel no inconsistency about postulating paroxysmal events for the more remote past. It was only to be expected, in his view, that agencies such as volcanoes and earthquakes would have been active on a larger scale in the earlier periods of the earth's history, and that they would have tended directionally towards quiescence.

So far I have discussed the directional character of this synthesis of the 1820's, and I have pointed out that saltatory events were only postulated where they seemed necessary, against a background of generally gradualistic changes. We now have to consider whether even this limited use of "catastrophe" was, as Lyell alleged, radically unscientific because it failed to be actualistic in method. Did it leave the door wide open to unrestrained speculation, by failing to interpret the past in terms of the present?

I think it is very difficult to substantiate Lyell's allegation. Most of the research I have been describing was certainly done with actualistic intentions, and with a good deal more methodological sophistication than Lyell was prepared to allow. Even Cuvier had postulated sudden *révolutions* only because he believed that no actual causes could account for the observed phenomena. But Cuvier saw no inconsistency in using actualism to interpret other phenomena, such as sedimentation; and of course he also used an actualistic argument to refute Lamarck's transformism, pointing out that since transmutation had demonstrably *not* occurred during recorded human history, no amount of geological time could justify postulating it for the more remote past. So even for Cuvier, comparison with the present was the proper fundamental method for geology, as indeed for palaeontology too; and non-actualistic events were only to be postulated where the method of actualism failed.

Comparison with the present was, moreover, not merely good geological method: it was a natural corollary of the general directionalist synthesis. This may seem paradoxical unless we remember the implications of Fourier's view of an *exponential* fall in the surface temperature of the earth. All geological research tended to confirm this general picture of a gradual slowing down of the rate of change of conditions on earth. Consequently it was entirely reasonable that

I

the span of human history should be, in general, a quite reliable key to the more recent epochs of earth-history.

But it did not follow that actual causes could be assumed *a priori* to be *totally* sufficient for the reconstruction of the past. For one thing, the further one penetrated into the past, the more conditions would have differed from the present, and therefore the more tenuous the analogy would become. Even if the *kinds* of processes had been the same, their *degree* of intensity might have been so much greater that their manifestations might have no close parallel at the present day.

For another thing—and this underlines again the immensity of geological time that was assumed—the period of human history was so short by geological standards that it could not be taken *a priori* as an altogether adequate sample of even the more recent past. For if there were in fact causes that only operated occasionally, such as epochs of elevation, they might never have come under human observation at all, and yet they might have been immensely important in shaping the earth's development.

So there was general agreement that actualistic comparison of past and present was the best policy for research, but that it could not be expected *a priori* to be totally adequate for geological explanation. It would be unhistorical to pick out a few figures in the 1820's as actualistic forerunners of Lyell, and to attribute a lack of scientific method to the rest. All geologists were actualists by policy; they varied only in the *extent* to which they believed the past could be interpreted by close analogy with present causes.

There was in fact a continuing lively debate on just this point. This is reflected in the Göttingen prize question which was won by Karl von Hoff, with his massive compilation of the historically authenticated effects of actual causes. Von Hoff emerges as a convinced believer in the power and efficacy of actual causes; but it should come as no surprise to find that he was nevertheless not convinced that *all* phenomena in the past could be explained in this way. His belief in the explanatory power of actual causes was perfectly consistent with his recognition that some more paroxysmal events might have occurred in the past. It was also of course consistent with his directionalism.

This can be seen equally well, for example, in Poulett Scrope's work. Scrope could argue for the heuristic value of actualism as strongly as Lyell was to do later; but this did not stop him advocating a directionalist volcanic theory according to which vulcanism had declined in intensity during earth-history. Moreover, when later he wanted to point to the sources of Lyell's actualism, he was content to couple his own work with that of Charles Daubeny, although Daubeny allowed an even greater role to paroxysmal events than Scrope himself wished to concede.

Finally, what of the theological status of this directionalist synthesis and of the causes it postulated? Were "catastrophes" unscientific, as Lyell again alleged, in that they opened the door to non-material causation?

Once again, I think this is difficult to substantiate. Even Buckland was prepared to suggest possible secondary causes for the diluvial episode, and the more typical "catastrophism" of von Buch and Élie de Beaumont was even more clearly naturalistic. If the saltatory events of earth-history were sometimes left with their causation unexplained, this is not because they were thought to be due to divine intervention, but simply because many geologists wisely concentrated on establishing that certain events *had* occurred, recognizing that their causation was a logically separate question that could best be left until there was better evidence to go on.

The same methodological point also explains in part the general reluctance to speculate on the origin of new species. Probably many naturalists *did* believe that the phenomena of adaptation placed this problem beyond reasonable hope of any explanation in secondary terms, but the example of Geoffroy shows once more that generalizations are hazardous on this point.

So with the important—but in a sense limited—exception of the origin of new species the geology and palaeontology of the 1820's were thoroughly naturalistic. But of course this did not prevent them being given a providentialist interpretation. In particular, the belief in a directional approximation to the present relatively stable state of things was readily assimilated to a belief that the earth had thus been gradually prepared until it was in a fit state to be the habitation of Man.

But while this may have been a powerful motive behind the work of some directionalists, it is possible that historical concentration on the English scene has led to an over-emphasis on the role of natural theology in the science of this period. The Enlightenment tradition in Continental Europe may perhaps have made natural theology a much less significant factor in the areas where most of the best science was being done.

To sum up this very tentative analysis of the synthesis into which Lyell's *Principles* was launched: it was a coherent synthesis of geology and biology, in which the earth and its inhabitants had undergone a series of *directional* changes, the rate of change generally decreasing as conditions approached those of the present day. Superimposed on generally *gradualistic* changes, occasional *saltatory* events were inferred where necessary; the last such event might have been geologically recent, but their frequency and intensity were usually assumed to have decreased in the course of time. *Actualistic* comparison with the present was agreed to be a heuristic policy for research, but could not be taken *a priori* to be a totally adequate means of explaining all events in the past, and particularly the more remote periods of earth-history. Finally, with the general exception of the origin of new species, all events were taken to be due to *natural* secondary causes, operating in complete conformity with the ordinary laws of nature, though their pattern might be interpreted in terms of providential design.

I have left myself with no time to attempt a corresponding analysis of Lyell's own synthesis. Briefly, however, we may note that although he was deeply committed to a thoroughgoing *naturalism*, he did not differ in this respect from other men of science. Even at the one point where many of them were inclined to fall back on non-material causation—namely the origin of new species—Lyell too was content to use the conventional language of "creation," leaving it to be inferred, as he later told Herschel, that he believed some unknown secondary cause to be responsible. It can be argued that Lyell could excuse his ambiguity on this point, because the exact mode of origin of species was peripheral to his task of establishing a more fundamental position, namely a steady-state terrestrial system.

Second, it is well known that a rigorous *actualism* is an important element in Lyell's synthesis. He insisted that past causes had been similar to present, not only in kind but also in degree. But this is only an extreme form of a methodological policy which was accepted in varying degrees by all his contemporaries. And in practice it seems to have been governed less by considerations of methodological rigor than by a concern to demonstrate that geological processes were not declining in intensity.

Thus, for example, causes that had never been observed in human history were still acceptable to him if they might possibly occur in the future. Likewise he was not averse to inverting his actualism on occasion, as when he postulated species-production as a rare but unobserved event, on the basis of the evidence that new species *had* been introduced in the past.

Third, Lyell is generally supposed to have been fundamentally committed to gradualistic explanation in opposition to any form of catastrophism. Yet even here, his position on the gradual/saltatory

I

continuum was in fact some way from extreme gradualism, and was governed more by the canons of actualism and hence by the requirements of a steady-state system. Thus he initially rejected the alleged rise in the Baltic shorelines, in spite of its extreme gradualism, because he doubted its actualistic authenticity and because it failed to fit his own (more saltatory!) theory of elevation by earthquakes. Similarly he rejected the extreme gradualism of Lamarckian transmutation in favor of an unspecified process which was, by implication, considerably more saltatory in character.

Fourth and last, Lyell's fundamental commitment is, I believe, to be found in the steady-state system he propounded. It is only at this level that the conflict between his synthesis and that of the directionalists bears the marks of a genuine confrontation of irreconcilable paradigms. Only here do we find the characteristic rival interpretations of the same phenomena, the viewing of the evidence through alternative spectacles.

I think this is reflected in the way in which the *Principles of Geology* was received by the scientific community: its attempt to extend the heuristic value of actual causes was generally acclaimed, but its use of actualism in the service of a steady-state or uniformitarian system was equally widely criticized. For the entire structure of the *Principles* is governed by the overriding strategy of advocating such a system, and this seemed to Lyell's contemporaries to involve the most serious straining of the evidence.

Throughout the 1830's, in fact, while Lyell saw his actualistic policy used increasingly in geological research, his steady-state system gradually seemed less and less tenable in the light of that research.

Edouard Lartet's discovery of the first fossil primates within the Tertiary epoch, Louis Agassiz's delineation of an age of fish preceding the age of reptiles, Roderick Murchison's description of a vast Silurian system largely pre-dating even the age of fish—these and many other discoveries seemed to confirm still more clearly the essential validity of the directionalist view of the history of life, rather than Lyell's steady-state view. Lyell's steady-state system for the earth itself fared little better, because he was unable to point to any indefinitely renewable source for the earth's internal heat, on which his system of balanced forces ultimately depended.

In conclusion, therefore, I suggest that however Lyell's work is evaluated—and I have scarcely touched on this question—it must be seen in the context of a rival synthesis of great scientific power and sophistication. Directionalism gave meaning and coherence to a vast range of empirical observations, and it was based on the best physics and chemistry of the time. Its theories were neither fanciful nor speculative: on the contrary they were consciously kept within the strict empiricist limits of "positive science." Lyell may have wished, as he said, to "establish the principle of reasoning in the science," but to his contemporaries the principles were already secure.

References

Brongniart, Adolphe, Nov. 1828. "Considérations générales sur la nature de la végétation qui couvrait la surface de la terre aux diverses époques de formation de son écorce," *Ann. sci. nat.*, *15*, 225–58.
————, 1828–[37]. *Histoire des végétaux fossiles, ou recherches botaniques et géologiques sur les végétaux renfermés dans les diverses couches du globe.* 2 vols. Paris (Dufour and d'Ocagne).
Buch, Leopold von, 1820. "Ueber die Zusammensetzung der basaltischen Inseln und über Erhebungs-Cratere" (Vorgelesen 28 Mai 1818), *Abh. Phys. Klasse, Königl.-Preuss. Akad. Wissensch.*, Berlin, *1818–19*, 51–68.
————, 1825. *Physicalische Beschreibung der Canarischen Inseln.* Berlin (K. Akad. d. Wiss.).
Buckland, William, 1823. *Reliquiae Diluvianae; Or, Observations on the Organic Remains Contained in Caves, Fissures, and Diluvial Gravel, and on Other Geological Phenomena, Attesting the Action of an Universal Deluge.* London (John Murray).
Cordier, [P.] L. [A.], 1827. "Essai sur la température de l'intérieur de la terre" [lu à l'Acad. Sci., 4 juin, 9, 23 juillet 1827], *Mém. Mus. d'Hist. Nat.*, *15*, 161–244; *Mém. Acad. Roy. Sci., Inst. France*, *7* (1827), 473–555.
Cuvier, G., 1812. *Recherches sur les ossemens fossiles de quadrupèdes, ou l'on rétablit les caractères de plusieurs espèces d'animaux que les révolutions du globe paroissent avoir détruites.* 4 vols. Paris (Deterville).
Daubeny, Charles, 1826. *A Description of Active and Extinct Volcanos; With Remarks on their Origin, their Chemical Phaenomena, and the Character of their Products, As Determined by the Condition of the Earth during the Period of their Formation.* London (W. Phillips).
Élie de Beaumont, L., 1829–30. "Recherches sur quelques-unes des révolutions de la surface du globe, présentant différens exemples de coïncidence entre le redressement des couches de certains systèmes de montagnes, et les changemens soudains qui ont produit les lignes de démarcation qu'on observe entre certains étages consécutifs des terrains de sédiment," *Ann. sci. nat.*, *18* (1829), 5–25, 284–416; *19* (1830), 5–99, 177–240.
Fourier, [J. B. J.], 1820. "Extrait d'un mémoire sur le refroidissement séculaire du globe terrestre," *Ann. chim. phys.*, *13*, 418–38.
————, 1827. "Mémoire sur les températures du globe terrestre et des espaces planétaires," *Mém. Acad. Roy. Sci., Inst. France*, *7*, 569–604.
Geoffroy St. Hilaire, É., 1825. "Recherches sur l'organisation des gavials; Sur leurs affinités naturelles, desquelles résulte la nécessité d'une autre distribution générique, *Gavialis, Teleosaurus* et *Steneosaurus*; et sur cette question,

si les Gavials (*Gavialis*) aujourd'hui répandus dans les parties orientales de l'Asie, descendent, par voie non interrompue de génération, des Gavials antédiluviens, soit des Gavials fossiles, dits Crocodiles de Caen (*Teleosaurus*), soit des Gavials fossiles du Havre et de Honfleur (*Steneosaurus*)," *Mém. Mus. d'Hist. Nat., 12*, 97–155, pls. 5, 6.

———, 1828. "Mémoire où l'on propose de rechercher dans quels rapports de structure organique et de parenté sont entre eux les animaux des âges historiques, et vivant actuellement, et les espèces antédiluviennes et perdues," *Mém. Mus. d'Hist. Nat., 17*, 209–29.

Hoff, Karl Ernst Adolf von, 1822–34. *Geschichte der durch Überlieferung nachgewiesenen natürlichen Veränderungen der Erdoberfläche*. 3 vols. Gotha (Justus Perthes).

Hooykaas, R., 1957. "The Parallel between the History of the Earth and the History of the Animal World," *Arch. intern. d'hist. sci.*, no. 38, 3–18.

———, 1959. *Natural Law and Divine Miracle. A Historical-Critical Study of the Principle of Uniformity in Geology, Biology and Theology*. Leiden (E. J. Brill).

Kuhn, Thomas S., 1962. *The Structure of Scientific Revolutions*. Intern. Encycl. Unif. Sci., *2* (2).

Lyell, Charles, 1830–33. *Principles of Geology, Being an Attempt to Explain the Former Changes of the Earth's Surface, by Reference to Causes Now in Operation*. 3 vols. London (John Murray).

Lyell, (Mrs.) [K.], 1881. *Life, Letters and Journals of Sir Charles Lyell, Bart.* 2 vols. London (John Murray).

Scrope, G. Poulett, 1825. *Considerations on Volcanos, the Probable Causes of their Phenomena, the Laws Which Determine their March, the Disposition of their Products, and their Connexion with the Present State and Past History of the Globe; Leading to the Establishment of a New Theory of the Earth*. London (W. Phillips).

———, 1827. *Memoir on the Geology of Central France; Including the Volcanic Formations of Auvergne, the Velay and the Vivarais*. 2 vols. (1 of plates). London (Longman, Rees, Orme, Brown, and Green).

Sedgwick, (Rev.) Adam, Feb. 1831. "Address to the Geological Society, delivered on the Evening of the 18th of February 1831," *Proc. Geol. Soc. Lond., 1*, No. 20, 281–316.

Whewell, William, Mar. 1832. [Review of *Principles of Geology*], *Quart. Rev., 47*, No. 93, 103–32.

———, 1837. *History of the Inductive Sciences from the Earliest to the Present Time*. 3 vols. London (John W. Parker).

Wilson, Leonard G., 1967. "The Origins of Charles Lyell's Uniformitarianism," *Geol. Soc. Amer. Spec. Pap. 89*, 35–62, 3 pls.

II

Lyell and the *Principles of Geology*

Abstract: Lyell's *Principles of Geology* is still treated more often as an icon to be revered than as the embodiment of a complex scientific argument rooted in its own time and place. This paper describes briefly the origins of Lyell's project, in the international geological debates of the 1820s and in his own early research; the structure of argument of the first edition (1830–33), and its relation to its intended readership; and the modification of the work in subsequent editions, and the transformation of its strategy in response to its critical evaluation by other geologists. The fluidity of the *Principles* (and its offshoot the *Elements of Geology*) reflects the ever-changing interaction between Lyell and his fellow-geologists, and between them and a much wider public, during one of the most creative periods in the history of geological science. Lyell's scientific stature is best appreciated if he is placed not on a pedestal but among his peers, in debates at the Geological Society and elsewhere.

Charles Lyell's *Principles of Geology* (first published in 1830–33) is one of the most significant works in the history of the earth sciences, but it is still more often cited than read, more often revered than analysed. Above all, it needs and deserves to be understood in the context of its own time, before being recruited to support modern arguments, or repudiated for its failure to support them.

Unlike many comparably important works in the history of art or literature, the *Principles* was not launched into the public realm as a once-for-all achievement, thereafter to be admired or criticized but not significantly modified. On the contrary, Lyell reissued it in successive editions over almost half a century, and during that period it changed radically in character. The *Principles* was in effect one side of a *dialogue* between Lyell and his contemporaries: not only the other leading geologists in Britain and abroad, but also much wider groups in British society. Like some other important scientific works from the same period – Charles Darwin's *Origin of Species* (1859) is a good example – Lyell's *Principles* was a work with multiple goals and a diverse range of intended readers (Secord, 1997). It was not only an original scientific treatise directed at other geologists; it was also at the same time a work in what the French recognize as a distinctive genre, that of *haute vulgarisation* or authoritative high-level popularization. The construction of the first edition, and its modification in subsequent editions, should therefore be seen as Lyell's lifelong contribution to continuing debates, not only about the changing profile of geological knowledge and its basic 'principles', but also about the place of such knowledge in a scientific view of the world.

While he was an undergraduate at Oxford, Lyell's adolescent interests in natural history were channelled into geology by the charismatic presence and famously entertaining lectures of William Buckland (Rupke, 1983; Wilson, 1972). When in 1819 Lyell graduated and moved to London to begin a legal career, Buckland sponsored him for membership of the Geological Society (itself little more than a decade old). This put Lyell at the heart of geological debate in Britain (Morrell, 1976), and in 1823 he became one of the Society's secretaries. He then took an early opportunity to visit Paris, making himself known in the world centre of scientific research and improving his fluency in the international language of the sciences. Back in London, as a member of the Athenaeum – a new social club for intellectuals of all kinds – he also became known far beyond scientific circles: some of his most important geological ideas were first published in essay-reviews for the influential Tory periodical *Quarterly Review*, which was widely read by the social, political and cultural elites in Britain.

It was a time of intense public interest in all the sciences; elementary books were selling well and proving profitable to their authors. Lyell therefore planned to use that well-tried format to propagate his own emerging conception of geology; like many young barristers he urgently needed other sources of income, in his case to supplement a modest allowance from his father. But as he began to put his ideas on paper, he found them expanding beyond the bounds of an introductory work. Eventually he realised they would need to be expounded in a full-length treatise, and he began to talk about writing a book that would establish the basic 'principles' of the science. In other words, Lyell's proposed work was modified profoundly before it was ever published. The word 'principles', when used at that time in a book title, carried connotations not of an elementary textbook but of Isaac Newton's monumental *Principia*; adopted by Lyell, a man barely into his thirties, it signalled substantial scientific ambitions. Nonetheless, Lyell's book never completely lost its initially intended character as a work of *haute vulgarisation*, and its later offshoot the *Elements of Geology* returned explicitly to that genre.

Resources for the *Principles*

During the later 1820s, Lyell formulated his own conception of geology, constructed the argument for his work, and assembled empirical material to support it. Occasionally he did so in the field, in direct contact with the natural world, but far more often it was in interaction with his scientific contemporaries, either face to face or through their publications. However, rather than being passively 'influenced' by them, he treated them as his scientific *resources*; like any great scientist, he actively selected from all the past and present research that was available to him, choosing what best served his own intellectual goals and reshaping it for his own purposes.

To help make sense of this rich body of empirical material, Lyell drew on two contrasting theoretical models for the study of the earth. The importance of one is well known, but the significance of the other has been widely overlooked. The first was the Huttonian theory of the earth. This was taken by Lyell, as it was by most of his contemporaries, not from James Hutton himself but from John Playfair's bowdlerized *Illustrations* (1802). Like the many other works using the title 'theory of the earth', Hutton's had been a hypothetical model (1788) followed – in his case, years later – by a mass of empirical 'proofs and illustrations' (1795). But Playfair had shorn it of its pervasive deistic metaphysics and anthropocentric teleology, and had recast it in an effort to turn geology into a branch of physics. As such, geology would be above all a science of *causes*, of natural processes the same yesterday, today and for ever. The earth would be treated as a theatre of ceaseless and essentially repetitive change, played out on a timescale left conveniently indefinite but treated in practice as virtually infinite. The earth was assumed to have been roughly the same kind of place at all times: as a mathematician, Playfair had found the cyclic stability of the Newtonian solar system an appealing parallel. Such theories were commonplace in Enlightenment ideas about the natural world in general, and Playfair's application of them to the then new science of geology was attractive to many in Lyell's generation. Specifically, Lyell followed Playfair in putting the Newtonian notion of a *vera causa* or 'true cause' at the heart of his method: phenomena should be explained only in terms of causal agencies that are observably effective, both in kind and in degree (Laudan, 1982, 1987). For a science like geology, faced with understanding the past as well as the present, that meant in practice that explanations should always be in terms of 'modern' or 'actual' causes ('actual' was used in a sense now obsolete in English, meaning current or present-day, as in the modern French *actualités* for the day's news). In geology, *verae causae* were necessarily also actual causes, since only the present (and, by extension, the reliably documented human past) could be directly observed and witnessed.

The other theoretical model, equally important for understanding Lyell's work, was that of the earth as an product of *history*. This was quite compatible with the physical model, since all the events in the history of the earth and its life – with the possible exception of the origins of new species, and particularly the human species – were assumed to have had natural causes of some kind. But on the historical model such causes were taken to have produced an earth that had *not* always been the same kind of place; it had had a complex contingent history that could not be predicted from any physical model, and that could only be discovered and reconstructed by attending to the presently observable traces of specific past events. The historical model had first been formulated, a century and a half before Lyell, within the framework of biblical history and its traditional short timescale. But by the time of Buffon's *Époques de la Nature* (1778) it had been fully secularized; Buffon's experiments with cooling globes gave it a quantified timescale that was modest by modern standards but already literally beyond human imagination (Roger, 1989). Just after the French Revolution, the anatomist Georges Cuvier had greatly deepened the historical model, and highlighted the 'otherness' of the earth's past, by using fossil bones to claim the reality of extinction as a perennial natural phenomenon (Rudwick, 1997). Cuvier had become convinced that no physical cause known to him could have wiped out fossil species as well-adapted as he believed them to have been. So he had adopted the common idea that some kind of natural but catastrophic agency must have been responsible for extinctions (e.g. Cuvier, 1812). He had remained uncertain about the physical character of these putative catastrophes; but others, for example Hutton's old friend James Hall (1814), had suggested they could have been mega-tsunamis, as it were, caused by sudden major bucklings of the oceanic crust. In other words, the historical model for the earth was here interpreted with resources drawn from the physical model: an observable kind of natural event, albeit of a magnitude unwitnessed in human history, was held responsible for otherwise inexplicable effects in the earth's history.

The historical model, and Cuvier's version in particular, was mediated to Lyell by his mentor Buckland. Buckland's fieldwork in the Oxford region convinced him that the puzzling deposits he termed 'diluvium' were quite distinct from the more recent alluvium along the present rivers: for example the 'diluvial' boulder clay and gravels had apparently been swept from the Midlands plain over the watershed of the Cotswolds into the Thames valley. Buckland's inference of a geologically recent mega-tsunami or 'geological deluge' gave concrete form to what Cuvier had inferred mainly from fossil bones. It then seemed to be confirmed by the chance discovery of Kirkdale cave in Yorkshire, which Buckland (1822, 1823) interpreted geohistorically as a den of extinct hyaenas that had been annihilated by the geological deluge. His 'diluvial' interpretation became a powerful scientific theory, which most geologists found highly persuasive as an explanation of some extremely puzzling phenomena; it won Buckland the Royal Society's Copley Medal, and Lyell defended it when he visited Paris in 1823.

Buckland's 'diluvialism' was presented, however, within a context that made it far more than a scientific theory; its reception was moulded by societal issues more prominent in England than in France, and more pressing in Oxford than in London. Cuvier had sought to link the new geohistory to the far briefer annals of human history, by searching the literature of *all* ancient civilizations for possible early human records of the most recent catastrophe; as a cultural relativist typical of the Enlightenment he had treated the biblical story of Noah's Flood as just one of many obscure folk-memories of some such event (Rudwick, 1997). Buckland, on the other hand, needed to prove to his colleagues that geology was a legitimate field of academic study in the intellectual centre of the Church of England, and that it would not undermine Oxford's traditional textual forms of learning (Buckland, 1820; Rupke, 1983). He therefore invoked the geological deluge as decisive evidence for the historicity of the biblical Flood in particular; he interpreted it as a violent event of global extent and colossal magnitude, although in fact this entailed taking substantial liberties with the literal meaning of the text (English scholars were still largely ignorant of the textual criticism pioneered in the German universities, which rendered biblical literalism obsolete, not least in religious terms). In other words, in Britain – though never in the same way on the Continent – an effective geological theory was entangled with issues of biblical interpretation, ecclesiastical authority and, above all, the political power of the established Church.

Once Lyell moved from Oxford to London, away from generally conservative dons and into the company of politically liberal lawyers, he became uneasy about this mixing of science and divinity, although (as mentioned already) he continued for some time to champion Buckland's diluvial theory as a good explanation for many otherwise puzzling phenomena. His shift away from Buckland's kind of theorizing was unquestionably powered by his growing hostility, not to religion in general, but to specific aspects of the contemporary Church of England, and particularly its virtual monopoly on higher education. But what eventually convinced him that the geological deluge was a chimera was the cumulative weight of specific empirical cases, in which the phenomena could be still better explained without recourse to any recent catastrophe. These cases were drawn from the published literature of the multinational geological community, backed up by Lyell's own first-hand confirmation of some of the more striking examples.

For example, when he visited Paris in 1823 Lyell was taken into the field by Constant Prévost, a former student of Cuvier's colleague Alexandre Brongniart. Cuvier and Brongniart's joint monograph on the Paris Basin (1811) was the outstanding exemplar of how a detailed geohistory could be inferred from a pile of formations and their fossils: it was a local history of tranquil sedimentation on a vast timescale, punctuated by occasional sudden alternations between marine and freshwater conditions (Rudwick, 1996, 1997). But Prévost was already arguing against such sudden changes, and claiming that the ancient environments were paralleled in every respect in the modern world, for example around the estuary of the Seine (Bork, 1990). Lyell was duly convinced by Prévost's arguments (Wilson, 1972), and the next year he followed the Frenchman's example by studying a recently drained lake near his family home in Scotland. In his first major paper to the Geological Society, Lyell (1826) argued that the modern sediments were identical in every detail to some of those that Prévost had shown him in France. For Lyell, it was a powerful vindication of the claim that actual causes might be adequate to explain *everything* in the geohistorical record.

In the wake of Cuvier's claims to the contrary, many geologists had realized that any such conclusion about actual causes depended on a better knowledge of just what effects such agencies were actually having in the present world. In 1818 the Royal Society of Göttingen had therefore offered a valuable prize for a monograph on the subject. The prize was won by the civil servant (and part-time geologist) Karl von Hoff of Gotha. Hoff's exhaustive search of the geographical and historical literature (1822–34) showed that ordinary agencies such as sedimentation and erosion, volcanoes and earthquakes, had indeed effected major physical changes, even within the two or three millenia of reliably recorded human history (Hamm, 1993). The implication was that in the far vaster tracts of geological time there might be no limit to what they could produce. Lyell quickly saw the significance of this for his own project. He learnt German specifically to exploit the empirical riches of Hoff's work (and thereby also gained access to a major new scientific literature). He also followed Hoff's example by searching through the accounts of the voyages and expeditions of all the scientific nations, finding copious further evidence of the power of actual causes within recorded history (Rudwick, 1990–91).

A third example of great importance to Lyell came, like Prévost's, from fieldwork studies of the geohistory of a specific region. George Scrope, a geologist who had joined Lyell as a secretary of the Geological Society, published a speculative 'theory of the earth' (1825) based like Buffon's on a cooling model of the earth's history. He backed it up, however, with a detailed study of the extinct volcanoes of Auvergne (1827), illustrated vividly by a set of geologically interpreted panoramas of the landscape there. Scrope's work borrowed extensively – with scant acknowledgement – from earlier French writers, but it presented anglophone geologists with persuasive evidence for a continuous geohistory of this famous region, uninterrupted by any diluvial event. The occasional episodes of volcanic activity, stretching far back into the geological past, served in effect as markers recording successive stages in the continuous slow fluvial erosion of the topography;

and Scrope claimed that what was needed for the correct interpretation of the region was simply 'Time! Time! Time!' (Rudwick, 1974).

Lyell was deeply impressed by Scrope's case study, and explained its significance to a wider audience in the *Quarterly Review* (Lyell, 1827). He suggested that it was a model for how geology ought to be done, using observable actual causes to penetrate from the known present back into the more obscure past. This was in fact just the strategy that Cuvier had famously advocated, but Lyell now intended to use it to undermine Cuvier's argument for a recent catastrophe. He extended Scrope's argument to cover the organic world as well as the inorganic, claiming that the fossils being found by local geologists in Auvergne proved that there had been no recent mass extinction or marine incursion there. When in 1828 he had an opportunity to do fieldwork on the Continent, his very first destination was Auvergne, where he duly confirmed Scrope's interpretation and his own extension of it.

A fourth and final example of Lyell's use of the rich contemporary geological literature strengthened that conclusion about gradual faunal change; it extended to Tertiary formations and their fossils what Scrope had claimed for the volcanic and erosional history of the same era. Cuvier had suggested that the formations on the flanks of the Apennines, with their superbly preserved fossil molluscs, might help to bridge the taxonomic and geohistorical gap between the Parisian fossil faunas and the living molluscs of present seas. Giovanni-Battista Brocchi of Milan had pursued that idea with a fine monograph on the Italian fossils (1814); but to interpret them he had formulated a model of piecemeal faunal change, which dispensed with the need for any Cuvierian catastrophes, at least for marine faunas (Marini, 1987). Extracts had been translated by Leonard Horner (Lyell's future father-in-law) in the *Edinburgh Review* (1816), the Whig counterpart of the *Quarterly*, and Lyell later learnt Italian, partly in order to read Brocchi's work more thoroughly; it became the basis for his own more extensive use of *all* the known Tertiary molluscan faunas in the reconstruction of Tertiary geohistory.

These four examples are sufficient to suggest how, when writing the *Principles*, Lyell appropriated a massive and mature body of contemporary geological literature, published not only as books but also in a growing number of scientific periodicals. Lyell exploited that literature to the full, by being as international in outlook – and as multilingual – as the leading geologists in the rest of Europe (the word 'Europe' is used in this paper in its proper pre-Thatcherite sense, to include the British Isles). It was no coincidence that when in 1826 Lyell resigned as a secretary of the Geological Society he was promptly appointed its Foreign Secretary.

The strategy of the *Principles*

Immediately after his fieldwork in Sicily, the culmination of his long expedition on the Continent in 1828–29, Lyell described his proposed book in a letter to Roderick Murchison (his companion on the earlier part of the trip). It was to vindicate his '*principles of reasoning*' in geology, which were that '*no causes whatever* have ... ever acted, but those *now acting*, and that they never acted with different degrees of energy from that which they now exert' (Lyell, 1881, vol. 1, p. 234). In other words, actual causes were wholly adequate to explain the geological past, not only in kind but also in degree. The first proposition was relatively uncontroversial, but with the second Lyell distanced himself from a virtual consensus among his fellow geologists (Bartholomew, 1979). Even those like Buckland who claimed there had been a drastic catastrophe in the geologically recent past thought it might have been analogous to the tsunami that had famously destroyed Lisbon in 1755; but they insisted that the physical evidence of its action proved that it must have been on a far larger scale. Likewise, even those like Prévost and Scrope who most strongly supported Lyell's claims for the power of actual causes believed that volcanic and tectonic events, for example, must have been on a far larger scale in the earth's hotter and more active infancy. But Lyell was now convinced that if geology were to be truly 'philosophical' (or in modern terms, scientific), the Newtonian *vera causa* principle would need to be applied rigorously, confining causal explanation to agencies observably active in the modern world, acting *at their present intensities.*

Lyell therefore had a major task of persuasion ahead of him, if he was to convince other geologists – let alone the general educated public – that everything in the geological record could be adequately explained in these terms. This is where Lyell's legal training stood him in good stead: as his contemporaries noticed at once, the *Principles* was a barrister's brief from start to finish, rhetorical in character throughout (using the word 'rhetoric' in its proper and non-pejorative sense). Lyell set out to present a good *case* for the explanatory adequacy of actual causes: as he put it in his carefully phrased subtitle, the work was 'an attempt to explain the former changes at the earth's surface, by reference to causes now in operation'. Lyell's *Principles*, as Darwin later said of his own book, was 'one long argument'.

The core of Lyell's argument was his exhaustive survey of the whole range of actual causes. Before embarking on that survey, he inserted important preliminary chapters. In a brief introduction, he stressed not only the causal goals of geology but also its close analogy with human history (Rudwick, 1977), thus showing that he proposed to combine the physical and the historical models for the science. He then launched into a long account of the history of geology, with much of the factual material taken straight from Brocchi (McCartney, 1976). But Lyell's history was – as he admitted privately – a discreet cover for criticisms of his contemporaries (Porter, 1976, 1982). He rewrote the history of geology to imply that those who now insisted on the reality of catastrophic events in the earth's past were following a tradition that had consistently retarded the progress of the science. It was, he claimed, a tradition that invoked catastrophes arbitrarily whenever the explanatory going got difficult, that failed to appreciate the sheer scale of geological time, and that concocted speculative accounts of the very origin of the earth as well as its subsequent history. In fact this was highly unfair to Buckland and others like him, as they were quick to point out once the *Principles* was published. But for Lyell it had the polemical advantage of associating them with the quite distinct genre of 'scriptural geology', which was currently enjoying great popularity among the general public. It implied that some of the most prominent English geologists were as wrong-headed as the popular writers (none of them members of the Geological Society) who were trying to reconcile the new science with a traditional and literalistic exegesis of Genesis. In effect, Lyell equated his own conception of 'the uniformity of nature' with being *scientific* in matters of geology, thereby branding his anticipated geological critics as unscientific.

Lyell then attacked those same geologists for what they regarded as an increasingly well-founded inference in geology. This was that the immensely long history of the earth had been broadly *directional*, from a hot and highly active origin to a cool and less turbulent present, with a broadly progressive history of life to match (Rudwick, 1971). As foreshadowed in his letter to Murchison, Lyell set against this his own Huttonian vision of a broadly stable, or at most a cyclic, history of both the inorganic and organic realms. He had reluctantly conceded that there was good evidence for major climatic change in the course of geohistory: the tropical-looking Carboniferous flora, for example, was difficult to explain away. At the eleventh hour, as it were, he devised a new theory that accounted for such climatic changes without conceding anything to the directional theory of a cooling earth (Ospovat, 1977). Drawing on the climatology of the great Prussian geographer Alexander von Humboldt (1817), Lyell argued that local climates in the distant past, as at present, were products not only of latitude but also of the configuration of continents and oceans, winds and ocean currents. If – as he argued – those configurations had changed continuously through the ordinary agencies of geological change, any given region might have had highly diverse climates at different periods of geohistory.

From this ingenious explanation Lyell drew consequences that his geological readers found nothing less than startling. Since organisms were closely tied to their environments, he claimed that the history of life was likewise in the long run directionless, so that organisms long extinct might recur in at least approximately the same forms: to illustrate the point he indulged in a speculative thought experiment about how the ecosystem of the Jurassic reptiles might reappear in the far distant future. This undercut the foundations of the historical model of the earth, as Lyell's contemporaries understood it: it denied the reality of the unrepeatedness of the fossil record. Lyell had to explain away that appearance, by stressing the extreme imperfection of the record, as a consequence of the extreme chanciness of preservation. The recent discovery of rare mammalian fossils in the (Jurassic) Stonesfield 'slate', for example, gave him welcome evidence that mammals were not after all confined to the Tertiary, and enabled him to argue that they might have been in existence as far back as the fossil record could then be traced (in modern terms, well into the Lower Palaeozoic). Such were the lengths to which the consistent application of his extremely rigorous 'principles' required Lyell to go.

After these preliminary discussions, Lyell embarked on his survey of actual causes both inorganic and organic, describing with a wealth of detailed examples how much change they had effected even within the short span of recorded human history. The inorganic causes alone took up the whole of the rest of the first volume (1830). Although Lyell borrowed extensively from Hoff's factual material, he rearranged it to illustrate his own theoretical model of the earth as a system of *balanced* physical agencies. Erosion was balanced by sedimentation, for example, elevation by subsidence, and so on; it was in fact the Newtonian model of the solar system applied to the earth, as it had been by Hutton and Playfair before him. This explains Lyell's surprising choice for the frontispiece of his opening volume (at that time the marketing equivalent of a modern dustjacket). Instead of a picture of, say, a violently erupting volcano or an idealised section through the earth's crust, Lyell chose a picture of a human artefact: a view of the famous Temple of Serapis near Naples, which he had seen on his travels (Fig. 1). It was in fact a highly effective epitome of his argument. It linked the human world with the natural, human history with geohistory, classical scholarship with scientific research; it illustrated the power of actual causes, and the reality of crustal movements even within the short span of time since the Roman period; and it was a vivid demonstration that those movements had been in both directions, both elevation and subsidence, thus illustrating his conception of the earth as a system in dynamic equilibrium.

Lyell's repertoire of actual causes continued in his second volume (1832) with a survey of processes in the organic world. As with inorganic agencies, his argument was that even the largest features in the geological record could be explained by

Fig. 1.   The frontispiece of the first volume of Lyell's *Principles of Geology* (1830). This engraving of a view of the ruined 'Temple of Serapis' (now thought to have been a market) at Pozzuoli near Naples was redrawn from one published by a local archaeologist (Jorio 1820). The zone bored by marine molluscs on the surviving pillars was well understood to be evidence of changes in relative sea-level since the Roman period; for Lyell, the site epitomised in miniature his model of the earth as being in a dynamic non-directional equilibrium expressed in repeated small-scale changes in the earth's crust, even within human history.

the summation of smaller effects over vast spans of time. So his discussion centred on the reality of species, and the processes by which their appearances and disappearances, origins and extinctions, might be explained. He rejected Lamarck's notion of the imperceptibly gradual 'transmutation' of one species into another, claiming instead that species were real and stable units of life (Coleman, 1962). Lamarck's evolutionary views, increasingly in vogue on the Continent, were widely suspect in Britain for their supposed implications of materialism, irreligion and even revolutionary politics (Corsi, 1978; Desmond, 1989). Lyell's explicit repudiation of such ideas signalled that his conception of geology, by contrast, should be acceptable to his intended readers, however conservative their opinions might be (Secord, 1997); privately, he also found deeply repugnant the implication that human beings had evolved from animal ancestors (Bartholomew, 1973). But Lamarck's kind of evolution – denying the reality of extinction altogether – was also unacceptable for strictly scientific reasons: Lyell's method of reconstructing geohistory was going to depend on the stability of species as natural units, each with a definable time of origin and of extinction; and above all, Lamarck's ideas embodied no observable *vera causa*.

In the course of a discussion of biogeography and the means by which organisms may be dispersed, Lyell tried to define the circumstances in which new species have originated; but he left the process itself unexplained. He told his friends privately that he thought it was a natural process of some kind, but publicly his reticence was prudent, deflecting any suspicion that he was a closet evolutionist (Secord, 1997). With extinction there were no such problems: Lyell was free to argue that species become extinct gradually and one by one, by equally gradual changes in their environments, without recourse to sudden catastrophes or mass extinctions. So he outlined a model of organic change, developed from Brocchi's, in which species originated and became extinct in piecemeal fashion, in both time and space.

Lyell's survey of actual causes concluded with those in which the inorganic and organic worlds interact. Much of his discussion was about taphonomy, analysing the circumstances in which organisms may – or may not – be preserved as potential fossils. This served to underpin his insistence on the extreme imperfection of the fossil record; but he also pointed out that some groups, such as the molluscs, were likely to have a less imperfect record than others, such as mammals. This foreshadowed his crucial use of fossil molluscs in the reconstruction of geohistory.

Such applications followed in the third and culminating volume (1833). Lyell emphasized that the first two volumes were 'absolutely essential' to the third: actual causes were merely the 'alphabet and grammar' of geology, making it possible to decipher nature's 'language', in which the past history of the earth had been recorded. The construction of geohistory and its causal explanation was the goal of geology; understanding actual causes was simply the means to that end. Lyell therefore began his final volume by discussing how the ordinary succession of formations and their fossils should be interpreted in order to yield reliable geohistory. Underlying this was not only his profound sense of the sheer magnitude – and therefore the explanatory power – of the timescale itself, but also his conviction that the continuous piecemeal change in faunas and floras could provide the basis for *quantifying* geochronology (Rudwick, 1978). On the assumption that the rates of species-origins and extinctions had been statistically uniform and constant, the age of any formation should be directly related to the proportion of its fossil species that were still extant. Lyell had therefore recruited (and paid for) the taxonomic expertise of Paul Deshayes in Paris, who arranged all the known Tertiary molluscan faunas according to their percentages of extant species. The faunas fell into three distinct groups, for which the Cambridge philosopher William Whewell supplied Lyell with the appropriately classical names 'Eocene', 'Miocene' and 'Pliocene', to express respectively the 'dawn', minority and majority of modern species in the successive faunas (Lyell split the Pliocene into 'Older' and 'Newer', making four groups). For Lyell, however, these were not four contiguous *periods* of time; they were merely random samples from the deep past, separated perhaps by far longer spans of unrecorded geohistory (Rudwick, 1978). So as he penetrated still further back from the present, the *total* disjunction between the Eocene fauna and that of the youngest part of the Chalk was attributed not to any sudden catastrophe or mass extinction but to an even longer span of gradual but wholly unrecorded faunal change. Such was the kind of conclusion to which Lyell was led, by his insistence on the vast timescale of geohistory, on the extremely gradual character of physical and organic change, and on the extreme imperfection of the natural record of geohistory.

The bulk of Lyell's last volume was devoted to a detailed study of the Tertiary era, presented as an exemplar of how the whole of geohistory should be reconstructed. As Cuvier had recommended, the rocks and fossils were analysed retrospectively, from the known present back into the unknown past. Lyell analysed each of the preserved fragments of Tertiary geohistory in turn, from the Newer Pliocene (which he later renamed the 'Pleistocene') back to the Eocene, in order to demonstrate that the same processes had been operative at each time, processes the same in kind as those in the present world, acting at much the same intensities. Throughout the unimaginable span of Tertiary time, the earth had been much the same kind of place. Lyell tried to demonstrate just how vast the timescale was, in the longest part of his discussion of the Newer Pliocene. He used his own fieldwork in Sicily to suggest a rough calibration of his mollusc-based geochronology in terms of the far briefer timescale of human history (Rudwick, 1969). He presented evidence to suggest that the many eruptions of Etna recorded in the past two millenia had added very little to its bulk, so that the huge cone as a whole must be extremely ancient in human terms. Yet it appeared to be built on formations that he identified as Newer Pliocene in age, with a molluscan fauna almost identical to that still living in the Mediterranean. So a volcano that was unimaginably ancient in human terms was also extremely recent on the timescale of even the Tertiary formations, let alone geohistory as a whole.

Having established his methods with respect to the Tertiary, Lyell could afford to treat the record of the 'Secondary' and 'Transition' eras (roughly, in modern terms, the rest of the Phanerozoic) briefly and programmatically. Finally, he knocked the bottom out of the consensual reconstruction of geohistory, by claiming that the 'Primary' rocks, so called because they appeared to be the most ancient (in modern terms, they were mostly Palaeozoic or Precambrian), were not primary at all, but either igneous or what he called 'metamorphic' (coining that term for the first time). In any case, geohistory was left without any recoverable beginning; so Lyell ended on an unmistakeably Huttonian note, by claiming that the earth was a complex physical and biological system in dynamic equilibrium, for which there were no traces of a beginning and no foreseeable end (Gould, 1987).

Transformations of the *Principles*

Lyell's *Principles* was published by John Murray, the leading British scientific publisher at the top end of the market (and also the publisher of the *Quarterly Review*); Murray's impeccable Tory credentials, and the handsome format and high price of the work, indicated both the readership to which it was directed and also its social and political respectability. The process of dialogue between Lyell and his contemporaries began as soon as the first volume was published, as it generated reviews in a wide range of periodicals, and of course much unpublished comment too. (It would have had a much wider impact beyond Britain, if Prévost had not become embroiled in the politics of the July Revolution, abandoning his plan to publish an edition in the international language of the sciences.) British geologists generally admired Lyell's demonstration of the surprising power of actual causes, but continued to doubt whether they were adequate to account for *all* the traces of apparent catastrophes – particularly the most recent – however generous the time allowed. Reviewers representing a broader public gave it a more mixed reception, depending largely on their position on religious and political issues (Secord, 1997). Some welcomed its application to geology of the same kind of naturalistic rationality that was proving so successful in other spheres. Others were alarmed at its apparent implications for traditional ways of understanding the relation between the human and the divine, and above all between the natural and the social; it should not be forgotten that the work was published at the height of the political unrest surrounding the passing of the great Reform Act of 1832.

Reviewing the second volume in the *Quarterly*, Whewell (1832) coined the terms 'uniformitarians' and 'catastrophists' for what he called 'two opposing sects' in geology, predicting correctly that Lyell would have a harder task than he imagined to persuade the sceptics on the other side. But in fact there were few if any strict uniformitarians apart from Lyell himself; and anyway those famous labels confused the issues. Lyell had subtly amalgamated several distinct meanings of 'uniformity', about some of which there was little argument; and his critics were less concerned at his rejection of catastrophes than at his denial of any direction to geohistory (Rudwick, 1971). It was the latter that aroused the most vigorous criticism among geologists. Henry De la Beche epitomised their opposition in his famous and widely circulated cartoon of a 'Professor Ichthyosaurus' lecturing to a reptilian audience on a human skull (Fig. 2). This ridiculed Lyell's ideas by turning his own verbal thought-experiment into visual form: De la Beche portrayed a distant post-human *future*, on the next round of Lyell's cycle of directionless geohistory (Rudwick, 1975, 1992). Lyell's attempt to explain away the increasingly solid evidence for an unrepeated sequence of distinctive geohistorical periods seemed utterly unconvincing to geologists, even to those such as Scrope who were most sympathetic to his project (Scrope, 1830). Unlike Lyell's neo-Huttonian model of a steady-state earth, however, his reconstruction of the Tertiary portion of geohistory was generally admired, and was prominent in the citation for the Royal Medal that the Royal Society awarded him in 1834.

Lyell began to respond to his critics promptly, even before the *Principles* was complete. In a polemical introduction to the third volume, he again set out the reasons for his rejection of any catastrophes greater in intensity than those authenticated by reliable human history. He accused his opponents of arbitrarily 'cutting the Gordian knot' when they had recourse to catastrophes, instead of engaging in the hard work of 'patiently untying' it; he thereby claimed the high ground morally as well as methodologically. But by this time he was mainly concerned to bring out a new edition of the whole work. The first edition (and the so-called 'second', reprints of the first two volumes) had sold well, by the standards of a large work of non-fiction, so Murray brought out the next (the 'third' edition, 1834) in the same cheap format as his series of popular scientific books, making it accessible to a much wider range of readers (Secord, 1997). It was full of small

Fig. 2. Henry De la Beche's lithographed cartoon (1830) showing a 'Professor Ichthyosaurus' lecturing an audience of other Liassic reptiles on the subject of a fossil human skull (Lyell had just been appointed part-time professor of geology at the new King's College in London). The cartoon was designed to ridicule Lyell's notion of a cyclic pattern of geohistory, as expounded in the *Principles*, in which the Jurassic past – or something much like it – would eventually recur in the distant post-human future.

but cumulatively important revisions, many of them made in the light of the criticisms the work had received both publicly and privately. In the subtitle, for example, Lyell's confident 'attempt' to explain everything in terms of actual causes became more modestly an 'inquiry into how far' such explanations could go. In this more popular format the work sold well enough for Murray to call for two more editions in quick succession, incorporating further modifications in the light of new research, including Lyell's own.

However, by the later 1830s Lyell's project was running into serious problems. He had failed to convince highly competent colleagues such as Scrope (1830) and De la Beche (1834) that the appearance of directionality in the geohistorical record was an artefact of differential preservation, and his neo-Huttonian steady-state theory was therefore threatened with a total collapse in plausibility

(Bartholomew, 1976). Even his model of uniform faunal change, on which his reconstruction of the Tertiary depended, was being undermined by the divergent opinions of the taxonomists, on whose expertise his quantitative estimates depended.

It was therefore no coincidence that just at this time Lyell decided to recast his work in a radically different form. In 1838 Murray published his *Elements of Geology* as a small and inexpensive volume, summarizing his interpretation of geohistory for the general public. Its diagrammatic frontispiece epitomized his claim that all the main classes of rocks had been produced – and all the agencies responsible had been active – at every period of geohistory (Fig. 3). The volume focussed on the rocks and fossils themselves, which Lyell, like his readers, regarded as the subject-matter of 'Geology proper'. Conversely, the next (sixth) edition of the *Principles* (1840) was transformed into a treatise on actual causes. In effect it was an enlarged and updated version of the first two volumes of the original work (and some theoretical chapters from the third); in the words of the subtitle – changed once more – it was on 'modern changes', now considered merely as 'illustrative of geology'. Lyell planned a parallel treatise on the Tertiary formations and their faunas, but that work never appeared; instead, the second edition of the *Elements* (1841) was greatly enlarged, partly to incorporate a more adequate summary of his interpretation of the Tertiary. So a decade after he launched his work, Lyell's argument had been split into two distinct parts: a treatise on actual causes and a more popular work on geohistory. The unitary conception of the original *Principles* had been lost.

In the years that followed, Lyell's insistence on the explanatory power of actual causes came increasingly to be taken for granted by other geologists: his repertoire of persuasive examples seeped into the ordinary practice of others, so that on many (though not all) issues most geologists became in effect Lyellians. On the other hand, his 'uniformitarian' geohistory remained deeply implausible. Indeed it became more so, as new discoveries cumulatively reinforced the sense of directionality (Bartholomew, 1976). For example, even allowing for the imperfection of the record – which geologists now conceded, though not in Lyell's extreme form – the resolution of the famous Devonian and Cambro-Silurian controversies (Rudwick, 1985; Secord, 1986) greatly clarified the distinctive character of the earlier part of the fossil record. So Lyell's continued denial of the directionality of geohistory left him isolated in his own community (Bartholomew, 1979). Underlying his dogged reluctance to concede directionality – as his private notebooks have revealed (Wilson, 1970) – was Lyell's continuing repugnance at the implications of evolution for human dignity; for a directional fossil record was now being used by others as the strongest evidence for evolution, and Lyell knew

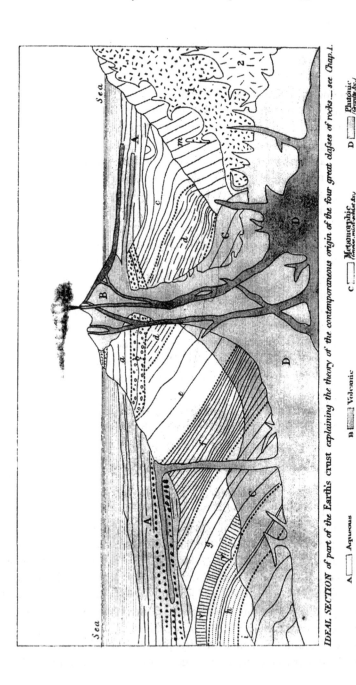

IDEAL SECTION *of part of the Earth's crust explaining the theory of the contemporaneous origin of the four great classes of rocks—see Chap. I.*

A ☐ Aqueous   B ▓ Volcanic   C ☰ Metamorphic &c,   D ☐ Plutonic; /Granite &c./

Fig. 3.   The frontispiece of Lyell's *Elements of Geology* (1838). This engraved and hand-coloured 'ideal section' through the earth's crust illustrated Lyell's claim that *all* the main classes of rock (A, Aqueous; B, Volcanic; C, Metamorphic; D, Plutonic) are still being formed in the present world; earlier products of the same processes are also shown (a–i, Aqueous [in retrospective order]; m, Metamorphic; v, Volcanic; 1, 2, Plutonic).

that Darwin, his closest geological ally, was working privately to establish evolution as a respectable and persuasive scientific theory. After Darwin published his theory in the *Origin of Species* (1859), he was deeply disappointed that Lyell failed to support him unambiguously. In the *Antiquity of Man* (1863), a work occasioned by the coincidental discovery of clear evidence of prehistoric human life in the Pleistocene, Lyell did review the case for an evolutionary interpretation of geohistory, but hardly announced his own conversion in the ringing tones that Darwin wanted (Bynum, 1984). Only in the next (tenth) edition of the *Principles* (1868) did Lyell at last concede the case for evolution and for the directionality of the fossil record. This entailed a major recasting of the argument of the book, at least in its treatment of actual causes in the organic world. But the work continued to sell well, and Lyell saw a twelfth edition through the press shortly before his death in 1875. The *Elements* ended its equally successful career as a textbook explicitly designed for students, having become a set book in the new regime of Victorian examinations.

Conclusions

In retrospect, Lyell's *Principles* may seem paradoxical. On the one hand, its eloquent argument for the adequacy of actual causes in geological explanation was largely successful, by the criterion most appropriate in the sciences, that of collective amnesia. The systematic use of physical and biological processes observable in the present world came to be taken for granted by geologists as the necessary first step towards interpreting the always fragmentary traces of past geohistorical events; and later generations of geologists were often unaware of Lyell's role. Likewise his persuasive explanatory use of a timescale of humanly unimaginable magnitude came to be taken for granted in the everyday work of geologists; Lyell showed them the implications in practice of what they already conceded in principle (and that practice was scarcely affected either by Kelvin's subsequent restriction of the timescale or by its still later expansion after the discovery of radioactivity).

On the other hand, Lyell's claim that his methods entailed a denial of any kind of directionality in geohistory became increasingly implausible, and was abandoned eventually even by Lyell himself. Likewise his principled rejection of any explanations invoking causal processes more intense than those observed in human history became increasingly a straitjacket on legitimate theorizing. Even in his own lifetime, for example, Lyell's stance made it difficult for geologists to recognize the full severity of the Pleistocene glaciations; and a century later Lyell was still being invoked to justify the dogmatic dismissal of any conjectures about episodes of mass extinction, let alone their possibly extra-terrestrial causation.

However, such a two-edged legacy is just what we should expect from even the greatest scientists: like lesser mortals, they do not get everything right, nor should we expect them to. In fact the history of Lyell's *Principles* is a good illustration of the way that true progress in the earth sciences is generally the product of a synthesis between initially opposed positions (Rudwick, 1985). In retrospect, we can see that our modern understanding of geohistory and its causal explanation is a synthesis between Lyell's approach and that of his critics. We inherit the product of nineteenth-century debates, not least the famously vigorous arguments that took place at the Geological Society.

References

Note. The first edition of the *Principles* (1830–33) is now easily accessible and inexpensive, both as a facsimile reprint of the entire three-volume work and in a sensitively abridged one-volume format. The full reprint (Lyell, 1990–91) has an introduction analysing Lyell's argument, and a bibliography of Lyell's own sources, listing in full the items that he himself often cited obscurely in abbreviated form (Rudwick, 1990–91). The abridged reprint (Lyell, 1997) has a valuable introductory essay, stressing particularly the context of the work in British intellectual and cultural history (Secord, 1997). Copies of editions from late in Lyell's life are still relatively easy to find on the antiquarian market, but it should be remembered that they are a radically different work.

BARTHOLOMEW, M. 1973. Lyell and evolution: an account of Lyell's response to the prospect of an evolutionary ancestry for Man. *British Journal for the History of Science,* **6**, 261–303.

—. 1976. The non-progression of non-progression: two responses to Lyell's doctrine. *British Journal for the History of Science,* **9**, 166–174.

—. 1979. The singularity of Lyell. *History of Science,* **17**, 276–293.

BORK, K.B. 1990. Constant Prévost (1787–1856): the life and contributions of a French uniformitarian. *Journal of Geological Education,* **38**, 21–27.

BROCCHI, G.B. 1814. *Conchiologia fossile subapennina con osservazione geologiche sugli Apennini e sul suolo adiacente.* 2 vols. Stamperia reale, Milano.

BUCKLAND, W. 1820. *Vindiciae geologicae; or the connexion of geology with religion explained, in an inaugural lecture...* W. Buckland, Oxford.

—. 1822. Account of an assemblage of fossil teeth and bones belonging to extinct species of elephant, rhinceros, hippopotamus, and hyaena, and some animals discovered in a cave at Kirkdale, near Kirkby Moorside, Yorkshire. *Philosophical Transactions of the Royal Society,* **122,** 171–236.

—. 1823. *Reliquiae diluvianae; or, observations on the organic remains contained in caves, fissures, and diluvial gravel, and on other geological phenomena, attesting to the action of an universal deluge.* Murray, London.

BUFFON, G. DE. 1778. Des époques de la nature. *In:* BUFFON, G. DE. *Histoire naturelle,* Supplément **5**. Imprimérie royale, Paris, 1–254.

BYNUM, W.F. 1984. Charles Lyell's *Antiquity of Man* and its critics. *Journal of the History of Biology,* **17,** 153–187.

COLEMAN, W. 1962. Lyell and the "reality" of species, 1830–1833. *Isis,* **53,** 325–338.

CORSI, P. 1978. The importance of French transformist ideas for the second volume of Lyell's *Principles of Geology. British Journal for the History of Science,* **11,** 221–244.

CUVIER, G. 1812. *Recherches sur les ossemens fossiles de quadrupèdes, où l'on rétablit les caractères de plusieurs espèces d'animaux que les révolutions du globe paroissent avoir détruites.* 4 vols. Déterville, Paris.

— & BRONGNIART, A. 1811. Essai sur la géographie minéralogique des environs de Paris. *Mémoires de la Classe des Sciences Mathématiques et Physiques de l'Institut Impérial de France,* vol. for 1810, 1–278.

DARWIN, C. 1859. *On the Origin of Species by Means of Natural Selection, or the Preservation of Favoured Races in the Struggle for Life.* Murray, London.

DE LA BECHE, H.T. 1834. *Researches in Theoretical Geology.* Knight, London.

DESMOND, A. 1989. *The Politics of Evolution: Morphology, Medicine, and Reform in Radical London.* University of Chicago Press, Chicago.

GOULD, S.J. 1987. *Time's Arrow, Time's Cycle.* Harvard University Press, Cambridge, Mass.

HALL, J. 1814. On the revolutions of the earth's surface. *Transactions of the Royal Society of Edinburgh,* **7,** 139–211.

HAMM, E.P. 1993. Bureaucratic *Statistik* or actualism? K.E.A. von Hoff's *History* and the history of geology. *History of Science,* **31,** 151–176.

HOFF, K.E.A. von. 1822–34. *Geschichte der durch Überlieferung nachgewiesenen natürlichen Veränderungen der Erdoberfläche: ein Versuch.* 3 vols. Perthes, Gotha.

HORNER, L. 1816. [Review of] *Conchiologia fossile Subapennina...* di G. Brocchi, Milano, 1814. *Edinburgh Review,* **26,** 156–180.

HUMBOLDT, A. von. 1817. Des lignes isothermes et de la distribution de la chaleur sur le globe. *Mémoires de la physique et de la chimie de la Société d'Arceuil,* **3,** 462–602.

HUTTON, J. 1788. Theory of the earth; or an investigation of the laws observable in the composition, dissolution and restoration of the land upon the globe. *Transactions of the Royal Society of Edinburgh,* **1**, 209–304.

—. 1795. *Theory of the Earth, with Proofs and Illustrations.* 2 vols. Creech, Edinburgh.

JORIO, A. DE. 1820. *Ricerche sul Tempio di Serapide, in Puzzuoli.* Società Filomatica, Napoli.

LAUDAN, R. 1982. The role of methodology in Lyell's science. *Studies in the History and Philosophy of Science,* **13**, 215–249.

—. 1987. *From Mineralogy to Geology: The Foundations of a Science, 1650–1830.* University of Chicago Press, Chicago.

LYELL, C. 1826. On a recent formation of freshwater limestone in Forfarshire, and on some recent deposits of freshwater marl. *Transactions of the Geological Society of London,* 2nd ser., **2**, 72–96.

—. 1827. [Review of] Memoir on the geology of central France, including the volcanic formations of Auvergne, the Velay, and the Vivarrais, by G.P. Scrope, London. 1827. *Quarterly Review,* **36**, 437–483.

—. 1830–33. *Principles of Geology, Being an Attempt to Explain the Former Changes of the Earth's Surface, by Reference to Causes Now in Operation.* 3 vols. Murray, London.

—. 1834. *Principles of Geology, Being an Inquiry How Far the Former Changes of the Earth's Surface are Referable to Causes Now in Operation.* 3rd edn., 4 vols. Murray, London.

—. 1838. *Elements of Geology.* Murray, London.

—. 1840. *Principles of Geology, or, the Modern Changes of the Earth and Its Inhabitants Considered as Illustrative of Geology.* 6th edn., 3 vols. Murray, London.

—. 1841. *Elements of Geology.* 2nd edn., 2 vols. Murray, London.

—. 1863. *The Geological Evidences of the Antiquity of Man with Remarks on Theories of the Origin of Species by Variation.* Murray, London.

—. 1868. *Principles of Geology, or, the Modern Changes of the Earth and Its Inhabitants Considered as Illustrative of Geology.* 10th edn., 2 vols. Murray, London.

—. 1881. *Life Letters and Journals of Sir Charles Lyell, Bart.* [ed. K. Lyell]. 2 vols. Murray, London.

—. 1990–91. *Principles of Geology* (Facsimile reprint of 1st edn.), 3 vols. University of Chicago Press, Chicago.

—. 1997. *Principles of Geology* (abridged from 1st edn.). Penguin, London.

MCCARTNEY, P.J. 1976. Charles Lyell and G.B. Brocchi: a study in comparative historiography. *British Journal for the History of Science*, **9**, 177–189.

MARINI, P. (ed.). 1987. *L'opera scientifica di Giambattista Brocchi (1772–1826)*. Città di Bassano del Grappa.

MORRELL, J.B. 1976. London institutions and Lyell's career, 1820–41. *British Journal for the History of Science*, **9**, 132–146.

OSPOVAT, D. 1977. Lyell's theory of climate. *Journal of the History of Biology*, **10**, 317–339.

PLAYFAIR, J. 1802. *Illustrations of the Huttonian Theory of the Earth*. Creech, Edinburgh.

PORTER, R. 1976. Charles Lyell and the principles of the history of geology. *British Journal for the History of Science*, **9**, 91–103.

—. 1982. Charles Lyell: the public and private faces of science. *Janus*, **69**, 29–50.

ROGER, J. 1989. *Buffon: un philosophe au Jardin du Roi*. Fayard, Paris.

RUDWICK, M.J.S. 1969. Lyell on Etna, and the antiquity of the earth. *In*: SCHNEER, C.J. (ed.) *Toward a History of Geology*, M.I.T. Press, Cambridge, Mass., 288–304.

—. 1971. Uniformity and progression: reflections on the structure of geological theory in the age of Lyell. *In*: D.H.D. ROLLER (ed.) *Perspectives in the History of Science and Technology*. Oklahoma University Press, Norman, Okla.

—. 1974. Poulett Scrope on the volcanoes of Auvergne: Lyellian time and political economy. *British Journal for the History of Science*, **7**, 205–242.

—. 1975. Caricature as a source for the history of science: De la Beche's anti-Lyellian sketches of 1831. *Isis*, **66**, 534–560.

—. 1977. Historical analogies in the early geological work of Charles Lyell. *Janus*, **64**, 89–107.

—. 1978. Charles Lyell's dream of a statistical palaeontology. *Palaeontology*, **21**, 225–244.

—. 1985. *The Great Devonian Controversy: The Shaping of Scientific Knowledge among Gentlemanly Specialists*. University of Chicago Press, Chicago.

—. 1990–91. Introduction [and] Bibliography of Lyell's sources. *In*: LYELL, C. *Principles of Geology* (Facsimile reprint). University of Chicago Press, Chicago, vol. 1, vii–lxviii; vol. 3, Appendices, 113–160.

—. 1992. *Scenes from Deep Time: Early Pictorial Representations of the Prehistoric World*. University of Chicago Press, Chicago.

—. 1996. Cuvier and Brongniart, William Smith, and the reconstruction of geohistory. *Earth Sciences History*, **15**, 25–36.

—. 1997. *Georges Cuvier, Fossil Bones, and Geological Catastrophes: New*

Translations and Interpretations of the Primary Texts. University of Chicago Press, Chicago.

RUPKE, N.A. 1983. *The Great Chain of History: William Buckland and the English School of Geology (1814–1849).* Clarendon, Oxford.

SCROPE, G.P. 1825. *Considerations on Volcanos ... Leading to the Establishment of a New Theory of the Earth.* Phillips, London.

—. 1827. *Memoir on the Geology of Central France, Including the Volcanic Formations of Auvergne, the Velay and the Vivarrais.* 2 vols. Longmans, London.

—. 1830. [Review of] *Principles of Geology* by Charles Lyell, London, 1830. *Quarterly Review,* **43**, 411–469.

SECORD, J.A. 1986. *Controversy in Victorian Geology: The Cambrian-Silurian Dispute.* Princeton University Press, Princeton.

—. 1997. Introduction. *In*: LYELL, C. *Principles of Geology* (abridged). Penguin, London.

WHEWELL, W. 1832. [Review of] *Principles of Geology* by Charles Lyell, vol. 2. London, 1832. *Quarterly Review,* **47**, 103–132.

WILSON, L.G. (ed.). 1970. *Sir Charles Lyell's Scientific Journals on the Species Question.* Yale University Press, New Haven.

—. 1972. *Charles Lyell, the Years to 1841: The Revolution in Geology.* Yale University Press, New Haven.

III

POULETT SCROPE ON THE VOLCANOES OF AUVERGNE: LYELLIAN TIME AND POLITICAL ECONOMY

> Every step we take in its [geology's] pursuit forces us to make
> almost unlimited drafts upon antiquity. The leading idea
> which is present in all our researches, and which accompanies
> every fresh observation, the sound of which to the ear of the
> student of Nature seems continually echoed from every part
> of her works, is—
>
> Time! — Time! — Time!
>
> (G. Poulett Scrope, 1827)[1]

I

EARLY in 1826, at the age of 28, Charles Lyell began writing the first
of a series of articles for J. G. Lockhart, the new editor of the *Quarterly
review*.[2] These articles gave him his first opportunity to express to the
educated public his views on the state of science in general, and of geology
in particular, in English society. According to the convention of the
Quarterly, each article was nominally a review of one or more recently
published works, but like other reviewers Lyell clearly chose them as
'pegs' on which to hang his own arguments. In content, the articles form
a kind of 'gradualistic' series, rather like his own later interpretations of
geological phenomena. At one end of the series (though published third)
was an essay on the place of science in general in English university
education.[3] Another article (the first to appear in print) focused on some
of the English institutions specifically devoted to science.[4] Here there was
a hint that the need for reform in the place of science in English society
was not unrelated to a similar need for reform in Lyell's chosen branch of
science. The next article enlarged on this hint by examining the publica-

* Centrum Algemene Vorming, Vrije Universiteit, De Boelelaan 1081, Amsterdam,
Netherlands. This article is based on historical fieldwork in the Massif Central in 1967 and 1968.
On the latter occasion I had the pleasure and benefit of the company of Dr Kenneth Taylor,
who helped me to understand the eighteenth-century background of the debate I discuss here.
I am also indebted to the University of Connecticut for inviting me as a Visiting Professor in
1972, when I was able to write a first draft of this article, and to my former colleagues at Cambridge for making possible the period of leave in which the work was completed. I am grateful
to Dr Jonathan Hodge and Mr Roy Porter for some very helpful comments on the article.
    [1] G. Poulett Scrope, *Memoir on the geology of central France, including the volcanic formations
of Auvergne, the Velay and the Vivarais* (London, 1827), p. 165; cited hereafter as *Memoir*. The
second edition, *The geology and extinct volcanoes of central France* (London, 1858), is substantially
altered.
    [2] Leonard G. Wilson, *Charles Lyell. The years to 1841: the revolution in geology* (New Haven
and London, 1972), pp. 143–73, gives an account of the writings of these articles.
    [3] [Charles Lyell], 'Art. VIII' [Review of works on university education], *Quarterly review*,
xxxvi (1827), 216–68.
    [4] [Charles Lyell], 'Art. VIII' [Review of works published by scientific institutions],
Quarterly review, xxxiv (1826), 153–79,

tions of the Geological Society of London, on which Lyell had recently served as Secretary.[5] This essay expressed for the first time in a general context Lyell's characteristic emphasis on the need for actualistic comparison between present and past. Finally, what he needed to complete the series was an article in which he could show in detail the positive explanatory advantages of following this method in geology. The 'peg' which he chose for this purpose was a single work, George Poulett Scrope's *Memoir on the geology of central France*.[6]

Scrope's work proved in fact to be far more than a mere 'peg'. It is clear that Lyell was profoundly impressed by its method and its conclusions, and perhaps particularly excited by its magnificent coloured panoramic illustrations.[7] In the following spring (1828) he went to the Massif Central to see for himself the phenomena that Scrope had described. That fieldwork was so fruitful that Lyell pursued his train of thought by extending the expedition far beyond his original plan, into Italy and Sicily. He later referred to it as 'my tour which made me what I am in theoretical geology'.[8] On his return to England he abandoned his earlier intention to write a short popular book on geology, and set to work on what became his great three-volume treatise, the *Principles of geology*.[9]

The place of the *Memoir* in the context of Lyell's research should be sufficient to suggest that Scrope's geology deserves more historical attention than it has hitherto received. In this article I shall not attempt any comprehensive analysis of Scrope's work. Instead, I shall concentrate on certain themes that seem particularly important for understanding the conceptual foundations of geological thought at this period. I shall suggest how they were related to the thought of one representative opponent of Scrope's (Charles Daubeny) as well as to that of his most powerful ally (Charles Lyell). In conclusion, I shall suggest that some of Scrope's most distinctive themes find their roots in the work that he was undertaking at the same period in an apparently unrelated field outside natural science altogether—namely, in political economy.

II

Scrope was Lyell's almost exact contemporary. He had gone first to Oxford, but had soon transferred to Cambridge, where his interest in geology was stimulated by Adam Sedgwick and John Henslow (who

---

[5] [Charles Lyell], 'Art. IX. [Review of] *Transactions of the Geological Society of London*, vol. i. 2d series, London. 1824', *Quarterly review*, xxxiv (1826), 507–40.

[6] [Charles Lyell], 'Art. IV. [Review of] *Memoir on the geology of central France* . . . By G. P. Scrope, F.R.S., F.G.S., London, 1827', *Quarterly review*, xxxvi (1827), 437–83.

[7] The illustrations form a separate volume of the *Memoir*. For his panoramic sketches Scrope used a system of *geological* colouring, which was later adopted by Lyell for the frontispieces of volumes 2 and 3 of the *Principles of geology*. In the second edition of the *Memoir* the illustrations are reduced in size, uncoloured, and much less striking.

[8] Mrs Lyell, *Life letters and journals of Sir Charles Lyell, Bart.* (2 vols., London, 1881), i. 355.

[9] Charles Lyell, *Principles of geology, being an attempt to explain the former changes of the earth's surface by reference to causes now in operation* (3 vols., London, 1830–3).

were to influence the young Charles Darwin in the same way a decade later). In 1817–18 he wintered in Naples with his parents and witnessed an eruption of Vesuvius. This turned his interests towards the geology of volcanoes, and in 1819–20 he returned to Italy and studied a wide range of volcanic phenomena at first hand. He was therefore well prepared to apply actualistic comparisons to the extinct French volcanoes when he travelled to the Massif Central in the summer of 1821.[10]

The volcanoes of central France had been 'classic' geological phenomena since the mid-eighteenth century, when Guettard discovered lava-flows and well-preserved volcanic cones—against all contemporary expectations—in the hills around Puy de Dôme, above the town of Clermont in the province of Auvergne.[11] Desmarest had later used evidence from the same area to argue forcefully that the common but puzzling rock basalt was volcanic in origin—a lava—and not sedimentary as Werner believed.[12] Faujas de St Fond soon afterwards gave a detailed account of similar volcanic phenomena in the provinces of Velay and Vivarais to the south.[13] On the eve of the French Revolution the Auvergnat naturalist Montlosier published an important account of the volcanoes of his native countryside, extending Desmarest's pioneer interpretation of them.[14] Yet, despite all this work, the spectacular nature of the volcanoes of central France was not well known in England before the outbreak of war made first-hand observation virtually impossible.

After peace returned to Europe in 1815, one of the first English men of science to visit the volcanoes was Charles Daubeny. While studying medicine, Daubeny had attended Buckland's geological lectures at Oxford and Jameson's at Edinburgh. In 1819, at the age of 24, he travelled extensively through the Massif Central; and on his return to Oxford he wrote up his results in the form of two long letters, emphasizing the importance of the area and the lack of any description of it in English. Daubeny's letters were published in Jameson's *Edinburgh new philosophical journal*—at that time one of the main media for geology in Britain—only a few months before Scrope went to France.[15] It is almost inconceivable

[10] *Memoir*, p. vii.

[11] Jean Étienne Guettard, 'Mémoire sur quelques montagnes de la France qui ont été des volcans', *Mémoires de l'Académie Royale des Sciences*, vol. for 1752 (1756), 27–59. Gavin De Beer, 'The volcanoes of Auvergne', *Annals of science*, xviii (1964), 49–61.

[12] Desmarest, 'Mémoire sur l'origine et la nature du Basalte à grandes colonnes polygones, déterminées par l'histoire naturelle de cette pierre, observée en Auvergne', *Mémoires de l'Académie Royale des Sciences*, vol. for 1771 (1774), 705–75. Kenneth L. Taylor, 'Nicholas Desmarest and geology in the eighteenth century', in Cecil J. Schneer (ed.), *Toward a history of geology* (Cambridge, Mass., 1969), pp. 339–56.

[13] Faujas de Saint-Fond, *Recherches sur les volcans éteints du Vivarais et du Velay* (Grenoble and Paris, 1778).

[14] De Montlosier, *Essai sur la théorie des volcans d'Auvergne* (Paris, 1789; new edn., Clermont, 1802). I have consulted the later edition, which is said to be a corrected re-impression of the original.

[15] Charles Daubeny, 'On the volcanoes of the Auvergne', *Edinburgh new philosophical journal*, iii (1820), 359–67; iv (1821), 89–97, 300–15. The last part appeared in the issue for April 1821; Scrope arrived at Clermont early in June.

that Scrope was unaware of them. Possibly they were the immediate cause of his decision to study the volcanoes for himself. Certainly his conclusions contrasted sharply with Daubeny's in certain fundamental respects, so that it seems likely that Daubeny's work provided at least the stimulus of conclusions against which Scrope could react.

Daubeny's main concern, arising from the earlier 'Neptunist-Vulcanist' debates, had been to check the alleged similarity between the undoubted lavas in Auvergne and the much older basalts or 'trap' rocks of Scotland.[16] Much of his article was therefore concerned with the chemical constitution of the French rocks, but he concluded that field-evidence (i.e. association with other volcanic phenomena), rather than chemical petrology, must be decisive in settling the origin of such rocks. In Auvergne some of the basalts were certainly volcanic, since, as Guettard had first discovered, they clearly emerged as lava-flows from unmistakable cones of volcanic debris. Others must also be volcanic, as Desmarest had first argued, although subsequent erosion had left them on the tops of hills, and no trace of their cones remained. Daubeny termed these two classes of volcanic rocks 'modern' and 'ancient' respectively, because they dated from after and before the erosion of the present valleys.

Taking examples which Daubeny derived from Montlosier (see Figure 1), one flow had emerged from two well-preserved volcanic cones (Puy de la Vache and Puy de la Solas) south of Puy de Dôme,[17] and had poured into an already existing valley, flowing down it to the east for another 10 km. Where it entered the valley it had dammed the river and impounded a new lake (Lac d'Aidat), which now drained over the 'parapet' of lava. The next valley to the north was likewise occupied by a long narrow flow of lava from an adjacent cone. Such volcanic rocks obviously dated from after the erosion of the present valleys. Between these two valleys, on the other hand, was a long narrow plateau (La Serre) which also sloped gently to the east. This was capped by a strip of basalt, which outcropped in small cliffs around the edge of the plateau. Further north was a still more isolated hill (Gergovia) similarly capped with basalt. These basalts obviously dated from before the erosion of the present valleys.[18]

Daubeny referred to the 'modern' lavas as being 'more correctly speaking, the post-diluvian lavas'.[19] When later he rewrote his article to

---

[16] His main direct source, which referred to earlier work in Auvergne, was Ami Boué, 'Short comparison of the volcanic rocks of France with those of a similar nature found in Scotland', *Edinburgh philosophical journal*, ii (1820), 326–32.

[17] In the dialect of the region, the word 'puy' denoted any steep-sided and more or less conical hill. Such hills include cones of volcanic débris (e.g. Puy de la Vache), 'plugs' of solid igneous rock (e.g. Puy de Dôme), and isolated outliers capped with basalt (e.g. Puy Girou).

[18] Both Figure 1 and Figure 2 have been re-drawn from Desmarest's 1823 map, op. cit. (29), because it is clearer and more detailed than Scrope's. There are no substantial differences between the two maps, within the areas shown here.

[19] Daubeny, op cit. (15), pp. 360–1.

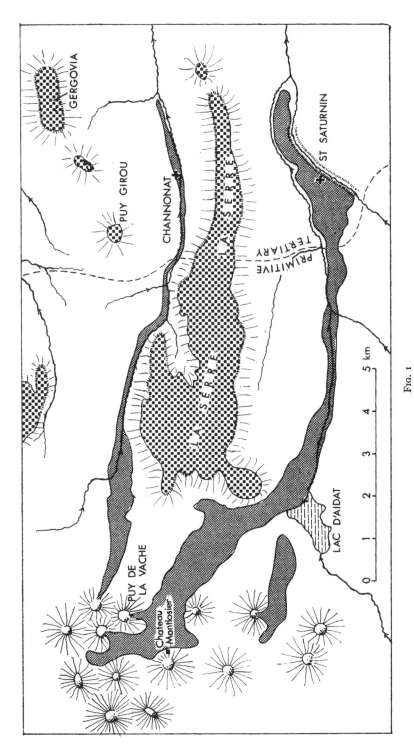

Fig. 1

Map of part of the eastern side of the chain of Puys, in Auvergne, to show the cones of volcanic debris (many with craters preserved), the 'modern' lava-flows in the valleys (finely stippled), and the 'ancient' basalts on the hilltops (coarsely stippled). Based on Desmarest's map (1823) with minor modifications from Scrope's map (1827).

incorporate it into his published lectures on volcanoes, he made this point more explicit by re-naming the 'ancient' and 'modern' eruptions of central France 'ante-diluvial' and 'post-diluvial' respectively.[20] It is not surprising that Daubeny assimilated his observations in this way into Buckland's diluvial theory, for he and Buckland were in close contact as members of the small circle of Oxford dons who were interested in the natural sciences. Before going to France, he may have heard Buckland enunciate the theory in his inaugural lecture as Reader in Geology in May 1819; and anyway the lecture was published in Oxford on his return, shortly before he wrote his letters to Jameson.[21] It is clear from a slightly later article that he agreed with Buckland at least to the extent of attributing the erosion of valleys (in general, not only in Auvergne) to 'the simultaneous operation of one general cause'. He may have gone further with Buckland, in identifying this violent event with the Flood of scripture; but if so, he was careful to point out that his use of diluvial terminology' was not dependent on that identification.[22] All that the terms implied was that there had been a 'Deluge' of some kind—a violent transient inundation that had scoured out the present valleys. A belief in such an event was of course a commonplace of geological theorizing at this time, and was not necessarily linked to a concern with the historicity of scripture.[23] Daubeny may have known, for example, that Dolomieu had approved the distinction between 'anciens' and 'nouveaux' lavas in Auvergne, although Dolomieu's conception of the event that had scoured out the valleys was even further than Buckland's from any literalistic reading of Genesis.[24]

Daubeny's confidence in the reality of the distinction between 'ancient' and 'modern' eruptions would have been strengthened still further by other authoritative precedents. Desmarest had first seen that even the older basalts must have flowed down what were *then* the floors of valleys, since any lava must necessarily flow down whatever slope is adjacent to its source, and continue down any valley there may be. But

---

[20] Charles Daubeny, *A description of active and extinct volcanoes: with remarks on their origin, their chemical phaenomena, and the character of their products, as determined by the condition of the earth during the period of their formation* (London, 1826). Daubeny was elected Professor of Chemistry at Oxford in 1822, in succession to John Kidd.

[21] William Buckland, *Vindiciae geologicae: or the connexion of geology with religion explained* (Oxford, 1820).

[22] Daubeny, 'Sketch of the geology of Sicily', *Edinburgh philosophical journal*, xiii (1825), 107–18, 254–69; see p. 264 n.: also op. cit. (20), p. 9.

[23] Leroy E. Page, 'Diluvialism and its critics in Great Britain in the early nineteenth century', in Schneer, op. cit. (12), pp. 257–71.

[24] 'Rapport fait à l'Institut national, par le citoyen Dolomieu, ingénieur des mines, sur ses voyages de l'an V et de l'an VI', *Journal des mines*, vii (an VI, 1798), 385–432. Dolomieu attributed the erosion of valleys to the sudden emergence of the present continents from beneath the sea, an event which, following Deluc, he regarded as geologically recent. But he emphasized that such a conclusion might be correct even if the religious opinions that had first suggested it (i.e. Deluc's version of a diluvial theory) were themselves 'absurdes'. See his 'Mémoire sur les pierres composées et sur les roches', *Observations sur la physique, sur l'histoire naturelle et sur les arts*, xxxix (1791), 374–407; xl (1792), 41–62, 203–18, 372–403; see xl (1792), 41 n.

after an eruption the stream or river flowing down that valley would continue its erosion, generally excavating through the bedrock to one side of the lava (which is generally harder). Given enough time, such an excavation would be deepened into a new valley, which would leave the lava as a terrace at a gradually increasing elevation above the river bed. If another stream was simultaneously eroding a valley on the other side of the lava, the latter would eventually be left as a hard resistant capping to a hill between two adjacent valleys. In other words, Desmarest had realized that the older basalts were identical to the modern lavas in mode of origin. Nevertheless, he considered that the distinction was a real one in *temporal* terms, and that the eruptions dated from two widely separated 'époques'.[25]

Montlosier had developed Desmarest's insight, terming the more recent lavas Nature's 'doctrine publique', since their volcanic origin was easy to see, whereas that of the older basalts was Nature's 'doctrine cachée', for it involved accepting a striking reversal of the present topography: hills had become valleys, and valleys hills.[26] Like Desmarest, he attributed the erosion of the valleys to the action of the present streams and rivers, continued over long periods of time, yet he too considered that the eruptions (and hence the lavas) fell naturally into two temporal classes, i.e. those on hilltops and those on the floors of valleys.

The description of the lavas of Auvergne in terms of two discrete classes was thus *not* a product of Daubeny's diluvial theory of the sudden erosion of the valleys. It seemed equally 'natural' to observers who held a quite different theory of erosion, i.e. as gradual. Daubeny's adoption of a division between 'ancient' and 'modern' lavas should therefore be seen in the first instance as articulating a *common-sense* interpretation of the volcanic rocks. Of course, his confidence in this interpretation would have been strengthened by its integration into a diluvial theory that regarded the two periods of volcanic activity as having been separated by an unusual episode of rapid erosion. But, conversely, it follows that this common-sense distinction between 'ancient' and 'modern' would be unlikely to be broken down except by someone with strong theoretical reasons for actively searching for *intermediate* instances that had not been obvious to previous observers. Scrope, it seems, was such a person.

III

In the preface to his *Memoir* Scrope asserted that when studying the geology of Auvergne he had 'laid down and adhered to a resolution not to open any author who had written on the subject' before seeing the

---

[25] Nicholas Desmarest, 'Extrait d'un mémoire sur la détermination de quelques époques de la nature par les produits de volcans, et sur l'usage de ces époques dans l'étude des volcans', *Observations sur la physique, sur l'histoire naturelle et sur les arts*, xiii (1779), 115–26. See also Taylor, op. cit. (12).

[26] Montlosier, op. cit. (14), p. 23.

phenomena for himself; 'I was, in short,' he said, 'thoroughly determined to form an opinion exclusively my own.'[27] It is hard to tell how genuine this claim may have been, or to what extent Scrope was merely conforming to the 'empiricist' norms that were dominant at that time in English geology. Certainly he borrowed some of his most striking examples and conclusions from Montlosier's book; and it is even possible that, like Lyell and Murchison a few years later,[28] he visited Montlosier in the château he had built in an inhospitable spot among some of the most spectacular volcanic cones (Figure 1). Similarly it is difficult to believe that he constructed his fine geological map of Auvergne single-handed in the course of one field-season, without drawing on Desmarest's maps, though his acknowledgement of Desmarest was less than generous.[29] In any case, by the time he wrote the *Memoir* he was certainly aware of most of the earlier published work. In particular, he seems to have been reacting against Daubeny's conclusions, for at the start of his description of the volcanic rocks he noted that he would reject the alleged distinction between 'ancient' and 'modern' eruptions.[30]

The structure of Scrope's *Memoir* is primarily geographical, but the descriptive sections are interspersed with frequent comments on the significance of the phenomena for a *temporal* reconstruction of events. Thus the varied state of preservation of the more recent volcanic cones, which Daubeny had described in terms of three 'classes' without temporal significance,[31] was interpreted by Scrope in terms of a smoothly graduated temporal *series*. They witnessed to successive eruptions 'often at distant periods' (i.e. 'distant' from each other). In an oblique criticism of the most recent of the three 'époques' that Desmarest had distinguished, he asserted that even these newer volcanoes were 'far from belonging to a single epoch'.[32] Puy Pariou, for example, to the north of Puy de Dôme (see Figure 2), was 'the product of one of the last eruptions' because it had the most perfectly preserved crater.[33] Cones with less well preserved craters, or with no trace of a crater left, were correspondingly older.

Scrope believed that still clearer evidence of relative ages could be derived from the lava-flows and their relation to the present valleys. Borrowing a striking example from Montlosier, he described how two

[27] *Memoir*, p. viii.
[28] Mrs Lyell, op. cit. (8), i. 188.
[29] *Memoir*, p. 36. Desmarest published some small maps to illustrate his 'époques', but the full version of his superb map of Auvergne was published posthumously by his son in 1823 (too late for Scrope to have used it for his fieldwork, but in time for him to have incorporated it into his own, closely similar map): Desmarest, *Carte topographique et minéralogique d'une partie du département du Puy de Dôme dans le ci-devant province d'Auvergne où sont déterminées la marche & les limites des matières fondues & rejettées par les volcans ainsi que les courants anciens & modernes pour servir aux recherches sur l'histoire naturelle des volcans* (Paris, 1823).
[30] *Memoir*, p. 40 n.
[31] Daubeny, op. cit. (15), p. 362.
[32] *Memoir*, p. 50.
[33] Ibid., p. 62, and Pl. IV. See also Pl. [VI] (unnumbered), for the Puy de La Vache crater and its flow, with Lac d'Aidat.

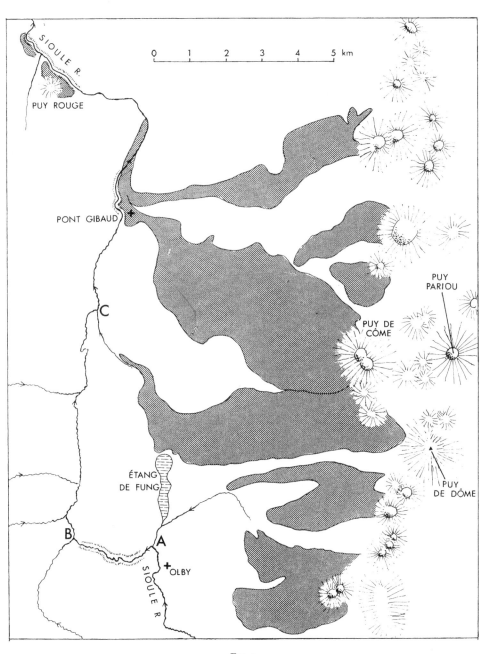

FIG. 2
Map of part of the western side of the chain of Puys, in Auvergne, to show the effects of some 'modern' lava-flows (stippled) on the course of the river Sioule ('ancient' basalts omitted). Based on Desmarest's map (1823) with minor modifications from Scrope's (1827).

FIG. 3

Part of Scrope's drawing (1827) of the cratered volcanic cone above Jaujac (Ardéche). The village is built on a thick lava flow which has been eroded into a deep gorge since its eruption. Examples such as this were crucial to Scrope's argument that the valleys of the Massif Central had been eroded by ordinary rivers over vast intervals of time, and not by any sudden 'diluvial' event.

huge flows from the fairly well-preserved volcanic cone Puy de Come had affected the course of the river Sioule (Figure 2).[34] The more northerly flow had blocked the valley at the site of Pont Gibaud, impounding a lake upstream (like Lac d'Aidat, but since filled in by natural silting). Yet even since the eruption the river had already eroded a small gorge along the edge of the lava. (Montlosier had also noted a similar gorge along the edge of another 'recent' lava at St Saturnin: see Figure 1.) The more southerly flow had blocked the original valley of the Sioule (Figure 2, A–C) so completely that the river had cut a new channel, not along the edge of the lava, but further upstream (from near Olby) through a hill of 'soft alluvial tufa' into a tributary valley to the west (Figure 2, A–B). Another lake (reduced by silting to a semi-artificial pond, the Étang de Fung) had been impounded in the abandoned stretch of the original valley. In other words, in both cases the Sioule had accomplished some erosion even *since* these geologically 'recent' eruptions. This implied that valley erosion had *not* been confined to a period before the 'modern' eruptions: it was still going on.

Having described many of the volcanoes in Daubeny's 'modern' or 'post-diluvial' class, and having indicated that they were not all of the same age, Scrope argued that there was no 'decided line of separation' between these and the 'ancient' or 'ante-diluvial' class.[35] 'One of the first connecting links' was Puy Rouge (top left of Figure 2), where the lava had evidently flowed down the floor of the valley of the Sioule as it then existed; but since that eruption a deep gorge had been eroded between one edge of the lava and the valley side.[36] It could not be argued that the eruption had been 'pre-diluvial' and the erosion of the gorge 'diluvial', because any general and violent inundation would surely have swept away all trace of Puy Rouge itself, which like all the cones was composed of small 'loose scoriae and ashes, which actually let the foot sink ankle-deep in them'.[37] Such a gorge, Scrope argued, therefore 'evinces the immeasurably long continuance of the erosive action, as well as its irresistible power'.[38]

He found even more striking examples of such 'connecting links' among the volcanoes of the former province of Vivarais to the south. Here the river Ardèche and its tributaries had eroded deep gorges through the sides of thick lava-flows and even through the underlying hard 'Primitive' rocks as well. Yet the volcanic cones associated with these eruptions were

---

[34] Ibid., pp. 58–61, and Pl. III. Compare Montlosier, op. cit. (14), pp. 29–37, who termed it 'un des plus beaux traits de géologie que le naturaliste puisse désirer'.

[35] *Memoir*, p. 84.

[36] Ibid., p. 85. Puy Rouge, also known as Puy Chalucet or Chaluzet from the hamlet at its foot, was apparently unknown to Desmarest and is not shown on his map. Dolomieu referred to it, and Scrope described it more fully, though it lay outside the area of his map. I have inserted it on Figure 2 from a modern map.

[37] Ibid., p. 163.

[38] Ibid., p. 85.

still well preserved with craters (Figure 3).[39] Once again, a diluvial interpretation seemed impossible. Scrope even traced the boulders of dark basalt in the river-beds downstream from these gorges, noting how they were gradually reduced in size and abundance by attrition against the harder (and light-coloured) boulders of 'primitive' rocks, until below Aubenas all trace of them disappeared. This was further evidence that the erosion was still continuing. Yet it was obviously very slow by human standards. It followed that the eruptions, although geologically so 'recent', must 'belong to an aera incalculably remote with respect to ourselves'.[40] In other words, although these volcanic cones looked so fresh, they must date from much further back beyond human history than Daubeny had allowed.[41]

The reality of Daubeny's distinction might still be defended, however, by saying that even these eruptions were clearly later than the main erosion of the broad valleys (though earlier than the erosion of the much narrower gorges incised into the valleys). Scrope therefore needed more 'connecting links' to enable him to move gradually from those lava-flows to the 'ancient' basalts that formed the tops of hills. He found 'one of those rare and valuable links' in the long narrow plateau of La Serre (see Figure 1).[42] He interpreted this as a lava-flow which had retained its linear form almost completely, yet which had been left above the valleys that had been eroded subsequently to the north and south of it. This much had been recognized by Desmarest long before. But Scrope saw the variation within the class of plateau-forming basalts as representing a *series* connecting them with the lavas on or near the floors of the valleys.

In Auvergne the main north-south zone of volcanic cones lies near the edge of an upland area composed of 'granite' and other 'primitive' rocks (in modern terms, metamorphic rocks). But to the east is a broad 'basin' (the Limagne) of soft Tertiary limestones. Many of the lavas that flowed eastwards from the cones had crossed the Primitive/Tertiary boundary, and their lower parts had flowed down valleys composed of soft sediments (see Figure 1). Subsequent erosion had therefore been (by geological standards) relatively fast. Even the plateau of La Serre showed at its lower end the first stages of disintegration. The hill of Gergovia to the north, with Puy Girou and another hill nearby, were seen by Scrope as an example of a much later stage in this 'slow but certain transformation'

---

[39] Figure 3 reproduces the right half of Scrope's panoramic Plate XIV. The left half, reproduced as Figure 27 in Wilson, op. cit. (2), shows the gorge more clearly but does not show the volcanic cone which gave the gorge its significance. See also *Memoir*, Pl. XV.

[40] Ibid., p. 154.

[41] It is of some interest to note that modern radiometric dating has established the age of the 'modern' lavas in Auvergne as only slightly greater than Daubeny's estimate, and far less than the age implied by Scrope: material under the Puy de la Vache flow at St Saturnin (Figure 1) has yielded a radiocarbon date of 7,650±350 B.P. (i.e. about 5,700 B.C.). See Aimé Rudel, *Les volcans d'Auvergne* (2nd edn., Clermont, 1963), p. 94. This book contains some striking aerial photographs of the volcanic cones.

[42] *Memoir*, p. 89.

into isolated basalt-capped hills (see Figure 4).[43] The spectacular isolated hills around the town of Le Puy in the former province of Velay were even more striking examples. They might 'yet outlast many of the generations of mankind', Scrope wrote, but they too would be eroded away 'after a time, comparatively brief in the calendar of nature', and all trace of the eruptions would be lost.[44]

Scrope had thus rejected Daubeny's (and other earlier writers') qualitative distinction between 'modern' lava-flows on the floors of valleys and 'ancient' ones on hilltops. He had interpreted varieties within both of Daubeny's 'classes' as members of an unbroken temporal *series*. Taking one limited area alone, as an example, he asserted that the isolated hill of Gergovia, the long plateau of La Serre, and the valley of Channonat between them (Figure 1) represented 'three distinct steps in the process of excavation' (or four, counting some slight erosion since the eruption of the lava in the valley), for each of the lavas must have flowed down what was the floor of the valley at that time. They were '*witnesses*' to different stages in the continuous process of erosion.[45] This contrasted with the diluvialists' assertion that there had been only two steps, distinct in kind, i.e. the diluvial erosion of the broad valleys themselves and the post-diluvial erosion of small narrow gorges within some of the valleys.[46]

Scrope's rejection of the diluvial interpretation was linked explicitly to a methodological viewpoint. It was philosophically unsound to attribute one part of the erosion to one cause (i.e. a Deluge) while accepting some later erosion (i.e. the gorges) as the work of the present rivers. Such recourse to 'a vague and unexampled hypothesis' was, he said,

> in defiance of all the laws of analogical reasoning, by strict adherence to which we can alone hope to obtain the least acquaintance with those operations of Nature's laboratory of which we have not been actual eye-witnesses.[47]

Scrope asserted that, contrary to the diluvialists' opinion, the process of erosion had been absolutely continuous. 'Every link in the chain of proofs is complete,' he wrote in his concluding chapter. Since no other 'gradual and progressive excavating forces' were conceivable, he inferred (like Desmarest and Montlosier before him) that the ordinary action of rain, streams, and rivers had been responsible throughout. This agency was adequate to explain all the observed erosion—'with an unlimited allowance of time'.[48]

With this concept of erosion as gradual and continuous, Scrope was able to replace the qualitative distinction between 'ancient' and 'modern'

43 Ibid., p. 90, and Pl. II.
44 Ibid., pp. 44–5, and Pl. XII.
45 Ibid., p. 46.
46 Ibid., pp. 161–2.
47 Ibid., p. 163.
48 Ibid., p. 162.

with a purely quantitative criterion of relative age: the degree of subsequent erosion, as measured by the present height of any given lava-flow above the nearest '*Thalweg*' or present valley-floor in longitudinal profile. This criterion of relative age was illustrated by one diagram of fundamental theoretical importance (see Figure 4). Seen diagrammatically as though looking south from near Clermont, the various lava-flows on the east side of the zone of volcanic cones could be seen to lie at many different degrees of elevation above the present river-beds, from almost zero in the case of the most recent lava-flows, up to more than 1500 ft in the case of hills like Gergovia. Scrope maintained that they therefore formed 'a natural scale for measuring the duration of the process' of gradual erosion. Only the calibration of this 'natural scale' in terms of years was still required.[49]

There is one important feature of Scrope's 'natural scale' which he did not make explicit but which can be inferred from the text of the *Memoir*, from the panoramic illustrations, and above all from his diagram. He claimed that the elevation of the lava-flows was directly proportional to their age. But this relationship was essentially probabilistic or stochastic in form. Although the rivers that flow eastwards from the Puy de Dôme uplands (i.e. those implied in Figure 4) are all 'graded' to join the north-ward-flowing river Allier (which passes through Pont du Château, Scrope's baseline on his diagram), the valleys in which they flow are not all at exactly the same height. Presumably this would also have been the case at any given time in the past, when a certain valley-floor might have been preserved by the eruption of a lava-flow. I think it is therefore clear that Scrope meant only to assert that in any 'population' of lava-flows in any given area, the *chances* are that age will be directly proportional to elevation.

The probabilistic character of Scrope's temporal sequence is even clearer if it is considered in terms of the 'morphology' of the lava-flows and their gradual disintegration. He had eliminated the ancient-modern distinction by demonstrating an unbroken series not only in terms of elevation but also in terms of the *form* of the lava-flows and of the volcanic cones. Each flow was extruded from a cratered cone in more or less linear form (depending on the underlying topography); but subsequent erosion would gradually eliminate the cone and break up the lava into one or more isolated 'outliers' capping hilltops; and these outliers would gradually be reduced in size and ultimately lost. But the rate at which a given cone and lava-flow passed through this 'morphological' sequence from initial eruption to final destruction would obviously depend on many contingent factors in addition to the sheer lapse of time: for example, the size and composition of the cone, the petrological character of the lava (which affects the rate of decomposition), the width of the flow, the size

[49] Ibid., pp. 46, 162.

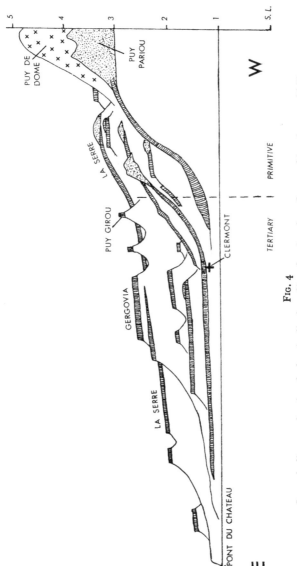

FIG. 4

Scrope's diagram showing the longitudinal profiles of various lava-flows and basalt-capped hills around Clermont, in Auvergne, to illustrate his conception of a 'population' of lava-flows that could be used as a chronological 'natural scale' of erosion. Lava-flows hatched; volcanic cones stippled. Scale of altitude in thousands of feet above sea-level (S.L.). Re-drawn from Scrope's coloured version (1827).

and erosive power of the river which erodes it, and so on. Scrope's diagram shows clearly that he was aware of this: for example, the plateau of La Serre, which is preserved almost entire, is much higher and hence presumptively older than other flows that are already more disintegrated into isolated basalt-capped hills (see Figure 4). Hence the stage of dis-integration of the lava-flows, like their elevation, would be proportional to their age as a probabilistic generalization that was valid for the whole 'population' of flows, but not necessarily or exactly for any individual flow taken in isolation.

Apart from this probabilistic component, Scrope's conclusions were less original than his acknowledgements suggest. Montlosier had regarded the ancient-modern distinction only as a convenient generalization. Indeed, he thought the importance of Auvergne lay precisely in the fact that it preserved a series of 'monuments' graded in age, with many 'nuances intermédiaires'.[50] He considered that ordinary streams and rivers were quite capable of eroding the valleys, given 'une infinité de siècles'.[51] He termed the various basalts '*témoins* de la nature' (compare Scrope's '*witnesses*') recording the stages of this slow process, and he thought their age was directly proportional to their elevation.[52] He asserted that the gorge at St Saturnin proved that this erosion was still continuing, and he even suggested that if the present rate of erosion in another similar gorge could be measured, the whole process could be calibrated in years.[53]

However, I am not concerned here with questions of priority. Scrope may have borrowed much from Montlosier, but his knowledge of the earlier work is not sufficient to explain his results. Daubeny too had read Montlosier, yet did not share his conclusions after seeing the volcanoes for himself. There must surely be deeper reasons why Scrope was sensitive to the *kind* of interpretation that Montlosier had suggested, and why he was able to enlarge on the earlier work and present it in a much more persuasive form.

IV

It is not surprising that Lyell reviewed Scrope's *Memoir* enthusiastically for the *Quarterly review*, as an almost perfect example of the style of reasoning that he himself wished to urge on geologists. Scrope had eliminated the ancient-modern distinction, and with it a major source of evidence for the diluvial theory; he had used an 'unlimited allowance of time' to do so; and he had stressed the heuristic value of the actualistic method of analogical comparison between present and past.

[50] Montlosier, op. cit. (14), pp. 112–20: 'on trouve à tous les âges des médailles frappées par la nature, pour attester toutes les gradations de ses travaux et de sa marche' (p. 118).
[51] Ibid., p. 95.
[52] Ibid., p. 118.
[53] Ibid., pp. 120, 43.

But, for all these 'Lyellian' themes, Scrope was far from holding the rigid and even dogmatic conception of 'uniformity' that Lyell was later to expound. Although he rejected any widespread Deluge as an explanation of valleys, he had no compunction about postulating 'catastrophic' events further in the past, when the circumstances suggested an analogy with sudden and violent events known in the present. For example, he interpreted the thick volcanic agglomerates derived from the great volcanic centre of Mont Dore as the products of 'aqueous deluges' or mud-flows like those known to occur occasionally in modern eruptions.54 Likewise he did not think the Massif Central itself had been elevated gradually, but by 'successive convulsions': the 'force' of elevation had acted violently but intermittently, like the volcanic activity to which he believed it was causally connected.55 One such sudden movement, he thought, had been responsible for the final rupture of whatever barriers had enclosed the lakes of the Massif Central, in which the Tertiary sediments (e.g. of the Limagne) had been deposited.

In his own eyes the *Memoir* was simply a detailed illustrative example, an 'appendix' or 'pièce justificative', for a more ambitious enterprise in geological explanation.56 This was his 'new theory of the earth', a preliminary outline of which formed the culmination of his *Considerations on volcanos* (1825).57 In this book be elaborated the viewpoint he had used in writing the *Memoir* in the service of a general interpretation of the history of the earth.

He criticized the common use of 'catastrophes' in geological explanation, not because they had never occurred, but on the methodological grounds that such an explanation generally 'stops further enquiry'. It was counter-heuristic because it prevented the possible discovery of non-catastrophic causes which might in fact be adequate. What had to be avoided in geological explanation was not sudden or violent events as such, but 'recourse to the gratuitous invention of vague and unexampled occurrences, referable to no known law of nature'. What Scrope found repugnant about the diluvial theory, on this level at least, was not that the alleged Deluge was violent but that it was 'unexampled' and had no known natural cause. 'The only legitimate path of geological enquiry,' he argued, 'was to try first to explain the past in terms of known geological processes, 'with the most liberal allowances for all possible variations [i.e. in intensity] and an unlimited series of ages'.58 The proviso about liberality was important, because it enabled him to make full use of violent events in the past wherever the phenomena seemed to demand

---

54 *Memoir*, pp. 101–4.
55 Ibid., pp. 166–8.
56 Ibid., p. x.
57 G. Poulett Scrope, *Considerations on volcanos, the probable causes of their phenomena, the laws which determine their march, the disposition of their products, and their connexion with the present state and past history of the globe: leading to the establishment of a new theory of the earth* (London, 1825).
58 Ibid., pp. iv–vi.

them (e.g. in the elevation of mountains), provided only that such events were similar in *kind* to those observable in the present (e.g. earthquakes). Thus there had been 'catastrophes', especially in the more distant past, but only because 'rare combinations of circumstances sometimes gave a prodigious force' to ordinary processes.[59]

This moderate and flexible actualistic methodology led, in Scrope's view, to a 'theory of the earth' that conceived the planet as having had a broadly directional history. Like other geologists in the 1820s, he attributed this directionality to the gradual cooling which the planet was believed to have undergone. The heat stored in the earth's interior, as a major source of energy for geological processes, had been gradually dissipated, with a consequent gradual diminution in their intensity. Of the process of erosion and deposition, for example, he wrote:

> It has proceeded generally by a lent [i.e. slow] and uniform process, gradually diminishing in energy from the beginning to the present day; but occasionally presenting partial [i.e. localized] crises of excessive turbulence, resulting from accidental combinations of circumstances favourable to the maximum of violence.[60]

Corresponding to this broadly directional history of the earth, Scrope believed that the history of life had also been directional, indeed progressive, in character. Although this aspect was only outlined briefly in the *Considerations*, it is clear that Scrope, like other geologists at the time, saw the history of life as having progressed in parallel, as it were, with the changing character of the environment. It had begun with 'organic beings of simple structure, and a constitution suited to the circumstances under which they were created', that is, while the early oceans were still at a relatively high temperature. Gradually 'new tribes of organised beings' had come into existence; and so the world had eventually arrived at its present state 'by gradual change, wrought by the slow refrigeration of the outer zones of the globe'.[61]

In terms of the analysis that I have outlined elsewhere for geological theories of this period, Scrope's two books (and his later papers and essay-reviews) form a coherent whole. His 'theory of the earth' was *naturalistic* in its metaphysics, *actualistic* in its methodology, *gradualistic* in the rates attributed to geological processes (while allowing that such processes might occasionally generate 'catastrophic' events), and *directionalist* in the overall pattern that it attributed to the history of the earth and of life.[62] It was original chiefly in its persuasive emphasis on the adequacy of actualistic 'causes' operating gradually over long spans of time.

---

59 Ibid., p. xxxi.
60 Ibid., p. 240.
61 Ibid., pp. 237–9.
62 M. J. S. Rudwick, 'Uniformity and progression: reflections on the structure of geological theory in the age of Lyell', in Duane H. D. Roller (ed.), *Perspectives in the history of science and technology* (Norman, Oklahoma, 1971), pp. 209–27.

V

The magnitude of past time had been, as we have seen, a frequent refrain throughout Scrope's *Memoir*. Even the most recent-looking volcanic cones and lava-flows dated from far back beyond all human records; yet their age was negligible, judged by his 'natural scale', compared with the age of a basalt like that of Gergovia. The latter in turn overlay and was younger than the hundreds of feet of finely laminated Tertiary lake-sediments in the Limagne; yet even these were relatively recent in geological terms, compared with the whole known succession of strata. It was with thoughts such as these that Scrope allowed himself the luxury of the purple passage I have quoted at the beginning of this article.

Yet other geologists, and in particular diluvialists such as Daubeny, were not averse in principle to the notion of such vast spans of time. It should no longer be necessary to emphasize that the nineteenth-century diluvialists' conception of the history of the earth was not constrained within the few thousand years of seventeenth-century chronology.[63] Nor was their use of sudden paroxysmal events in geological explanation due primarily to an inadequate appreciation of the magnitude of past time; for in the most recent and therefore clearest case, i.e. the Deluge itself, it was the nature of the phenomena themselves that seemed to demand explanation in terms of some extraordinary event.[64] Only at one point could the diluvial theory be said to have incorporated a constricted time-scale: namely, for the period *since* the Deluge.

Daubeny followed Dolomieu and more recent diluvialists such as Buckland in believing that the most recent 'catastrophe' had happened only a few thousand years ago. But this would have seemed to be confirmed by the astonishingly fresh appearance of some of the volcanic cones in Auvergne. For example, Puy Pariou had a sharp-lipped crater and was almost as well preserved as any of the cones on the flanks of Etna, some of which were known to date from eruptions in historic times. Yet all the eruptions in Auvergne had apparently been pre-historic, since there was no human record of their eruptions. Caesar had fought with Vercingetorix, who was encamped on the natural fortress of Gergovia, but his writing made no mention of any volcanoes in the region. More striking still, the poet Sidonius Apollinaris had had his episcopal palace by Lac d'Aidat, which clearly owed its very existence to one of the most

---

[63] As an early example, Dolomieu, op. cit. (24), thought the most recent 'revolution' (which had excavated the valleys) dated from not more than 10,000 years ago, but that the preceding time had been so immeasurably long that millions of years had been like minutes in a human life. Some such proportion was obvious to all (including, for example, Buckland and Daubeny) who were aware of the vast thickness of geological formations, many of which had clearly been deposited extremely slowly under tranquil conditions.

[64] It was therefore not until the advent of the glacial theory, postulating an extraordinary event in the form of an Ice Age, that the diluvial phenomena began to receive a satisfactory explanation. See M. J. S. Rudwick, 'The glacial theory', *History of Science*, viii (1970), 136–57.

recent eruptions.[65] It was therefore not unreasonable for Daubeny to conclude that the 'modern' eruptions must date from the relatively short period between the Deluge and the Classical period.

This compression of all the 'post-diluvial' eruptions in Auvergne into a very short interval might be regarded as a substantive effect of his diluvial interpretation. It could be argued that this accounts for his failure to see that the varying states of preservation of the 'modern' volcanic cones and lava-flows might represent a temporal *series* of varying degrees of decay and erosion. But no such explanation is adequate to account for Daubeny's precisely analogous 'failure' to see that the varieties of 'ancient' basalts might also represent a similar temporal series, for he was under no limitation with regard to 'ante-diluvial' time.[66]

I suggest, therefore, that this double 'failure' had deeper roots, in Daubeny's understanding of the qualitative nature of geological time. Daubeny often referred to events before and since the Deluge as belonging respectively to the 'former' and 'present' 'order of things'.[67] The casual use of these phrases is as revealing as the phrases themselves. They imply that he saw the history of the earth before and since the Deluge as periods that were distinct in an almost ontological (or, at least, cosmological) sense. Metaphorically speaking, Daubeny regarded the history of the earth as having been like a vast drama, the successive acts of which were separated by the fall of the curtain, bringing one 'order of things' to a close before the start of the next. His belief that the Deluge could be dated (at least approximately) by its attachment to human chronology in no way reduced that cosmological difference, just as the fixing of dates for the providentially significant periods of human history in no way reduced the qualitative distinction between, for example, the Old 'Covenant' and the New.

By drawing that parallel, I have indicated the most probable source of Daubeny's 'periodized' concept of geological history, in the similarly 'periodized' concept embedded in the Christian view of human history.[68] I suggest therefore that Daubeny's theistic belief in a God whose activity had given *meaning* to the successive periods of human history disposed him to see the much longer time-scale of geological history as a similar succes-

[65] Daubeny, op. cit. (20), p. 14.

[66] In Auvergne he would have seen the several hundred feet of finely laminated Tertiary sediments that underlie (and are therefore older than) even the 'ancient' basalts; and he was certainly aware from general geological knowledge that these sediments in turn are all very 'young' by geological standards.

[67] Daubeny, op. cit. (20 and 22). The phrases were of course traditional.

[68] The literature on this point is vast. For an introduction that stresses its application to geology, see Francis C. Haber, *The age of the world. Moses to Darwin* (Baltimore, 1959), and 'The Darwinian revolution in the concept of time', *Studium generale*, xxiv (1971), 289–307; also G. J. Whitrow, 'Reflections on the history of the concept of time', ibid., xxiii (1970), 498–508.

sion of cosmologically distinct periods.[69] In other words, history itself was divided inhomogeneously into periods characterized by different modes of divine activity.

Of such periods, those before and after the creation of Man as a rational moral being would have seemed the most significant for geology; and the Deluge was the nearest event in time to that great divide, of which geological traces could reasonably be anticipated. But this periodization into 'ante-diluvial' and 'post-diluvial' epochs might have been extended in a weaker form to the long ages before Man appeared. Those ages, whatever their length in years, could be seen as 'preparing the earth's surface for the abode of the existing races of animals'[70] and above all for Man, just as the long centuries of history recorded in the Old Testament had had meaning in the light of the New, as periods of preparation for the coming of Christ.

If this interpretation of Daubeny's viewpoint is valid, we might then expect him to have seen events *within* each period in terms of characteristic 'classes', rather than temporal *series*. This is precisely what his interpretation of the Auvergne volcanoes demonstrates. Obviously he believed that time itself had flowed 'homogeneously' in a Newtonian manner from one period to the next—even through the diluvial episode—but this was not incompatible with a tendency to see history as broken into discrete periods of contrasting character.

In attributing the deep structure of Daubeny's thought to his adherence to a traditional Christian interpretation of history, it would be easy to fall into the trap of making Scrope, by contrast, into a Huxleyan agnostic or even a modern sceptic. Clearly he was neither. In a footnote to the passage I have quoted at the beginning of this article, Scrope commented perceptively on the *imaginative* barrier to an acceptance of the vast time-scale that he proposed. But he followed this with a more explicitly theological comment:

> There are many minds that would not for an instant doubt the God of Nature to have existed *from all Eternity*, and would yet reject as preposterous the idea of going back a million of years in the History of *His Works*. Yet what is a million, or a million million, of solar revolutions to an eternity?[71]

Scrope's reference to 'the God of Nature' cannot be dismissed as a conventional concession to the opinions of his readers. When a few years later

[69] Daubeny was an Anglican layman of Broad Church, anti-Tractarian persuasion: see his *Miscellanies: being a collection of memoirs and essays on scientific and literary subjects, published at various times* (Oxford and London, 1867), pp. xiii–xvi; also his reprinted review of Lecky and defence of Baden Powell, ibid., part IV, pp. 3–40. Although these writings date from much later in his life, I know of no evidence that he did not hold the same beliefs in the 1820s.

[70] Charles Daubeny, 'On the diluvial theory, and on the origin of the valleys of Auvergne', *Edinburgh new philosophical journal*, x (1831), 201–29 (203). This was Daubeny's defence of his earlier conclusions, provoked by Lyell's use of Scrope's interpretation in the first volume of the *Principles*.

[71] *Memoir*, p. 165 n.

he reviewed the first volume of Lyell's *Principles* for the *Quarterly*, he went out of his way to begin his essay by emphasizing how geology displayed in its progressive designfulness 'the sure marks of a First Cause, acting by invariable laws'. There is no good reason to doubt the sincerity of such remarks. Scrope did believe in the divine origin of the world studied by geology; but he saw its designfulness displayed most clearly in the sheer scale of its history under natural law. Just as astronomy extended the spatial bounds of the created universe, so geology extended its temporal bounds: both sciences were equally sublime in the views they disclosed.[72]

It would be easy to dismiss such a sentiment as a commonplace of natural theology. Instead of applying that intellectualistic label, however, it might be more illuminating to note its religious function.

Scrope's work in Auvergne involved a rejection of what I have characterized as the 'common-sense' division of the volcanic eruptions into two separate epochs. In order to interpret the varieties of volcanic phenomena as members of a single temporal series, Scrope had to break down a division that most other observers had felt was objectively 'given' and overwhelmingly natural. The driving force behind his determination to blur the common-sense distinction between 'ancient' and 'modern' seems to have been provided by his evident antipathy to the diluvial theory. When Lyell later supported his interpretation of the erosion of valleys in the Massif Central, Scrope branded their opponents at the Geological Society as 'Mosaic geologists', so making no distinction between the diluvialists and the scientifically worthless 'scriptural geologists' of the time.[73] Indeed, it is clear that Lyell regarded Scrope as his ally in his campaign to purge the science from 'ancient and modern physico-theologians'.[74]

In a historiography that is still dominated by the literature of 'Conflict', that campaign is too readily interpreted in positivist terms as a struggle to free geology from theological constraints.[75] While it is beyond the scope of this article to discuss this question fully, it is worth suggesting that the attitude of both Scrope and Lyell was underlain by theological concerns as important as (though different from) those of the diluvialists.

If, as his phraseology suggests, Scrope believed in a deistic 'First Cause' or 'God of Nature', rather than the more active God of Christian theology, a canopy of human meaning could be stretched across the spatio-temporal vastness of the universe only by emphasizing that vastness

---

[72] [G. P. Scrope], 'Art IV. [Review of] *Principles of Geology* . . . By Charles Lyell, F.R.S., 2 vols, Lond. 1830', *Quarterly review*, xliii (1830), 411–69 (412–13).
[73] Scrope to Lyell, 23 December 1828, American Philosophical Society MSS., Philadelphia.
[74] Mrs Lyell, op. cit. (8), i. 268–71, 276.
[75] See, for example, Andrew D. White, *A history of the warfare of science with theology in Christendom* (2nd edn., New York, 1901), chapter 5, 'From Genesis to geology'. The same positivist assumptions are still apparent in more recent and more sophisticated studies: for example, Charles Coulston Gillispie, *Genesis and geology. A study in the relations of scientific thought, natural theology, and social opinion in Great Britain, 1790–1850* (Cambridge, Mass., 1951), chapter 4, 'Catastrophist geology'.

to the utmost.[76] If there had been no divine 'drama' manifest within the successive *periods* of human or geological history, then meaning could be found only by pointing to the magnificent *scale* of the universe within which the 'invariable laws' of nature had operated. In other words, history for Scrope was not 'periodized' qualitatively at all. Furthermore, this vast dimension of time was that of an homogeneous and 'neutral' medium, standing as it were 'outside' the events themselves. Only with this concept of time as a purely quantitative dimension could Scrope have constructed his 'natural scale' of erosion in Auvergne.

Perhaps his opponents such as Daubeny failed to treat geological time in the same manner precisely because they did not have this existential need to find human meaning in the grandeur of the time-scale. For Christians the meaningful character of geological history was not dependent on its quantitative magnitude; meaning was already incarnate in history, and could be extended relatively easily to the pre-human periods described by geology, whatever their length in years.

I suggest therefore that just as Daubeny's tendency to see geological time as 'periodized' was underlain by the traditional Christian conception of history, so Scrope's emphasis on the vast scale of geological time was underlain by the concerns of a deistic (or rather, Unitarian) theology.[77] Both theologies can be seen to have 'done work' on the mundane level of geological interpretation: Daubeny's, by reinforcing the 'common-sense' distinction between the 'ancient' and 'modern' lavas into a sharp periodization; Scrope's, by leading him to look for intermediates in order to blur that distinction into a uniform sequence of events on a vast time-scale.

VI

Scrope and Lyell had met each other by 1825, if not before;[78] they were both elected Fellows of the Royal Society in 1826 and were joint Secretaries of the Geological Society at the same period. It could therefore be argued that any similarities between their scientific ideas are as likely to be due to Lyell's influence on Scrope as to Scrope's on Lyell. More realistically, it might be suggested that their common concepts emerged in dialogue. But if we can trust Scrope's veracity, he had written the *Memoir* (or at least the bulk of it perhaps) by April 1822, less than a year

---

[76] I borrow the expressive term 'canopy' from Peter L. Berger's sociological work, *The social reality of religion* (London, 1967).

[77] I do not claim to have demonstrated the Unitarian character of Scrope's religious beliefs, but only that many clues in his writing suggest such an inference, which would be worth testing on further Scrope material. The vocabulary of 'First Cause', etc., was of course common to deists and Christians, but I see no sign in Scrope's work of any theistic concept of God as active throughout history.

[78] A letter to his sister dated 4 December 1825 seems to suggest that he had only recently met Scrope; see Mrs Lyell, op. cit. (8), i. 163. But they had been joint secretaries of the Geological Society since February 1825, and Lyell had been one of Scrope's sponsors for membership of the Society—probably though not necessarily from personal knowledge—as early as March 1824, soon after Scrope's return from the Continent (information from the Geological Society archives).

after his fieldwork and while he was still on the Continent;[79] and the more explicitly theoretical *Considerations*, to which the *Memoir* was an 'appendix', was probably written about the same time, and anyway not later than 1824 (since it was published in the following year). It was only at the end of 1824 that Lyell read his first scientific paper (to the Geological Society) with its 'trial run' of an actualistic argument.[80] Recent biographical research utilizing new manuscript sources has yielded little from earlier dates that would lead us to anticipate Lyell's later originality.[81] Indeed, at any time before 1830 a geologist sympathetic to their viewpoint might well have judged that Scrope's geology held more originality and promise than Lyell's; it is only in retrospect, in view of their later achievements, that one is revered as a founder of modern geology while the other's work has been left in obscurity. It is therefore plausible *prima facie* to suggest that Lyell may have been influenced in important ways in the later 1820s by work which Scrope had already completed and even published.

Lyell's public opinion of Scrope's work was significantly ambivalent. Scrope, as we have seen, regarded the *Memoir* as a detailed illustration of the theoretical views he developed in the *Considerations*. But when Lyell reviewed the *Memoir* for the *Quarterly*, he drew a sharp distinction between the two books. He referred disparagingly to the *Considerations* as having 'revealed to us a new system of cosmogony', and he contrasted it rather patronizingly with the 'maturer judgment' of the *Memoir*.[82] At first sight this is puzzling, because Lyell had privately praised the *Considerations* on its first appearance as a 'very creditable' work,[83] and he had been indignant at the shabby treatment it had received in the *Westminster review*.[84]

Furthermore, there are striking parallels between Scrope's explicit methodology and Lyell's later *Principles*. Scrope criticized those who failed to study present 'causes' before interpreting the past, on the grounds that they were having recourse to 'what *might be* rather than *what is*'. Lyell later criticized them likewise for 'guessing at what *might be*, rather than in inquiring *what is*'. Scrope attributed this to the common human 'love of the marvellous'; Lyell saw in it 'the ancient spirit of speculation revived'. Scrope criticized the unthinking use of terms such as 'catastrophe'

[79] *Memoir*, pp. ix–x.
[80] Charles Lyell, 'On a recent formation of freshwater limestone in Forfarshire, and on some recent deposits of freshwater formations; and an appendix on the Gyroconite or seed-vessel of the Chara', *Transactions of the Geological Society of London*, 2nd ser. ii, Part 1 (1828), 72–96. The paper was read on 17 December 1824 and 7 January 1825.
[81] Wilson, op. cit. (2). The judgment about originality is of course my own.
[82] Lyell, op. cit. (6), p. 439.
[83] Mrs Lyell, op. cit. (8), i. 63.
[84] [John MacCulloch], 'Art III. [Review of] *Considerations on volcanoes* . . . By G. Poulett Scrope, Esq., Sec. Geol. Soc. 8vo. 1825', *Westminster review*, v (1826), 356–73. MacCulloch dismissed Scrope's book in his opening page and devoted the rest of his review to expounding his own ideas on volcanoes. For Lyell's reaction, see Mrs Lyell, op. cit. (8), i. 170.

on the grounds that 'it stops further inquiry'; Lyell asserted that it was 'directly calculated to repress the ardour of inquiry'. Scrope described the actualistic method as 'the only legitimate path of geological inquiry', while Lyell asserted that 'history informs us that this method has always put geologists on the road that leads to truth'.[85] These parallels suggest that Lyell owed more to Scrope's *Considerations* than his public dismissal of that book would lead us to suppose.

On the other hand, Lyell's critical remarks might be explained, in the light of his later work, as a characteristic expression of his aversion to what he termed 'cosmogony'. If Lyell had confined this pejorative term to earlier works in the style of Burnet and Whiston, his attitude would be simple to understand in terms of the 'empiricist' norms of his own generation; but it is puzzling at first sight to find it used for a work by one of his contemporaries and colleagues who was explicitly trying to base his conclusions on the actualistic method that Lyell himself was advocating. Scrope's 'theory of the earth', available in published form as early as 1825, provided in outline a mode of interpretation of the history of the earth that was naturalistic in its metaphysical stance and consistently actualistic in its methodology. On the question of 'catastrophes' it was, as we have seen, reasonable and flexible. Scrope stated explicitly that his 'theory' was probably premature and would need much improvement, but he maintained that as a basis for further research it had substantial explanatory merits.[86]

It might have been expected that such a research programme would have appealed to Lyell, since it incorporated his own conviction that geology needed a more thoroughgoing application of the actualistic method. In his review of Scrope's work Lyell did indeed praise the *Memoir* for its use of the actualistic method coupled with the acceptance of an almost unlimited time-scale. Echoing Scrope's emphasis, Lyell remarked that 'the imagination is overpowered by the effort' of conceiving the intervals of time involved.[87] He approved Scrope's rejection of the ancient-modern distinction, and noted perceptively 'how prone we are to imagine strong lines of demarcation' and to overlook intermediate instances.[88] He also approved Scrope's methodological use of the 'comparatively recent' volcanoes: 'the knowledge acquired by studying the effects of time in these, leads us gradually on to the interpretation of what is most obscure in the characters of the more ancient volcanoes'.[89]

It might be argued, however, that although Lyell could wholeheartedly approve Scrope's use of geological time in the *Memoir*, he could not accept the more speculative elements in Scrope's *Considerations*. At

[85] Scrope, op. cit. (57), pp. iv–v, 242; Lyell, op. cit. (5), p. 518; op. cit. (9), iii. 2–3, 6.
[86] Scrope, op. cit. (57), p. vii.
[87] Lyell, op. cit. (6), p. 468.
[88] Ibid., p. 488.
[89] Ibid., p. 453.

certain specific points Scrope did indeed lay himself open to the charge that his explanations were purely speculative. For example, his physico-chemical argument about the nature of volcanic lava utilized a notion of a fluid 'vehicle' to explain the discrepancy between the observed behaviour of lava on extrusion and the melting-points of its constituent minerals.[90] But Lyell characterized the whole work as a mass of 'sweeping generalisations'. It may be significant that in doing so he explicitly rejected Scrope's 'elastic vehicle' for being as useless in geology as Lamarck's 'nervous fluid' was in biology.[91]

Lyell had read Lamarck's *Philosophie zoologique* early in 1827. It has been suggested recently that Lyell immediately recognized that Lamarck's transmutation theory posed a serious threat to the dignity of Man, and that this caused him to adopt a steady-state theory for the history of the earth and of life. Such a theory, to which he remained tenaciously committed for over a quarter of a century, provided the uniqueness of Man with a defensive bulwark by denying that there had been any overall progress in the earlier history of life and its environment.[92]

This interpretation, although admittedly tentative on the evidence at present available, receives some further circumstantial support from Lyell's treatment of Scrope's work. Lyell certainly emphasized the significance of Scrope's work in Auvergne for Man's self-understanding. 'These enlarged conceptions of the earth's antiquity', he wrote, would tend 'to derogate from the dignity of man'—but only 'if man be confounded with the inferior animals'. If, on the other hand, 'we hold mind to be something distinct from matter, it must be acknowledged that we assert its superiority more clearly by enlarging our dominion over time'. Through the contemplation of these 'myriads of years' we could include them all 'within the compass of our rational existence'.[93]

Whatever sources there may have been for Lyell's characteristic dualism of mind and matter, this argument indicates that the 'dignity of man' was indeed in Lyell's mind at this time. So also, I suggest, was the steady-state interpretation of the history of the earth that he was later to expound fully in the *Principles*. In his summary of Scrope's work in Auvergne Lyell used the retrospective component of actualistic comparison not just as a methodological necessity but as a deliberate descriptive idiom. Instead of trying (as Scrope did) to synthesize the geological evidence into an *historical* reconstruction of the sequence of events, Lyell summarized

---

[90] Scrope, op. cit. (57), pp. 19–25. Scrope realized that dissolved volatiles, particularly water vapour, must play an important role in determining the petrological character of both intrusive and extrusive igneous rocks. Though expressed in terms of contemporary caloric theory, his suggestions are far less 'speculative' than Lyell's critical comments might suggest.

[91] Lyell, op. cit. (6), pp. 439 and 440 n.

[92] Michael Bartholomew, 'Lyell and evolution: an account of Lyell's response to the prospect of an evolutionary ancestry for man', *The British journal for the history of science*, vi (1972–3), 261–303.

[93] Lyell, op. cit. (6), p. 474.

Scrope's conclusions in the form of four periods, described *retrospectively* from the most recent (i.e. since the last eruptions) backwards to the oldest (i.e. the period of the Tertiary lakes).[94]

At first sight, this might seem to suggest that his concept of history was nearer to Daubeny's than to Scrope's. But the similarity to Daubeny's 'periodized' history is more apparent than real. If an anachronistic metaphor is excusable at this point, both Scrope and Lyell saw geology as providing temporally limited human beings with a kind of Wellsian time-machine. Scrope saw this as a device from which the development of the earth could be watched on a vastly accelerated time-scale—but watched as a process of historical development from the remotest past continuously down to the present. Lyell, on the other hand, used it to travel *backwards* from the familiar present into the more remote past, stopping occasionally to inspect the state of the earth in detail. This notion is clearly evident (even if the metaphor would not have occurred to him) in his vivid word-picture of the scenery and life of Auvergne as it would have looked to a human observer in the earliest of his four periods.[95] Unlike Daubeny's periods, Lyell's were thus more or less arbitrary *samples* taken from a continuous sequence that was cosmologically as homogeneous as Scrope's.

Scrope's work shows that an actualistic emphasis on the need to refer back from the present in order to understand the past was perfectly compatible with a strictly historical presentation of the resultant reconstruction of the past. Lyell's descriptive idiom of sampling the past in retrospective order therefore seems to be related to his concern to establish the steady-state nature of that past. By retrospective sampling he could hope to demonstrate the essentially *unchanging* circumstances of the earth and of life.

This underlying strategy is in fact hinted at in Lyell's review. He argued that the common assumption that the earth was now in a state of relative quiescence after a more active past had only gained plausibility through a failure to appreciate the magnitude of past time.[96] Once the geological time-scale was properly appreciated, the apparent intensity of past events would be reduced. Conversely, a fuller knowledge of present processes would show that they were *more* effective than was commonly assumed. By this two-pronged reform in geological understanding, the present and the past would converge into conformity with each other, and the steady-state stability of the world would thereby be demonstrated.[97]

This programme is evident in Lyell's own evaluation of the phenomena

[94] Ibid., pp. 465–73.
[95] Ibid., pp. 472–3.
[96] Ibid., pp. 468–9.
[97] For the working out of this reform in the *Principles*, see Martin J. S. Rudwick, 'The strategy of Lyell's *Principles of geology*', *Isis*, lxi (1970), 4–33.

that Scrope had described. When he and Murchison studied the geology of the Massif Central in the following summer, he was clearly interested in seeing for himself the examples that would demonstrate the great *power* of ordinary river erosion within the geologically short interval since the more recent eruptions, as well as the evidence for the vast time-scale of earlier geological history. Hence he made a special study of the post-eruption gorges of the Sioule (Figure 2) and the Ardêche and its tributaries (Figure 3), as well as the evidence for Scrope's 'natural scale' (Figure 4) in the differential erosion of the lava-flows to the east of Puy de Dôme (Figure 1). He was also deeply impressed by the thinly laminated Tertiary sediments of the Limagne and other 'basins' within the Massif, for these demonstrated the vast time that had elapsed even before the earliest of the eruptions.

Lyell's subsequent paper on the 'hitherto-not-in-the-least-degree-understood subject' of valley erosion gave strong support to Scrope's conclusions.[98] He clearly believed (with Scrope) that the more recent eruptions, although prehistoric, were very young by geological standards. They were the most recent in a long sequence that stretched back to the time of the massive volcanic centre of Mont Dore, which was now deeply eroded. Mont Dore in turn was younger than any of the lake sediments, which by their fossils could be correlated (in the stratigraphical sense) with the well-known freshwater Tertiary formations of the Paris 'basin'.

In the final and culminating volume of the *Principles* Lyell used this evidence of the vast time-scale of geological history to argue that the state of the earth had been broadly constant throughout Tertiary time. For this purpose he adopted the same retrospective mode of description that he had used when reviewing Scrope's *Memoir*, in the service of a theory of the steady-state uniformity of the globe.[99]

VII

We have seen that Lyell rejected the directionalist theory in Scrope's *Considerations* as mere 'cosmogony'. This pejorative term should not be equated simply with large-scale theorizing, for Lyell himself was perhaps the most imaginative system-builder of his generation, and he significantly excluded one previous 'system' (Hutton's) from his general criticism of

[98] Charles Lyell and Roderick Impey Murchison, 'On the excavation of valleys, as illustrated by the volcanic rocks of central France', *Edinburgh new philosophical journal*, xii (1829), 15–48, Plates 1–3. Although nominally joint in authorship, it seems to have been mostly, if not entirely, Lyell's work. See also Mrs Lyell, op. cit. (8), i. 197; and, for the contrast with Murchison's views, ibid., i. 199. For Scrope's further development of the same topic, see 'On the gradual excavation of the valleys in which the Meuse, the Moselle, and some other rivers flow', *Proceedings of the Geological Society of London*, i, no. 14 (1830), 170–1.
[99] Rudwick, op. cit. (97). In the *Principles* Lyell changed his interpretation of the Auvergne volcanoes with an 'extremely rash' conjecture that even the most recent eruptions dated from the Miocene, far back in the Tertiary period. He seems to have been trying to protect still further his rejection of any diluvial interpretation of the area, in the face of Daubeny's defence: see Lyell, op. cit. (9), iii. 268; also notes 41 and 70.

such theorizing. 'Cosmogony' was applied specifically to large-scale interpretations that saw the history of the earth in broadly directional or developmental terms, from a 'beginning' to the present (Hutton's steady-state 'system' was thus excluded). Scrope's theory incorporated just such an overall directionality to the history of the earth. Furthermore, this was explicitly linked to a consequently directional history of life as well. It was this, I suggest, which caused Lyell to reject Scrope's *Considerations* as mere 'cosmogony'. Although it was based impeccably on naturalistic explanations and actualistic method, it led perilously towards a 'Lamarckian' history of life. Scrope himself (like Lyell in the *Principles*) used the conventional and vague terminology of 'creation' and 'coming into existence', but Lyell may have seen how easily such a gradually progressive history of life could be turned into Lamarckian form, to become the empirical basis of an explanation that would rob Man of his unique position in Nature. Only by separating the directionality of the *Considerations* from the detailed application of the actualistic method in the *Memoir* could Lyell deflect this threat to the dignity of Man.

Lyell's underlying concern with the *biological* implications of Scrope's work is also suggested by the fact that he began his review of the *Memoir* by noting its possible implications in a direction that Scrope had not mentioned at all. There was no fossil evidence of any marine incursion into Auvergne since before the Tertiary lake sediments had been deposited. Hence the Massif Central had presumably been undisturbed by whatever changes in land and sea might have affected more low-lying areas during that interval of time. The Cuvierian explanation of faunal change was therefore inapplicable: the faunas on the Massif could not have become extinct through sudden marine inundations, and then have been replaced by migration from elsewhere. Yet the faunas had undoubtedly changed as much as they had in other areas: Auvergnat naturalists had recently discovered (near Issoire) a rich fossil fauna of mammals belonging to extinct species usually regarded as 'pre-diluvial'.[100] Lyell therefore wrote that he looked forward to the discovery in Auvergne of more of these 'lost links in the great chain that unites the present with the past'. He realized that this area might yield significant evidence to test whether the faunal change had been sudden (as diluvialists like Buckland believed) or gradual (like Scrope's view of valley erosion); and also whether it had happened by Lamarckian transmutation or, as 'more numerous and more distinguished physiologists' believed, by some other means.[101]

[100] See preliminary report in *Bulletin des sciences naturelles et de géologie*, iii (1824), 330–1; Dévèze de Chabriol and Bouillet, *Essai géologique et minéralogique sur les environs d'Issoire, Dépt. du Puy-de-Dôme, et principalement sur la montagne de Boulade avec la description et les figures lithographiées des ossemens fossiles qui y ont été recueillis* (Clermont-Ferrand, 1825–7); Croizet and Jobert, *Recherches sur les ossemens fossiles du département du Puy-de-Dôme* (Paris, 18[26]–28). The sites were on Mont Perrier, above the village of Pardines and near the farm of Boulade. In modern terms the deposits are early Villefranchian (Pliocene) in age.
[101] Lyell, op. cit. (6), pp. 442–3.

It is possible that this hope was one of Lyell's principal reasons for deciding to visit Auvergne himself in the following summer. Much of his subsequent paper, as has already been mentioned, was devoted to confirming Scrope's conclusions on the erosion of valleys. But even where, in the bulk of the paper, Lyell was merely dotting the i's and crossing the t's of Scrope's account, the biological direction of his thoughts is suggested by his interest in finding ancient river-gravels preserved in places below the various lava-flows. These gravels confirmed that each lava had indeed flowed down what had been an ordinary river-bed at that time, thus refuting the diluvialist assertion that the valleys had been scoured out by an exceptional Deluge. But in addition, ancient river-gravels were the most hopeful sites for the preservation and discovery of the fossil remains of terrestrial animals. Scrope had argued that a 'natural scale' of the gradual erosion of the valleys had been preserved fortuitously by the accident of intermittent volcanic eruptions. Lyell seems to have hoped that this might be matched by a corresponding scale of faunal change, provided by the same accidents, which had preserved river-gravels of various ages beneath the lava-flows.

In the event, Lyell discovered no new faunas. But he was able to clarify a disputed point about the geological position of the fossil fauna already mentioned, and to confirm that it came from an ancient river-gravel now embedded in a hillside high above a present valley floor. It had been preserved not by a lava-flow but by a massive agglomerate derived from Mont Dore. But the implication was the same: on Scrope's 'scale' its age was equivalent to some of the relatively ancient basalts. In other words, an almost unimaginably long interval of time separated this fauna of extinct species from the related mammals living at the present day.[102]

In his paper, Lyell did not use this result explicitly to answer the question he had posed in his review of Scrope's work. But implicitly it suggested that the transition between the fossil fauna and the present (across the alleged 'diluvial' episode) could have been as slow, gradual, and continuous as the erosion that Scrope had demonstrated for the same interval of time.

The gradual character of faunal change could only be confirmed as a general proposition either by finding more intermediate mammalian faunas, in which there were mixtures of extinct and living species, or by finding a corresponding gradualistic series among fossil faunas of some other kind. Lyell had failed in the first option. But as soon as he left the Massif Central and began to study the marine Tertiary strata near the Mediterranean and in northern Italy, he found the material necessary for the second option, for it was well known that the molluscan faunas in the marine Tertiary strata contained both extinct and living species.

[102] Lyell, op. cit. (98), pp. 44–5.

Lyell's concept of gradual and piecemeal faunal change incorporated —and even required—a negative answer to the other question he had posed in his review: namely, whether faunal change, if gradual, had been by Lamarckian transmutation or not. I have already referred to his deeper reasons for rejecting the Lamarckian solution. The method he later developed for estimating the relative age of Tertiary faunas depended on the reality of *species* as discrete units, each of which came (somehow) into existence at one point in time and then endured until at a later point in time it became extinct.

This concept of discrete species, which (for any given biological group) had a certain statistically average longevity, is strikingly parallel to Scrope's analysis of discrete lava-flows. Just as lava-flows had been extruded sporadically into the valleys of Auvergne, so Lyell believed that species were (somehow) produced in piecemeal manner at appropriate space-time locations. Once extruded, each lava-flow would immediately be attacked by the remorseless powers of erosion; it would endure for a longer or shorter time, depending on many contingent circumstances; it would gradually be reduced in size and even broken up into isolated parts; and eventually, with the final destruction of the last remnant, all trace of the eruption would be lost. In the same way, once a new species had been produced, its adaptive constitution would enable it to survive within its initial physical and biotic environment; but sooner or later, with the contingencies of geological change, unfavourable circumstances would reduce its geographical distribution (perhaps to isolated areas) and gradually diminish its numbers; and, with its final elimination from the last of these areas, the whole species would have become extinct. At any given moment (e.g. the present), the 'population' of lava-flows would include 'individuals' dating from many different points of origin in the past, from the very ancient (and perhaps almost completely destroyed) to the most recently erupted. Likewise, the 'population' (in the statistical sense) of species comprising the total (say, molluscan) fauna at any given moment in time would include 'individuals' (i.e. species) dating from many different points of origin in the past, from the very ancient (perhaps surviving only as isolated relicts) to the most recently 'created'.

At this point the close analogy breaks down, because the 'population' of lava-flows that existed at any time in the past is not preserved in that form, whereas a 'population' of species may be, in fossil form. Where any such assemblage of fossils has been preserved, a comparison between its specific composition and that of a comparable fauna at the present would yield a measure of its age. The percentage of extinct species would be inversely proportional to the age of the fauna (though the relation would not be linear, because the proportion of species still extant would depend on rates of survivorship). Hence Lyell developed a natural 'chronometer' for Tertiary time, derived from the percentage of extinct species in the various molluscan faunas, and based on the assumption of a uniform rate of piecemeal faunal change.[103] This was closely parallel to Scrope's 'natural scale' of time in the geological history of Auvergne, derived from the degree of elevation of lava-flows and based on the assumption of a uniform rate of fluvial erosion.

It was not until the final volume of the *Principles* that Lyell formalized this concept of a natural 'chronometer' based on faunal change. But the first traces of it are evident in his correspondence soon after he left the Massif Central. From Nice he reported finding a molluscan fauna with species 'eighteen in a hundred of which are *living Mediterranean species*'; and in a letter from Milan he used this discovery explicitly in a context of *dating* the Tertiary formations.[104] This suggests that he may have derived the inspiration for his faunal scale, at least in part, from the analogy with Scrope's 'natural scale' of erosion. In the *Principles* he developed the 'chronometer' in exactly the same way as he had summarized the history of Auvergne in his review of Scrope's work, namely by describing four 'sample' periods of Tertiary time in retrospective order (Newer Pliocene, Older Pliocene, Miocene, and Eocene; the names indicate their probabilistic faunal basis).[105] As in his review, his analysis of events at these periods in the past was directed towards demonstrating their normality and similarity to the present. In this way he was able to assert the steady-state uniformity of the world throughout Tertiary time.

I therefore suggest that important and characteristic themes in Lyell's greatest work can be understood as developments of conclusions which he reached under the impact of Scrope's work in central France.

VIII

Scrope had two major intellectual interests throughout his life. Up to this point I have referred to him solely as a geologist with a particular interest in volcanoes. But he was also widely known in political circles as a political economist.[106] He was a Member of Parliament for thirty-five years; and though he is said to have never made a speech, he was active enough with the printed word to earn the nickname of Pamphlet Scrope. This was a play on his full surname of Poulett Scrope. In 1821, the year

[103] Rudwick, op. cit. (97).

[104] Mrs Lyell, op. cit. (8), i. 199, 201.

[105] Lyell, op. cit. (9), iii. 45–61. The names, suggested to him by Whewell, denote 'mostly recent [species]', 'few recent', and 'dawn of recent'; the four 'samples' were defined as having respectively 90–95 per cent, 35–50 per cent, c. 18 per cent, and c. 3 per cent species still extant. Lyell stated explicitly that he expected the 'gaps' between these would eventually be filled by new discoveries, i.e. that they were *samples* from a continuum and not sequential *periods*; see Rudwick, op. cit. (97), and *The meaning of fossils. Episodes in the history of palaeontology* (London, and New York, 1972), pp. 179–85, and, especially, Figures 4 and 5. For a modern statistician's appreciation, see Sir Ronald Fisher, 'The expansion of statistics', *Journal of the Royal Statistical Society*, series A, cxvi (1953), 1–6.

[106] Redvers Opie, 'A neglected English economist: George Poulett Scrope', *Quarterly ournal of economics*, xliv (1929), 101–37. This excellent article seems to have been neglected as much as its subject.

of his fieldwork in Auvergne, he had married the heiress of William Scrope, changed his surname from Poulett Thomson, and settled at his wife's family seat in Wiltshire. There he had had to undertake the duties of a local magistrate. This had impressed upon him the acute urgency of the social problems of rural poverty. In 1829 he published the first of a long series of pamphlets on the reform of the Poor Law. He contested the constituency of Stroud in Gloucestershire unsuccessfully in the first election under the Reform Act, but was elected the following year (1833). In the same year he published his most substantial work on economics, the *Principles of political economy.*[107]

As with some more famous figures in the history of science, Scrope's apparently separate areas of interest pose an historiographical problem. Like Newton the natural philosopher and scriptural chronologist, or Wallace the biogeographer and spiritualist, Scrope the geologist and economist was a single human being with a single self. The problem of reconstructing and understanding that unified selfhood is made only marginally easier by the intellectual respectability (by our current standards) of both his areas of interest.[108] Nevertheless, it seems worth exploring, however tentatively, the possible connexions between his geological thinking and his economic thinking. Such connexions are *prima facie* plausible, since he maintained the two interests in conjunction rather than in succession. Even if we take his geology to have been the earlier in inception, and note his later preoccupation with political affairs, there was still an important period of overlap in the 1820s and early 1830s when he was engaged actively and creatively in both enterprises.[109]

First, there are some straightforward analogies. Just as the causation of geological events, however unusual, was referable to the ordinary laws

[107] G. Poulett Scrope, *Principles of political economy deduced from the natural laws of social welfare and applied to the present state of Britain* (London, 1833).

[108] But compare, for example, the hostility of some Darwinian scholars towards the suggestion that Darwin's *science* may have been influenced substantially by the socio-political concerns of the Malthusians. For example, Gavin De Beer, *Charles Darwin. Evolution by Natural Selection* (London, 1963), p. 100: 'The view that Darwin was led to the idea of natural selection by the social and economic conditions of Victorian England is devoid of foundation.' For a critique of this view, see Robert M. Young, 'Malthus and the evolutionists: the common context of biological and social theory', *Past and present*, no. 43 (1969), 109–45; 'The historiographic and ideological contexts of the nineteenth-century debate on man's place in nature', in M. Teich and R. M. Young (eds.), *Changing perspectives in the history of science* (London, 1973), pp. 344–438, especially section 4.

[109] A parallel example is Montlosier (1755–1838), who continued an interest in geology throughout his life alongside active political involvement and historical research. As a constitutional monarchist, he entered national politics as a member of the Constituent Assembly, went into exile after the declaration of the Republic, returned to Paris under the Consulate and Empire but devoted himself to historical work; see his *De la monarchie française* (3 vols., Paris, 1814). He retired to Auvergne under the Restoration, campaigning in many publications against the growing power of clericalism (he was a strongly Gallican Catholic), and was called to the Chamber of Peers under Louis-Philippe. A. Bardoux, *Le Comte de Montlosier et le Gallicanisme* (Paris, 1881), includes a bibliography (pp. 385–91). Montlosier's geological use of historical metaphors of monuments and medals (see note 50), although fairly commonplace in French geology in the late eighteenth century, would have had added force to him as a serious historical scholar.

of nature, so Scrope believed that the most effective political economy would be one based on 'the natural laws of social welfare'. Just as he deplored 'the gratuitous invention of vague and unexampled occurrences' in geology, so he denied that there was anything mysterious in the 'laws of social economy'. Just as the 'general invariability' of the laws of nature was in no way contravened by demonstrating that the earth had 'passed through several progressive stages of existence' in the past,[110] so the detection of laws of social behaviour was compatible with an optimistic outlook for the future progress of mankind. And at the deepest level, just as Scrope wished his geology to demonstrate the overall providential design behind the law-bound progressiveness of the earth's history, so also he wished his political economy to vindicate the providential character of the world of human affairs by refuting the gloomy predictions of the Malthusians.

Scrope's *Political economy* was above all a polemic against the Malthusians, for he believed that their pessimism would prove unjustified if society adopted the practical economic 'principles' that could be 'deduced' from 'the natural laws of social welfare'.[111] But although his argument was anti-Malthusian, it was conducted within the same terms of reference as that of the Malthusians themselves. In other words, it involved the same kind of *probabilistic* analysis of population trends in relation to food supplies, even though Scrope used his analysis to reach opposite conclusions from those of the Malthusians.

I have already emphasized the similarly probabilistic character of Scrope's analysis of the sporadic eruption and gradual erosion of lava-flows in Auvergne. Although his *Memoir* was published several years before his *Political economy*, it seems likely that the ideas behind both enterprises were forming in his mind at about the same time during the 1820s. Probabilistic arguments were, of course, the commonplaces of the Malthusian debate, but in geology they were very unusual at this period. I therefore suggest that Scrope's probabilistic analysis of Auvergne geology may have been derived from his probabilistic thinking at the same period in the realm of political economy.

It has been argued forcefully, and I believe correctly, that the application of 'population thinking' to living organisms (as opposed to 'essentialist' or 'typological' thinking) was a major element in the 'Darwinian revolution'.[112] In seeking for possible sources of this mode of approach

[110] Scrope, op. cit. (72), p. 465.
[111] Scrope, op. cit. (107), title and subtitle. He believed that more efficient agriculture, together with emigration to still underpopulated parts of the world, would be more than sufficient to keep food supply ahead of population growth. See, for example, the frontispiece of *Political economy*. In the previous year Scrope had written a strongly critical review of Thomas Chalmers's Malthusian work *On political economy* (Glasgow, 1832) for the *Quarterly review*, xlviii (1832), 39–69; see also Young, 'Malthus', op. cit. (108), pp. 119–25.
[112] See, for example, Ernst Mayr, 'The nature of the Darwinian revolution', *Science*, clxxvi (1972), 981–9; David L. Hull, *Darwin and his critics. The reception of Darwin's theory of evolution by the scientific community* (Cambridge, Mass., 1973), chapter 5

to organisms, we should not be blinkered by the anachronistic application of twentieth-century divisions of science. 'Individuals' do not have to be self-reproducing in order to exist in statistical 'populations'. Darwin's shift of interest from geology to biology was a shift *within* the contemporary category of 'natural history'. Hence he may have derived his 'population thinking' about organisms, at least in part, from Lyellian biogeography.[113] But I would suggest in addition that Lyell may have derived *his* 'population thinking' about lava-flows, mountains, islands, *and* organisms from the geology of Scrope.

I have already pointed to the close analogies between Scrope's probabilistic analysis of the lava-flows in Auvergne and Lyell's similar analysis—developed immediately after seeing Scrope's evidence for himself—of gradualistic faunal change in Tertiary molluscan faunas. The suggestion that Lyell derived this insight into the history of life from Scrope's geology, and that Scrope in turn derived his insight from the world of political economy, is reinforced by the analogy employed by Lyell when he came to expound his concept of piecemeal faunal change. The analogy was that of census returns of population—a comparatively recent innovation and a matter of great public interest. This makes it clear that Lyell himself regarded an assemblage of fossil molluscan species as directly and validly analogous to a human population. Just as the individuals within the population would have changed continuously in a piecemeal manner by births and deaths, so the species within the fauna would have changed continuously by piecemeal 'creation' and extinction. And just as successive censuses would record discontinuous 'samples' from this continuous process of populational change, so the successive Tertiary formations preserved similar discontinuous 'samples' from the continuous process of faunal change.[114]

In the *Principles* Lyell extended probabilistic thinking into the whole realm of geology. His vision of the earth in a state of constant flux was based on an understanding of the complex dynamic interaction of all the varied processes of geological and biological change. I suggest that this characteristic Lyellian vision may have been derived, to an important degree, from Scrope's earlier work applying a similar probabilistic analysis to the geology of Auvergne; and I suggest that that in turn was possibly derived from the thought that Scrope was giving at the same period to the *Principles of political economy.*

One further point of contact between Scrope's geology and his economic thinking is more elusive but perhaps equally important. I have

[113] See Michael T. Ghiselin, *The triumph of the Darwinian method* (Berkeley and Los Angeles, 1969); Camille Limoges, *La sélection naturelle* (Paris, 1970); M. J. S. Hodge, 'On the origins of Darwinism in Lyellian historical geography', paper read at the summer meeting of the British Society for the History of Science, in Edinburgh, July 1971.

[114] Lyell, op. cit. (9), iii. 30–4: 'The hypothesis of the gradual extinction of certain animals and plants, and the successive introduction of new species' is also referred to as 'the gradual birth and death of species'.

emphasized how Scrope's use of a virtually unlimited time-scale enabled him to undermine Daubeny's interpretation of Auvergne geology. I have suggested that he had theological reasons not only for rejecting the diluvial theory but also for wishing to replace a concept of time as historically 'periodized' by divine activity with a concept of time as cosmologically homogeneous and vastly extended. I have also suggested, with Daubeny's use of contrasting 'orders of things', how an apparently casual phrase may be deeply revealing. It might be objected that, if such a phrase was casual, it was likely to have been used unthinkingly as part of the common idiom of the age (or, more accurately, of the writer's social group), and that it could not therefore be made to bear any great weight of historical significance. On the other hand, as with significant expressions in everyday discourse, such 'slips' may reveal aspects of the writer's thought that were either unconscious or else so 'taken for granted' that they would not normally have been articulated in explicit form. If an analysis of their meaning throws light on, and is consonant with, the more explicit components of a writer's thought, their use as an aid to historical understanding is surely justified.

There is a surprising and apparently casual phrase at the heart of Scrope's purple passage about the time-scale of geology, with which I began this article. The geologist, he wrote, is forced 'to make almost unlimited *drafts* upon antiquity' (my italics). The metaphor was never taken up explicitly, and Scrope himself may hardly have paused to consider why he had used it. But I suggest that it may be deeply revealing of one source of his characteristic attitude to time.

The metaphor comes of course from the world of banking. In one of his earliest pamphlets, *On credit currency*, published only three years after the *Memoir*, Scrope argued for the superiority of a sound banking system, with proper regulation of credit, over an economy dominated by coinage.[115] An efficient system of banking, he believed, would avoid the social evils caused by the fluctuating values of the precious metals, which had been so disastrous during the 1820s. Much of Scrope's political economy, both here and in his later book, is essentially a development of Ricardian economics and is not significantly original. What I suggest may have been radically novel was the way in which he seems to have translated his concept of money in economics into a concept of time in geology. It is this analogical translation that is suggested by his casual use of a banking metaphor to illustrate the geologist's use of time.

The metaphor might be expanded as follows. In order to 'finance' the work of geological explanation, the geologist needs to draw almost unlimited drafts or 'funds' of time from Nature's 'bank'. There is no

---

[115] G. Poulett Scrope, *On credit currency and its superiority to coin, in support of a petition for the establishment of a cheap, safe and efficient circulating medium* (London, 1830). This 'pamphlet' is in fact nearly 100 pages in length.

intrinsic limit to the supply of time, any more than there is to the supply
of money. More 'credit-currency' can be created at any time by printing,
if the economy warrants it; money is not a 'substantial, intrinsically
valuable' material, but a neutral 'circulating medium' of exchange.[116] So
likewise the geologist need not feel limited by the time-scale within which
his explanations must be framed; there is always more time available in
Nature's 'bank', if the explanation warrants it. 'The key to the riddle
[of rapid fluctuations in prices] is not that the supply of goods has been too
great, but *that of money too small*'.[117] In the same way, the riddle of geological
explanation would be solved not by ascribing a greater power and intensity
to geological processes in the past, but by attributing the observed effects
to the action of those processes over much greater periods of time. In
effect, the 'funds' of time are issued to 'finance' an explanatory project;
if the project is successful, the 'credit' (in the banking sense) will have
been justified. Hence although there is no intrinsic limit to the 'funds',
their issue as 'credit' is not uncontrolled. Furthermore, although the
project would be 'speculative' (in the banking sense), the enterprise
could hardly fail. If the phenomena remained refractory even when
injected with vast 'funds' of time, nothing was really lost, since the
geologist would then at least know that they could not (at least for the
present) be explained by reference to ordinary processes acting through
long periods of time. So the explanatory enterprise was more like a sound
investment. It was always worth *trying* to explain phenomena by making
the assumption that unlimited supplies of time were available.

But if time has no intrinsic 'value' within a geological explanation,
just as money has no intrinsic value within the economy, then its effective
value must be determined by some external standard of reference. In his
political economy Scrope proposed that an economic '*Tabular standard*',
based on an 'index of general prices', could serve as a generalized reference
point for the 'real' value of money at any time—what goods could actually
be exchanged for others through a given amount of the neutral 'circulating
medium'.[118] So likewise Scrope's geology shows that he believed that a
knowledge of 'actual causes' and their efficacy on the time-scale of human
history could serve as a reference point for the 'real' value of the 'neutral'
medium of time in geological explanation—what effects could have been
produced by the same processes over the vast intervals of time that the
geologist was being encouraged to invoke.

Thus Scrope's banking metaphor for geological time can be translated
directly into the prescriptive methodology of actualism—which he
explicitly recommended. If time is 'neutral', and external to actual events,
then geologists should begin by assuming that unlimited amounts of time

116 Ibid., p. 52, and subtitle.
117 Ibid., p. 69.
118 Scrope, op. cit. (107), p. 407.

are available, and should then explore the explanatory value of this 'neutral' medium with reference to the standard of processes observable in the present. Of course, Scrope was not the first to apply the actualistic method to geological problems. But he was perhaps original in finding a new form of conceptual underpinning for the method—persuasive and heuristic at least to himself—by seeing that it was implicit in a concept of geological time that was directly analogous to his concept of money in political economy.

Much of Scrope's geology, as I have expounded it in this article, will seem familiar to all who have read Lyell's *Principles*. I will end with a brief comment on the differences and the similarities. First, the differences. The example of Scrope's work should serve to emphasize the strange conceptual isolation of Lyell in the geological world of the 1820s and 1830s. No other geologist, perhaps, was a closer ally of Lyell's in his campaign to eliminate the last traces of scriptural concerns and to urge the adequacy of ordinary 'actual causes' in geological explanation. No ally had a more vivid appreciation of the possible magnitude of geological time and of the potential explanatory power of such a time-scale. Yet while sharing so much with Scrope, Lyell nevertheless isolated himself decisively by rejecting Scrope's directionalist view of the history of the earth as mere 'cosmogony'. This, I believe, gives further support to the suggestion that Lyell shifted into his distinctive position of arguing for a rigidly steady-state system, not for any intrinsic geological reasons, but as a reaction to his fears for the dignity of Man engendered by reading Lamarck.

Secondly, the similarities. I have suggested at many points the breadth and depth of Scrope's apparent influence on Lyell's thinking, at a crucial phase in the latter's development. The evidence suggests that the direction of influence was indeed from Scrope to Lyell and not *vice versa*. I therefore conclude that a full understanding of the Lyellian concept of geological time, which was so crucially important for the later development of geology and for Darwin's work in biology, must take into account its possible origin (at least in part) in the work of Scrope, who in turn may have derived it (at least in part) from his concern with the social problems of political economy. If further research confirms this suggestion, then the work of George Poulett Scrope may come to count, along with that of Malthus, as an important point at which the arguments of political economy impinged substantively on the scientific content of the 'Darwinian revolution'.

IV

Lyell on Etna, and the Antiquity of the Earth

Martin J. S. Rudwick   The significance of Charles Lyell in the history of geology has never, perhaps, been summarized more perceptively than it was by Darwin, when he said that "the great merit of the *Principles* was that it altered the whole tone of one's mind, and therefore that, when seeing a thing never seen by Lyell, one yet saw it partially through his eyes."[1] Lyell's purpose in the *Principles of Geology* was to reinterpret the whole range of geological knowledge as persuasively as possible, within a new conceptual framework. The key to this new viewpoint on the natural world was his extension of the time scale of earth history; or rather, his imaginative grasp of the full implications of such a time scale for the interpretation of geology. It was this that enabled others to see the phenomena of geology through new eyes.

But the source of Lyell's own sure grasp of the importance of geological time deserves closer and more critical study. It is fairly clear that his thinking was influenced profoundly by his reading of Playfair's exposition of Hutton, and that, perhaps as a result, he reacted strongly against the outlook of his teacher, Buckland.[2] But to such "literary" influences must be added the direct impact of Lyell's own firsthand experience of geological phenomena. Recently, while following the course of his journey through France and Italy in 1828 and 1829, I became convinced that one of the most decisive events in the formation of his viewpoint was his firsthand study of the volcano Etna. This paper is an attempt to justify that belief.[3]

[1] Darwin to Horner, August 29, 1844: *More Letters of Charles Darwin*, ed. F. Darwin and A. C. Seward (London: John Murray, 1903), Vol. 2, p. 117.
[2] L. G. Wilson, "The Development of the Concept of Uniformitarianism in the Mind of Charles Lyell," *Actes Xme Congr. int. d'Hist. Sci., Ithaca* (1964), Vol. 2, pp. 993–996.
[3] This analysis is based chiefly on Lyell's letters from France and Italy, published in *Life, Letters and Journals of Sir Charles Lyell, Bart.*, ed. K. M. Lyell (2 vols.;

IV

When Lyell crossed the Channel in May 1828 at the start of his first major scientific expedition, an outline for his most influential book was already clear in his mind, and he had even completed the first few chapters.[4] Unlike Darwin embarking on the *Beagle*, Lyell set out already convinced that he possessed a revolutionary conceptual viewpoint; and the object of his journey was to gain firsthand experience of those specific phenomena that would form the most persuasive examples in the service of his argument.

Therefore, he went first to see the extinct volcanoes of Auvergne, which George Poulett Scrope had interpreted persuasively as evidence both for the uniformity of present and past and for the immensity of the geological time scale.[5] Lyell's first major essay on the principles of reasoning in geological science had been built around a review of Scrope's book;[6] in going to see the volcanoes for himself, he was clearly intending to use them for his own extension of Scrope's central theme.

Following Scrope's work closely from Auvergne southward to the Vivarais, Lyell felt increasingly confident that all the phenomena of geology, and not only those of vulcanism, could be interpreted in terms of processes still in operation, and that no differences of kind need be postulated between the geological processes of past and present. The freshwater Tertiary strata of the Limagne and the Cantal were closely comparable to the modern lake-marls he had studied near his home in Scotland,[7] and they clearly represented an extremely long period of deposition. Yet although geologically quite recent, these strata antedated even the earliest volcanic rocks of the Massif Central; and these, in turn, were far older than the fresh but still prehistoric cones and lava flows of Auvergne. This extended sequence of events clearly implied an immensely long time scale.

Even more significantly, however, the freshwater strata seemed to Lyell to have been elevated most strongly where they directly underlay

---

London: John Murray, 1881), Vol. I, pp. 182–251; on the use he subsequently made of his observations when writing *Principles of Geology* (3 vols.; London: John Murray, 1830–1833) (all citations are to the first edition of each volume); and on my own field work in France and Italy in 1966 and 1967. I would like to thank those present at the New Hampshire Conference for their valuable comments on this paper, which was first read in a preliminary form to the British Society for the History of Science in July 1967.

4 *Principles*, Vol. III, pp. v, viii.

5 G. P. Scrope, *Memoir on the Geology of Central France, Including the Volcanic Formations of Auvergne, the Velay and the Vivarais* (2 vols.; London: W. Phillips, 1827).

6 *Quarterly Review*, XXXVI (1827), 437–483.

7 C. Lyell, "On a recent formation of freshwater limestone in Forfarshire and on some recent deposits of freshwater marl." *Trans. Geol. Soc. Lond.*, II (1826), 72–96.

the huge volcanic center of the Cantal.[8] This seems to have suggested to him that the connection between vulcanism and tectonic disturbance might be more closely causal than was generally admitted. If the spectacular uplift and folding of strata in mountain regions could be ascribed simply to the long-continued action of ordinary earthquakes, of the kind that commonly accompanies volcanic action, then there would be no grounds for postulating occasional paroxysmal convulsions of the earth's crust, and the keystone of his opponents' model of earth history would be removed.

The full implication of these observations seems to have emerged in Lyell's mind only after he and Murchison had descended from the Massif Central and moved on through Provence to Nice. While Murchison recovered from illness, their enforced rest from field work gave them an opportunity to write up their work for publication and to reflect on what they had seen. "The whole tour has been rich," Lyell wrote, "as I had anticipated (and in a manner which Murchison had not), in those analogies between existing Nature and the effects of causes in remote eras, which it will be the great object of my work to point out."[9]

Within this general program it was now essential to demonstrate the efficacy of ordinary earthquakes as the main agent of elevation. Lyell therefore decided on a radical change of traveling plans. His hypothesis of a causal connection between elevation and vulcanism could only be tested by a firsthand study of active volcanoes, and more particularly by a search for a direct correlation between the degree of recent uplift and the intensity of recent volcanic action. In studying an area of active vulcanism, he wrote confidently, "I scarcely despair of *proving* the positive identity of the causes now operating with those of former times."[10] Therefore, as Murchison told him, "Sicily is for your views the great end: there are the most modern analogies, volcanic, marine, elevatory, subsiding, &c."[11] Regretting that he had not originally planned this extension to the expedition, Lyell wrote to Herschel, Buckland, Daubeny,

---

[8] C. Lyell et R. S. Murchison [sic], "Sur les Depots lacustres tertiaires du Cantal, et leurs rapports avec les roches primordiales et volcaniques," *Ann. Sci. Nat., XVIII* (1829), 173–214, Pls. XII, XIII; read at Geological Society of London, April and May 1, 1829 (abstract in *Proc. Geol. Soc., I,* (1829), 140–142.

[9] *Life,* Vol. I, p. 199. The style and content of their joint papers on French geology strongly suggest that the inspiration for them came almost entirely from Lyell, although Murchison had of course shared the field work. This conclusion is supported by the comment quoted here; Murchison was evidently less ardent than Lyell, even at this stage, in believing in the adequacy of existing causes, and later his viewpoint became more explicitly catastrophist.

[10] *Life,* Vol. I, p. 199.

[11] *Life,* Vol. I, p. 200.

IV

and Scrope, all of whom had been to Sicily, asking for advice on where to go and what to see.

He and Murchison continued their original plan together as far as the Vicentin, to study the richly fossiliferous Tertiary strata and their interbedded volcanic rocks; for it was already clear to Lyell that any demonstration of the uniformity of present and past would need to take the Tertiary epoch as its chief sample of the past. Even before they began this field work, however, Lyell was already using another key hypothesis, namely, that the percentage of extant species in a Tertiary fauna was some measure of its relative age. This now combined with his tectonic hypothesis; for if Tertiary faunas had changed gradually and in piecemeal fashion into that of the present day, as he believed they had, then the crucial faunal links between present and past were most likely to be uplifted above sea level in the vicinity of active volcanoes.[12]

At Padua, Lyell left Murchison to return to England via the Tyrol, and he himself turned south towards Sicily. From Naples, he climbed Vesuvius and confirmed Scrope's observation that the active cone was similar in structure to the ancient partial cone of Monte Somma.[13] He also visited the "modern" volcano of Monte Nuovo (formed in 1538) and the celebrated half-submerged columns of the so-called Temple of Serapis at Pozzuoli, both vivid illustrations of the action of geological processes within historic times. But the observation that excited him most was his discovery of elevated strata, containing the shells of extant molluscan species, on the volcanic island of Ischia.[14] If Ischia had risen some 2,600 feet in the geologically recent past, how much more elevation, he argued, might be expected in the vicinity of the far greater volcano Etna: "I have little fear of bringing a great part of Trinacria [Sicily] into our own times, as it were, in regard to origin."[15]

After receiving letters of advice from Buckland and Daubeny, he wrote, "Between Messina and Syracuse [i.e., on and around Etna] lies all that is of the greatest importance to me."[16] Thus, even in anticipation, before he actually reached Sicily, Etna had come to have a dominant place in his thinking.

On his arrival there, he approached the mountain first from the east, climbing through the orchards and vineyards into the Val del Bove, a huge natural amphitheater five miles across, floored with recent lava

---

[12] *Life*, Vol. I, p. 201.
[13] G. P. Scrope, "On the Volcanic District of Naples," *Proc. Geol. Soc., I,* (1827), 17–19.
[14] *Life*, Vol. I, pp. 209–213.
[15] *Life*, Vol. I, p. 213.
[16] *Life*, Vol. I, p. 215.

flows and volcanic cones. This, as Buckland had told him, was a sight more worth his attention than any other in Sicily or perhaps in all Europe.[17] Buckland's interest in it apparently arose from his belief that it supported Leopold von Buch's elevation-crater hypothesis, implying that the whole mountain might have been uplifted in one great paroxysmal event. But in the cliffs surrounding the Val del Bove, Lyell found natural sections that did not at all support this hypothesis. They showed, on the contrary, a series of lava flows and layers of volcanic tuff ("ash"), sloping uniformly outward from the center of the mountain, except where the dip was locally disturbed by buried parasitic cones. This layered structure, cut by basalt dykes that had presumably fed the successive eruptions (Fig. 1), seemed to him to imply a gradual and hence prolonged growth for the whole mountain.

FIG. 1.   Lyell's sketch of basalt dykes cutting layered volcanic tuff, exposed on the side of the Val del Bove and accentuated by differential weathering (Principles, Vol. III, p. 90). The tuffs are dipping toward the observer at this point and therefore appear horizontal.

Soon afterwards, he found the kind of evidence of recent uplift that he had come to expect — marine mollusk shells interbedded with lava flows several hundred feet above sea level; and he hoped that the identi-

[17] Principles, Vol. III, p. 83, footnote. In his letters from Sicily, Lyell uses the Sicilian form "Val del Bue."

IV

fication of the species present would "fix the zoological date of the old-est part of Etna."[18]

He was now ready to climb the mountain itself (Fig. 2). Setting

FIG. 2.   *Lyell's sketch of Etna (probably from near Catania), showing the Val del Bove on the right (east) flank, the summit cone (a), and the parasitic cones (f, g) around the base (Principles, Vol. III, p. 83). The height shown in this sketch is about 10,000 feet, the breadth about 10 miles.*

out from Nicolosi with the local geologist Carlo Gemmellaro as his guide, he at once encountered some of the most recent lava flows, still barren and cindery on the surface but revealing, where they were quarried, a compact stony center indistinguishable from the most ancient volcanic rocks of Scotland. Nearby was the 500-foot double cone of Monti Rossi, thrown up during the great eruption of 1669 (Fig. 3);

FIG. 3.   *Lyell's sketch of "Minor cones on the flanks of Etna" (Principles, Vol. I, p. 364). Number 1 is Monti Rossi.*

this was only one of some 80 major parasitic cones encircling the mountain, in Daubeny's phrase, "like a court of subaltern princes waiting upon their soverign."[19] Each lava could be seen to have flowed down

18 *Life,* Vol. I, pp. 216–217.
19 C. Daubeny, *A Description of Active and Extinct Volcanoes* (London: W. Phillips, 1826), p. 203.

the mountain in a broad stream a mile or two across, often flowing around either side of an older volcanic cone and partly burying it. As they climbed higher, through the forest zone and into the zone of grassland, the main string of volcanic cones was reduced by the sheer scale of the mountain to the appearances of small pimples; and on climbing still higher, the view to the south opened out beyond the plain of Catania to the low hills of the Val di Noto toward Syracuse. Finally, in the desert zone of volcanic ash near the summit crater, the whole of eastern Sicily and the tip of Calabria were spread out around them like a map, while almost directly below them the broadly conical form of the mountain was gashed by the Val del Bove (Fig. 4).[20] Later, when

FIG. 4.   Lyell's "View from the summit of Etna into the Val del Bove" (Principles, Vol. III, p. 93). Note the basalt dykes projecting from the cliffs on the far side.

he came to describe this scene in the *Principles*, it evoked perhaps the most lyrical passage in the whole book,[21] leaving no doubt of the intensity with which his imagination had been stirred. He was, in fact, only just in time to make the ascent; for the following day Etna was

[20] The ascent is briefly described in Lyell's letter to his sister Marianne (*Life*, Vol. I, pp. 217–218).
[21] *Principles*, Vol. III, pp. 88–90.

IV

covered with the first snow of winter, and the weather held him for three days "a willing prisoner" in Gemmellaro's house, with one of the best geological libraries in Sicily at his disposal.

Lyell's ascent to the summit of the greatest volcano in Europe was the imaginative climax of his whole expedition. No other single experience, perhaps, could have convinced him so persuasively of the essential validity of his whole approach to geology. His ascent of Etna was also the intellectual climax of his journey, but only in retrospect, for its full significance could not be grasped until he could relate it to the geology of the surrounding area. The relation was not easy to make out, for some of the lava flows had extended so far from the foot of Etna that they obscured the foundations on which the mountain had been built up. But beyond the limits of such flows, at La Motta, Lyell found strata with marine fossils, which seemed to underlie the mountain itself (Fig. 5). These strata did not dip uniformly away from Etna, as the elevation-

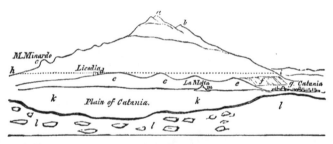

FIG. 5.   Lyell's sketch of Etna from the far side of the Plain of Catania (Principles, Vol. III, p. 64), to illustrate his belief that the whole mountain had been built up on a foundation of marine strata similar to those exposed in the foothills (e) around La Motta and in the cliffs (f) on the seaward side. The line h–i indicates "the highest boundary along which the marine strata are occasionally seen." The foreground (1) is the edge of the Val di Noto hills.

crater hypothesis would require, and he believed that their interbedded volcanic rocks were also marine in origin and unconnected with the later vulcanism of Etna itself.

Further south, beyond the alluvial plain of Catania, the low hills of the Val di Noto were also formed from marine strata with interbedded volcanic rocks, and Lyell was able to confirm Daubeny's impression that they were Tertiary in age.[22] The most conspicuous rock, through which quite deep valleys had been cut, was a massive limestone with

[22] C. Daubeny, "Sketch of the Geology of Sicily," Edin. Phil. Journ., XIII (1825), 107–118, 254–269, Pl. IV.

moulds of marine shells; but when Lyell reached Syracuse he found that this was underlain by a more rubbly limestone and then by a blue clay with much better preserved fossils. He was then astonished to discover that most of these shells seemed to belong to species still living in the nearby Mediterranean. According to his hypothesis, therefore, all the thick strata of the Val di Noto were not merely Tertiary, but very recent Tertiary in age; yet in general appearance they resembled more closely the "Secondary" strata of Britain. "All idea of attaching a high antiquity to a regularly stratified limestone, in which casts and impressions of shells alone were discernible, vanished at once from my mind."[23] Yet although these strata were geologically so recent, Lyell now believed that they underlay the huge mass of Etna, which in human terms was of immense antiquity.

From Syracuse, he continued to follow Daubeny's recommended itinerary around the island,[24] though significantly in reverse direction, having begun rather than ended with Etna. Daubeny had noted with regret his inability to discover the relation between the strata of the Val di Noto and those further west; and as Lyell travelled southward along the south coast, it was for evidence on this point that he searched particularly. When he reached Girgenti (now Agrigento), he found the famous Greek temples built on an escarpment of limestone that, as Daubeny had said, was "replete with shells not far, if at all, removed from existing species."[25] Daubeny had conjectured that such strata were equivalent to the Subappenine beds in the north of Italy;[26] but Lyell now recognized that they contained a far smaller proportion of extinct species, and he therefore believed they were "indubitably far, very far more recent" in origin.[27] This limestone was not merely uplifted but gently folded at Girgenti;[28] and Daubeny had reported that at Castro-giovanni (now Enna) in the interior of the island, it reached nearly 3,000 feet above sea level.[29] To Lyell, this alone was remarkable enough, since it meant that central Sicily must have been elevated by that amount

[23] *Principles*, Vol. III, pp. x, xi.
[24] In an Appendix to his paper on Sicily, pp. 268–269.
[25] Daubeny, *Sketch*, p. 117. Daubeny's interests were mainly in volcanic geology, and he did not pretend to have given more than cursory attention to the paleontology of Sicily; nevertheless he knew enough conchology to recognize a recent assemblage when he saw one.
[26] *Ibid.*, p. 259. Daubeny was chiefly concerned to argue for their *Tertiary* age, and he evidently had no thought of being able to subdivide that period.
[27] *Life*, Vol. I, p. 232.
[28] The folding can be seen most clearly from the east of the town, the direction from which Lyell approached it; the temples are built on the southern limb of a syncline and the modern town on the northern.
[29] Daubeny, *Sketch*, pp. 254–255.

since the molluscan species now living in the Mediterranean had come
into existence. But this striking evidence of the power of geological
agents in the recent past would still be susceptible of a paroxysmal interpretation unless it could also be shown that such a degree of elevation
could have occurred by very gradual stages. This required some demonstration that there had been ample time for gradual elevation, although the strata were geologically so recent. It was at this point that
the correlation between Girgenti and the Val di Noto was so crucially
important (Fig. 6). Lyell thought he could recognize a similarity be-

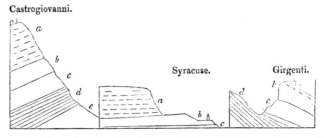

FIG. 6.   Sections showing Lyell's correlation of "Newer Pliocene" strata
in Sicily (Principles, Vol. III, p. 64). The "great limestone" of Val di Noto
(a) rests on the rubbly limestone of Girgenti (b), which, in turn, overlies
blue clay with well-preserved shells (c).

tween the rubbly limestone of Girgenti and the stratum underlying the
"great limestone" of the Val di Noto, and both were underlain by a
similar blue clay with well-preserved fossil shells. If this correlation was
correct, its implications, from Lyell's point of view, were extremely far
reaching, for it established that the time required for the elevation of
the strata in the center of the island had also been time enough for the
accumulation of the huge mass of Etna. He believed that this time could
be shown to have been immensely extended by human standards, and
hence amply long enough for the elevation to have occurred gradually
by a succession of ordinary earthquakes. "I got so astounded by the
results I was coming to," he wrote to Murchison later, ". . . that I
began to doubt them; and . . . I struck back again to Val di Noto right
through the centre of the isle."[30] But this only confirmed the correlation
he had made tentatively at Girgenti; and when he crossed the interior
once more on his way to Palermo, he was able to study the section at
Castrogiovanni and confirm Daubeny's report on the elevation of the
strata there.

[30] *Life*, Vol. I, p. 232

Back in Naples, he got his fossil collections identified by Costa; this confirmed his own provisional estimate of the age of the Sicilian strata, and therefore reinforced his confidence in his general conclusions. "The results of my Sicilian expedition exceed my warmest expectations by way of modern analogies," he wrote to Murchison; and significantly, he followed this remark with a brief but revealing statement of the aims and methodological principles on which his book was to be based.[31] It is therefore not surprising to find his Sicilian observations playing a crucial role in the over-all strategy of the *Principles of Geology*.

If it seems questionable to speak of an "over-all strategy," it can only be because the *Principles* have suffered in modern times from a failure to view the work as a whole, and indeed from being more venerated than read. All too many geologists admire it uncritically as the first "modern" textbook of the science, while perhaps remaining puzzled by its curious order of topics. Among historians of science, Lyell is duly applauded for his emphasis on the gradual action of existing causes, which by implication are often assumed to be the only causes that could be "natural" or "scientific" and so escape the "contamination" of theological overtones. But most commonly, Lyell features only as a forerunner of Darwin; attention is concentrated on half of the second volume, and the remainder of the work is dismissed as of merely technical geological interest.

All these views fail to recognize the significance of Lyell's own remark, in his letter from Naples, that the detailed geology in his book would not merely illustrate his dual principles of actualism and uniformitarianism, but would be presented "as evidence strengthening the system necessarily arising out of the admission of such principles." The word *system*, which reappears frequently in the book itself, is crucially important, for the *Principles of Geology*, far from being the first "modern" textbook of geology, can best be regarded as the last major all-explanatory system in earth science, the last important representative of a class of work that Lyell himself strongly condemned.[32] As might be expected of an all-explanatory system, there is indeed an underlying strategy to the book, which makes any part unintelligible without the rest.[33]

---

[31] *Life*, Vol. I, p. 234.
[32] Lyell, of course, contrasted his own work with most earlier "systems" on the grounds that his (like Hutton's) excluded "questions as to the origin of things"; but he was able to define his subject in this way precisely because he was championing a uniformitarian and not a developmental model of earth history.
[33] Martin J. S. Rudwick, "The Strategy of Lyell's 'Principles of Geology,'" *Isis* (1970), in press.

IV

Thus, the work opens with a "history of geology," which is designed to demonstrate that only those who have adhered prophetically to Lyellian principles have contributed to the progress of the science.[34] This polemical introduction is followed at once by a discussion of the "Progressionist" model of earth history that the *Principles* were designed to refute. After these preliminaries, the remainder of the first volume describes the inorganic or physical "causes now in operation" on the earth's surface; this is designed to show that the effects of such processes, even in historical times, have been far greater than has been generally admitted, and hence that far more of the phenomena of the past can be explained simply by their agency, without invoking any former differences of degree.

The second volume continues the argument on the same lines, dealing with "causes now in operation" in the organic realm. The preparatory nature of the whole of this volume is made quite explicit in the preface, in which Lyell states that it "will be found absolutely essential to the understanding of the theories hereafter to be proposed,"[35] that is, in the third volume. Lyell's opposition to transmutation, and his discussion of the piecemeal extinction of species, cannot be understood except in this context. At this point in the strategy of the work, he simply needs to establish two points: the reality of the species as a discrete unit in natural history, and the probability that these units disappear and — although more mysteriously — appear, in essentially piecemeal fashion, in response to the physical changes he has already described. The remainder of the second volume (after the point at which Darwinian forerunner-hunters give up the chase) continues to set the scene for the third volume, with a discussion of the interaction of organic and inorganic processes, and of the chances that organisms have of being preserved in the fossil record.

The third volume, the climax of the work, begins by summarizing the preceding argument and showing its relevance to the problem of reconstructing the past history of the earth. It is this problem, the reconstruction of earth history, that is the fundamental object of the whole work. The study of present physical geological processes, as set out in the first volume, is necessary for the light it can shed on the interpretation of the past, and for its demonstration that "actual causes" are sufficient to explain all the phenomena of the past if only an adequate time

[34] Lyell's historiographical myth shows remarkable vitality; having been transmitted through Geikie, Zittel, Adams, and others, it is still regularly reproduced in the historical introductions of most modern textbooks of geology.
[35] *Principles*, Vol. II, p. v.

scale is conceded. Likewise, the arguments about species in the second volume are all important simply as the theoretical justification for the *geological clock* that Lyell proposes to apply to the record of the past. The quantitative precision of the physical sciences may be unattainable in a historical science like geology; but Lyell argues, in effect, that at least an approximation to a quantitative time scale can be achieved. Given the continual flux of the physical features of the earth's surface, and the consequent flux in ecological conditions, there should be a statistically uniform rate of change in the organic world. Therefore, it should be possible to calculate the relative age of any fossil-bearing stratum by estimating the ratio between the number of extinct species and the number of still-surviving species that it contains. And since marine organisms stand the greatest chances of preservation as fossils, it is by the abundant marine Mollusca that the geological clock should be calibrated, at least for the Tertiary epoch.[36]

The bulk of the third volume is devoted therefore, to a reconstruction — on the basis of this geological clock — of the whole of Tertiary earth history, divided into four arbitrary periods based on the percentage of fossil molluscan species still extant. The exposition is retrospective, from newer Pliocene to older Pliocene and back through Miocene to Eocene, precisely because, "in the present state of our science, this retrospective order of inquiry is the only one which can conduct us gradually from the known to the unknown."[37] Only by starting with the present state of the earth can the adequacy of present causes be demonstrated for increasingly remote periods of earth history. For the same reason, it is sufficient for his strategy to show the validity of uniformitarianism for the Tertiary epoch; and he is content to end the work with a very brief discussion of the pre-Tertiary history of the earth, simply "to show that the rules of interpretation adopted by us for the tertiary formations, are equally applicable to the phenomena of the secondary series."[38]

[36] The consistency with which he proposed to apply this technique is shown by his conjecture (*Principles*, Vol. III, p. 328) that the interval of time represented by the Cretaceous-Eocene faunal discontinuity may be even greater than that comprised by the entire Eocene-Recent series of strata! The significance of this passage seems to have been generally overlooked.
[37] *Principles*, Vol. III, p. 62.
[38] *Principles*, Vol. III, p. 324. The brevity of this section is not due to lack of knowledge of pre-Tertiary stratigraphy. Although the Transition and Primary rocks were still poorly understood in 1833, Secondary stratigraphy (i.e., Carboniferous to Cretaceous, inclusive) was *better* known than Tertiary. Indeed, Lyell's work was more generally valued for bringing Tertiary stratigraphy up to the standards of Secondary, than for its theoretical argument.

Lyell's experience of Sicilian geology can be seen to fit crucially into this general strategy. Thus, in the first volume, the comparison between the Sicilian Tertiary faunas and those of the present Mediterranean is presented as an *experimentum crucis* establishing the reality of climatic change in the geologically recent past; although the species are the same, the fossil specimens are, on the average, distinctly larger than those now living.[39] This evidence, added to the more debatable evidence from more remote periods of earth history, obliges Lyell to devise a climatic theory that will account for such changes on strictly uniformitarian lines, before beginning the main exposition of his system.

Later in the volume, Etna supplies him with valuable examples of the present efficacy of both aqueous and igneous "causes." Some of the lava flows produced within historical times had already been cut into small gorges by the existing rivers, and cut, moreover, through compact basalt.[40] He had, in fact, seen much more spectacular examples of such gorges in the Vivarais, cut through basalts erupted from fresh-looking volcanic cones; but these basalts, although perhaps geologically recent, were certainly prehistoric.[41] The Etnean examples were more valuable because they unquestionably belonged to what Daubeny (for example) regarded as "the present order of things."[42]

Etna also provided an important argument for the undiminished potency of igneous processes.[43] Its activity within historical times had been more fully recorded than any other volcano except Vesuvius; but it was far larger than Vesuvius, and the scale of some of its eruptions in recent centuries, judged by the extent of the lava flows produced, had been quite as great as any in its entire history. With an example of modern vulcanism on such a scale, it would be difficult to argue that similar processes could not explain such vast ancient volcanoes as that of the Cantal, or to maintain that the intensity of volcanic activity had diminished in the course of earth history. With a prospective eye to his

[39] *Principles*, Vol. I, p. 93.
[40] *Principles*, Vol. I, pp. 177–179.
[41] C. Lyell and R. I. Murchison, "On the Excavation of Valleys, as Illustrated by Volcanic Rocks of Central France," *Edin. New. Phil. Journ.*, *VII* (1829), 15–48, Pls. I–III. Curiously, by the time he wrote Volume III of the *Principles*, Lyell had greatly increased the age he assigned to even the most recent of the volcanoes of Central France, ascribing them to the Miocene period (*Principles*, Vol. III, p. 268).
[42] Daubeny, *Sketch*, *passim*. The undermining of this widespread assumption of a contrast between the present and the past "order of things" was, of course, a major object of Lyell's whole work.
[43] *Principles*, Vol. I, pp. 361–371.

use of Etna later in the *Principles*, Lyell also describes its internal strúc-
ture and mode of growth in some detail, briefly refuting von Buch's
elevation-crater hypothesis.

In the second volume, Etna figures only briefly, in a vivid thought
experiment, in which Lyell argues that even a sudden change in the
climate of Sicily would merely shift the ecological zones down the
slopes of Etna, and could not produce transmutation of species.[44]

In the third volume, however, the Sicilian Tertiary strata figure
prominently as the essential "type" for his newer Pliocene period,
dated on his geological clock by their very high proportion of molluscan fossils belonging to species still living in the nearby Mediterranean. These strata demonstrate how "the line of demarcation between
the actual period and that immediately antecedent, is quite evanescent."[45] With this role of eliminating the conceptual gap between
present and past, it is not surprising to find Lyell devoting more space
to the newer Pliocene than to any of the other three periods of Tertiary
time, nor to find that the first two-thirds of that space are devoted to
Sicily. It is within these four chapters[46] that the crucial role of Etna
becomes apparent.

Lyell first describes the thick strata of the Val di Noto (see Figure
6). He stresses the insignificant proportion of extinct species that they
contain, their astonishing degree of elevation in the center of the island,
and the evidence that they accumulated extremely slowly. These phenomena, he concludes, whetting the appetite of his readers, "suggest
many theoretical views of the highest interest in Geology."[47]

Without immediately explaining this remark, he goes on to describe
the profile of Etna as seen from the Val di Noto, arguing that these
very recent Tertiary strata probably underlie the huge mass of the
volcano, which must therefore be later in date (see Figure 5). He has to
admit that the reasoning at this point is unavoidably circumstantial; but
he is at pains to argue that the volcanic rocks interbedded with these
strata represent *submarine* eruptions unconnected with Etna, and therefore do not affect his conclusion.[48]

He then analyzes the internal structure of Etna, showing how the
great chasm of the Val del Bove allows that structure to be seen (see
Figures 2, 4). He reminds his readers how Etna seems to have been
built up during historical times by successive lavas flowing down the

[44] *Principles*, Vol. II, p. 174.
[45] *Principles*, Vol. III, p. 22.
[46] *Principles*, Vol. III, Chs. VI–IX.
[47] *Principles*, Vol. III, p. 75.
[48] *Principles*, Vol. III, pp. 81–82.

slopes of the mountain and gradually burying the older volcanic cones; he then shows that the sections on the sides of the Val del Bove reveal precisely the internal structure that would be expected (see Figure 1). There is no evidence to support von Buch's theory that a chasm like the Val del Bove might be a huge "crater of elevation" formed by the sudden upheaval of the whole mountain. On the contrary, the sections show that Etna has been built up gradually in the manner in which it is still growing, and there is nothing to suggest that its rate of growth has been significantly greater in the past.

Lyell asks, therefore, "whether any estimate can be made of the length of the period required for the accumulation of the great cone" of Etna.[49] He prepares his readers for his conclusion by reminding them of his belief "that confined notions in regard to the quantity of past time, have tended, more than any other prepossessions, to retard the progress of sound theoretical views in Geology."[50] He argues that Adanson's method of estimating the age of a large tree, by taking a sample of its outer growth rings, can be applied by analogy to a volcano, although with less precision. A volcano does not grow with such regularity as a tree, nor are its growth rings (i.e., the successive lava flows) continuous around its circumference. Nevertheless, they can suggest a quantitative estimate of age.

Thus, Etna has a circumference of some 90 miles, so it would need 90 flows, each a mile wide, to raise the sides of the mountain by even the thickness of one flow. Yet the flows recorded in historical times do not even amount to this much change. Moreover, out of some 80 major cones on the flanks of Etna, only Monti Rossi has been erupted in recent centuries. If even a quarter of these 80 cones were formed in the 30 centuries since the first human records of the mountain, that would still imply no less than 12,000 years for all 80. Yet these 80 are only the most recent cones, not yet buried by accumulating lava flows; far more cones must be buried out of sight within the mountain, as the Val del Bove sections demonstrate.

However the facts are viewed, Lyell concludes, "we cannot fail to form the most exalted conception of the antiquity of this mountain." Etna "must have required an immense series of ages anterior to our historical periods for its growth; yet the whole must be regarded as the product of a modern portion of the newer Pliocene epoch."[51] He thought his readers might become "reconciled" to this "staggering" in-

49 *Principles*, Vol. III, p. 95.
50 *Principles*, Vol. III, p. 97.
51 *Principles*, Vol. III, pp. 99, 101.

ference, if they also considered the degree of uplift that the newer Pliocene strata had undergone within the same period of time; for he attributed the uplift to the injection of an even vaster mass of subterranean "lava."[52] This was the logical outcome of the neo-Huttonian tectonic theory that had first brought him to Sicily.

This, then, was the key position of Etna in the strategy of the *Principles*. Lyell had expounded the sufficiency of actual causes to interpret the phenomena of the past if only a vast enough time scale were conceded. He had devised a geological clock, based on the continuous changes in the organic world, by which to show that the Tertiary period alone represented an immense interval of time, even by accepted geological standards. What was needed, to link these aspects together, was a means of calibrating the geological clock, however approximately, by the time scale of human history; this calibration was what Etna could supply.

But such an illustration of the disparity between the human time scale and the geological needed to carry not only intellectual cogency but also imaginative power. Lyell seems to have realized that he had to persuade the imagination of geologists, far more than their intellect, into recognizing the signs of a time scale that was, indeed, almost beyond imagination. It is significant, therefore, that for a frontispiece to his reconstruction of earth history he chose part of his own panorama of the Val del Bove.[53] If anything could effect in his readers the necessary conversion of the imagination, it would be the contemplation of the massive grandeur and immense antiquity of Etna, coupled with the thought of its insignificant place in the whole history of the earth.

[52] *Principles*, Vol. III, p. 107.
[53] Having decided at a late stage to publish the *Principles* in three volumes instead of two, Lyell had to leave this frontispiece incongrously in Volume II, although its relevance is wholly to Volume III (see Vol. II, pp. 303–304). It is significant that he chose the same view of Etna as one of his principal "visual aids" for his lectures at King's College in 1832 (before Volume III was published), commissioning the scene painter Scharf to make a 9-foot enlargement (*Life*, Vol. I, pp. 317, 365). Like the frontispiece to Volume I (the half-submerged "Temple of Serapis" at Pozzuoli), it was carefully chosen for its imaginative impact.

V

HISTORICAL ANALOGIES
IN THE GEOLOGICAL WORK
OF CHARLES LYELL

Introduction

Hooykaas's work on the 'principle of uniformity' has helped to make it a commonplace among historians of science that geology occupies a special position in the differentiation of the sciences[1]). When geology emerged in the early nineteenth century as a new and coherent discipline, it was distinctive because it was the first area of natural science in which the dimension of historical time became constitutive. By the 1830s, this characteristic was clear enough for the English polymath William Whewell to choose geology as his 'type' example of the sciences he termed 'palaetiological'[2]). By this he meant that geology was centrally concerned with the *causal* analysis of *past* events. For geologists, therefore, the concept of the 'uniformity of nature' became far more than an abstract metaphysical foundation for their science: it was the immediate basis for their everyday practice of the actualistic method. Once this method of reasoning from present to past had been shown to have heuristic value in geology, it was quickly absorbed by other areas of natural science, notably by biology; and it became a taken-for-granted component of the normal practice of many sciences as we know them today.

Like all good pioneer works, Hooykaas's analysis of the principle of uniformity raised as many problems as it tackled. It charted the conceptual positions of many important individuals scattered through history, and displayed the variety of meanings and uses that the idea of uniformity had had. But like a pioneer triangulation of unexplored territory, the analysis fixed the relative positions of key individuals in a rather abstract manner; the character of the topography itself, or in other words the development of the concepts within real individual minds in real historical situations, was only sketched in outline. Paradoxically, therefore, the very success of Hooykaas's project has highlighted some important historical puzzles about the circumstances in which the first 'palaetiological' science developed. I want today to discuss one such puzzle.

In surveying the history of science to find an authentic ancestor-figure, geologists have continually returned to the Anglo-Scottish lawyer (as he was originally), Charles Lyell (1797-1875). Even when nationalistic pride has caused geologists to champion a different hero of the actualistic method — for example Constant Prévost, Karl Ernst Adolph von Hoff, or of course the ubiquitous Lomonosov — the claims of such rivals have had to be measured against Lyell's. Historians of science, on the other hand, have tended in recent years to develop a 're-visionist' interpretation of Lyell and his 'uniformitarian' geology. With the exception of Wilson, the current biographer of Lyell, we have built on Hooykaas's conceptual analyses and on Cannon's historical insights, and we have tended to stress the validity and coherence of the geological work of Lyell's so-called 'catastrophist' opponents³). Hence we have tended to minimise the originality of Lyell and to blend him into his background. This 'demythologising' of Lyell has now gone so far that some participants in the international Lyell Centenary Symposium in 1975 were left wondering what precisely they were supposed to be celebrating!⁴)

I think the time has now come to revise this revisionism, but not by returning to a tradition of hagiography that made Lyell into a hero-figure. Instead, we should accept two pieces of evidence as being important for historical understanding: firstly, that Lyell's geological contemporaries of many nationalities regarded him as a theoretician and field-geologist of first-rate importance, even if they found him arrogant or thought some of his ideas wrong-headed; secondly, that Lyell himself sincerely *believed* that he was demonstrating for the first time the correct principles of reasoning in geology. Taking these contemporary judgements as a starting point, I want to look at certain analogies or models that crop up again and again in Lyell's early work, both private and published, to see whether they offer us a clue to the nature of his own self-perceived originality. All the analogies I shall discuss are in a broad sense *historical* analogies: they were drawn from various areas of contemporary study and debate about human affairs, but all of them involved some measure of historical or diachronic comparison.

V

The analogy with human historiography

The basic analogy I want to explore is the analogy between geology and human historiography itself. Lyell began his great three-volume work the *Principles of Geology* (1830-33) with a short introductory chapter, the summary of which began with the words 'Geology defined — Compared to History'[5]). Here he used the analogy with the study of history to outline his belief that geology should be concerned with *changes* and their *causes,* and not at all with ultimate *origins.* He concluded:

> 'Geology differs as widely from cosmogony, as speculations concerning the creation of man differ from history'.[6])

The same sentiment can be found in his private notebooks soon after he first began to think about writing a book. But here the phrase 'History of a nation'[7]) suggests that the model that Lyell had in mind was a reformed historiography in which grandiose theorising about ultimate origins was replaced by the more limited goal of a critical reconstruction of the history of particular nations.

Where did Lyell find such a historical model for a reformed geology? In his survey of the history of geology itself, Lyell argued that *no* earlier geological tradition had gained more than partial insights into the true method that the science should follow[8]). He ended this long introduction to the *Principles of Geology* by claiming for himself, and for the followers he hoped to attract, that in geology 'the charm of first discovery is our own'. I believe that one reason for this sublime confidence in his own originality can be detected in the rest of that concluding sentence:

> 'As we explore this magnificent field of inquiry, the sentiment of a great historian of our time may continually be present to our minds, that "he who calls what has vanished back again into being, enjoys a bliss like that of creating"'.[9])

This 'great historian' was Barthold Georg Niebuhr (1776-1831), and Lyell was quoting from the introduction to Niebuhr's *Römische Geschichte.* The rest of Niebuhr's last sentence is, like Lyell's, worth quoting in full (in the translation that Lyell read):

> 'It were a great thing, if I might be able to scatter, for those who read me, the cloud that lies on this most excellent portion of

(hi)story, and to spread a clear light over it; so that the Romans shall stand before their eyes, distinct, intelligible, familiar as contemporaries, with all their institutions and the vicissitudes of their destiny, living and moving'.[10])

It would be difficult to improve on this passage from Niebuhr's history, as a summary — in analogical terms — of all that Lyell hoped to achieve in geology: to 'scatter the cloud' of poetic mystery that had been spread over the earth's past, and to throw it into 'clear light'; to 'call what has vanished again into being', re-creating the earth's former inhabitants 'living and moving' within their original environments, making them 'distinct, intelligible, familiar as contemporaries'. Niebuhr's history of the Romans provided a perfect model for Lyell's reformed method for recovering the history of the earth.

The first part of Niebuhr's *History of Rome* was not published in English translation until 1828, by which time Lyell's theoretical position in geology was already well established (and he could not have read the original before 1829, when he first learnt German). In general terms, however, he was probably familiar with Niebuhr's work at an earlier date: the newly enthusiastic Germanist Thomas Arnold reviewed the *Römische Geschichte* in the influential *Quarterly Review* in 1825, only a year before Lyell contributed his own first article to it.[11]). Anyway Lyell, as a competent classical scholar, would surely have taken an interest in the new German *Altertumswissenschaft:* at that time it was still novel among English intellectuals, and would have been normal matter for discussion in the circles in which Lyell moved.[12]).

In a more limited sense, of course, the analogy with human history was already so well established as to be a commonplace of geological rhetoric. The metaphors of coins and medals, monuments and inscriptions, can be traced back even into the seventeenth century. For our present purpose, however, we need go no further back than Lyell's hero-figure the French anatomist Georges Cuvier. In the third edition of his great *Recherches sur les Ossemens fossiles,* published in 1825 during the most creative period of Lyell's life, Cuvier recalled his earlier pioneer work:

'Antiquaire d'une espèce nouvelle, il me fallut apprendre... à restaurer ces monumens des révolutions passées'.[13])

Yet 'antiquary' was precisely the right metaphor for Cuvier to use of himself. His anatomical reconstructions of extinct mammals from their fragmentary bones were analogous to an antiquary's reconstruction of (say) a Greek temple, by piecing together the fallen pillars and architraves. Perhaps Cuvier's work went somewhat further, since he did try to re-create his extinct animals as living beings that had been adapted to particular modes of life. Yet they were not set within a total reconstructed environment. William Buckland — Cuvier's follower and Lyell's teacher — did later extend Cuvier's approach by reconstructing *one* episode in its entirety. He used impeccable actualistic reasoning to recover the total ecological environment of the pre-diluvial cave-hyaena — its eating habits, even its defecating habits — with such vivid immediacy that his friends hardly knew where careful inference ended and playful fantasy began.[4])

I see no good reason to doubt that Lyell's admiration for Cuvier's work was sincere, or that Buckland's vivid reconstruction of the pre-diluvial scene was a model for Lyell's own word-pictures of past environments — this is just where contrasting labels such as 'catastrophist' and 'uniformitarian' can be most misleading.[5]). Yet what we find in Lyell's work, far more clearly than in that of any of his elders or contemporaries, is the consistent attempt to turn such isolated 'antiquarian' reconstructions into a truly historical *sequence* of reconstructions in continual flux.

I have already suggested that the model for this causal and sequential concept of the earth's history came from a deliberate and conscious analogy with human historiography. This is expressed clearly in a short essay entitled 'Analogy of Geology and History' which Lyell wrote while he was working in the Massif Central in 1828. Here he compared the human history of Auvergne with its earlier geological history of volcanic eruptions: in both cases the observable phenomena failed to make sense unless one recognised 'that the present state of things has grown out of one extremely different'. Lyell believed that to neglect this causal connection of past and present would be to repeat the mistake of earlier historians:

'We err as much as when we judge of a political constitution without considering the pre-existent state of the laws from which it has grown'.[16])

Although Lyell did not here refer to him by name, Niebuhr had combatted precisely this error, by tracing in particular the gradual development of the Roman legal system — a point that Lyell, as a lawyer himself, must surely have appreciated. As Thomas Arnold put it in his review article:

> 'Upon principles such as these Niebuhr has proceeded, and in doing so has adopted the only method by which a real knowledge of Roman history can ever be obtained'.[17])

Likewise, I suggest, Lyell realised that the correct 'principles' for geology, the only 'method' that would lead to 'real knowledge' of the earth's history, would be derived from a strictly analogous attention to the sequence of successive states through which the 'system of nature' had passed from epoch to epoch. The causal connection of past and present, in geology as in history, was indeed to be expected; as Lyell expressed it at the end of his little essay on the analogy:

> 'it belongs to the whole constitution of things, and if natural causes operate as they do, from day to day continually, one epoch must be affected by the events of another'.

This conclusion may seem obvious to us or even naive, yet it is important to see how Lyell took this insight from human history seriously, and applied it analogically to geology with a throughness that is unmatched among his contemporaries. He was not original in using actualistic reasoning to reconstruct episodes from the past; but only the analogy with human history could underpin his distinctive vision of the earth's past as a continuous causal chain, yielding a dynamic flux of total reconstructed *history.*

The analogy with linguistics
The second analogy I want to mention is the analogy with linguistics. In general terms, of course, the notion of a 'language of nature' is a very old one. Moreover, like the metaphors of the antiquary and his 'monuments', it was also well established by Lyell's day in the more specific sense of a means of deciphering the *past* history of nature. To illustrate this, I will extend a quotation I have already used. When Cuvier termed himself a new kind of antiquary, he described his task as

V

'à la fois à restaurer ces monuments des révolutions passées, et à *en déchiffrer le sens'.*[18])

In other words, the 'coins' and 'monuments' of nature could yield little insight unless the 'inscriptions' on them could be 'deciphered' and 'read'.

Lyell adopted this metaphor enthusiastically. For example, shortly after his essay on the history of Auvergne, he wrote as follows about his work with Roderick Murchison near Vicenza:

'The volcanic phenomena were just Auvergne over again, and we read them off, as things written in a familiar language, though they would have been Hebrew to us both six months before'.[19])

Lyell's adoption of the metaphor of linguistic decipherment is important, because it suggests how he was able to escape from the stultifying empiricism and aversion to theorising that characterised his main professional milieu, the Geological Society of London. For the metaphor implied a recognition of the essentially interpretative or hermeneutic task of the geologist. On this view, 'Baconian' fact-collecting was impossible in geology, or anyway useless, for the 'documents' of the past history of the earth had to be 'read' in a 'language' that had to be learnt.

Another example may illustrate this point more clearly. Lyell's work in Italy in the winter of 1828-29 convinced him that fossil molluscs held the key to an understanding of the relatively recent, or 'Tertiary', epochs of the earth's history. On his way back to England, he therefore stayed in Paris to study conchology under the guidance of Paul Gérard Deshayes. Writing home about his work with fossil shells, he said:

'It is the ordinary, or as Champollion says, the demotic character in which Nature has been pleased to write all her most curious documents'[20])

Here the immediate reference was to the newly successful decoding of Egyptian hieroglyphs, using the demotic script of the Rosetta stone as a linguistic 'bridge' to the Greek text. But Lyell's use of this linguistic analogy emphasises how he saw fossil molluscs as a natural 'language' which had to be deliberately learnt; at the same time, he implied that this language was easier to decipher, and perhaps a more reliable guide, than the spectacular but problematic 'language' of the fossil vertebrates that Cuvier had reconstructed.

Later, when writing the *Principles of Geology,* Lyell repeatedly used the metaphors of the 'documents' or 'inscriptions' on nature's 'monuments', as a means of illustrating the fragmentary character of the 'records' or 'annals' that nature had left. He stressed the need to learn the full vocabulary of the 'living language' of nature, or in other words the full extent of geological 'causes now in operation', before the records of past events could be correctly interpreted.[21]) Even Cuvier himself, Lyell suggested, had been inconsistent in concluding prematurely that 'the thread of induction was broken' between present and past, for Cuvier's successful reconstructions of extinct animals had been

'an acknowledgement, as it were, that a considerable part of the ancient memorials of nature were written in a living language'.[22])

At one point, Lyell applied the linguistic analogy in a significantly different way. Like many of his contemporaries, he was deeply struck by the rediscovery of Herculaneum under the foundations of Portici in the shadow of Vesuvius. He imagined an antiquary of a future period discovering a still fuller sequence of *three* buried cities, one above the other, in which inscriptions might show that the language had been Greek in the oldest city, Roman in the second and Italian in the youngest — a plausible possibility in Campania. Lyell used this thought-experiment to suggest that:

'the catastrophes, whereby the cities were inhumed, might have no relation whatever to the fluctuations of the language; and that ... the passage from the Greek to the Italian may have been very gradual, some terms growing obsolete, while others were introduced from time to time'.[23])

The primary purpose of this illustration was to emphasise by analogy the fragmentary nature of the geological record, and the need to distinguish occasional accidents of preservation from the original sequence of continuous processes. But there are two incidental features about Lyell's thought-experiment that are worth noting.

First, it shows even more clearly than my other examples that Lyell was aware of human languages as changing dynamic systems. Like his awareness of German historiography, his familiarity with recent developments in linguistics needs no special explanation: it was a subject of general interest in the intellectual circles in which he moved, and Lyell himself as a classical scholar would naturally have shared this interest. Certainly he was familiar enough with Sir William Jones's pioneer research on Sanskrit to use Jones's edition of the *Institutes of Hindoo Law* as the first item in his review of the history of geology.[24]) He may also have known of the more recent German philology of Grimm, Bopp and others, which became well-known in England after Rask published a review of Grimm's work in 1830.[25])

The idea of language as a continually changing system was thus available to Lyell as a conceptual resource from his general intellectual environment. Yet it is significant — and this is my second point — that he applied it only in a limited way. He does not seem to have grasped one essential feature of the new linguistics, namely the idea that words themselves have undergone imperceptibly gradual changes in the course of time. In his geological analogy, Lyell seems to have taken words as the basic units of language, units which could be 'introduced' or 'grow obsolete' but not themselves change. The language as a whole could change very gradually, but only by piecemeal change in the constituent units, the words. This model is strikingly parallel to Lyell's image of piecemeal change in faunas and floras, whereby the totality of organisms could change very gradually, but only by the introduction and extinction of the constituent units, the species. This parallel suggests that Lyell's insistence on the reality of unchanging units in biology may have had other reasons than a desire to avoid or forestall evolutionary explanations, and may have been rooted more deeply in the structure of his thinking.

The analogy with demography

The third analogy I want to mention is the analogy with demography or populational analyses. One essential component of Lyell's use of this in his geology was his concept of a biological species as an analogue of an individual organism. The initial value of this analogy was that it suggested a possible explanation of the phenomenon of extinction, the reality of which Cuvier's research had demonstrated for the first time beyond reasonable doubt. If species, like individuals, had intrinsically limited life-spans, then extinction might be explained without recourse to mere chance or accident. Lyell was using this analogy even in his first published essay on geology in 1826. Here he suggested that extinction was:

> 'a phenomenon perhaps not more unaccountable than one with which we are familiar, that successive generations of living species perish, some after a brief existence of a few hours, others after a protracted life of many centuries'.[26])

More than ten years previously Giovanni Battista Brocchi had written in the same context about the:

> 'disuguale misura dispensato agl'individui delle specie diversi: l'efemero non campa che poche ore, mentre il cervo, come si vuole, si mantiene per qualche secolo'.[27])

The almost identical phrasing strongly suggests that Lyell had read Brocchi's 'Riflessioni sul perdimento delle specie' even before he wrote this early essay.

This is not a matter of trivial *Quellenforschung:* it is important because it implies that this crucial analogy between species and individuals was available to Lyell at an early date in a context that linked it explicitly with the problems of Tertiary geology, for Brocchi's essay on extinction concluded the introduction to his great *Conchiologia fossile subappenina* (1814). Certainly Lyell was applying the analogy by the time he was working in the Massif Central in 1828, for he wrote in his notebook about 'the laws which regulate the comparative longevity of species'; and later, in the *Principles of Geology,* he wrote quite casually about the 'birth and death of species'.[28])

I cannot here go into the highly important and much-debated question of Lyell's early views on the origin of species. But I think it is worth noting that the analogy between species and individuals did give him an outline model for organic change that would be strictly naturalistic and yet non-Lamarckian.[29]) If species were in some sense 'born', they must have arisen by natural means from some pre-existing species; yet the analogy also implied that species might appear at definite points in time, with an individuality that would distinguish them from other species and characterise them throughout their life-span.

For our present purpose, however, what is more important is the way that Lyell extended this essentially biological analogy between species and individuals into a more complex model of organic change. In doing so, he made significant use of an analogy with demography. The stages by which he developed this analogy, and its role in his creative thinking, are still obscure; but in the last volume of the *Principles of Geology* it formed the core of his interpretation of the scattered Tertiary strata of Western Europe, which in turn was presented as a model example of his method of reconstructing the past history of the earth as a whole.

Lyell introduced the analogy simply as an illustrative device:

'Let the mortality of the population of a large country represent the successive extinction of species, and the births of new individuals the introduction of new species'.

Lyell then imagined itinerant 'commissioners' visiting the different provinces of the country in succession, taking a series of censuses and then leaving behind them a series of 'statistical documents' for each province. The recorded changes in the population of a given province would then be proportional to the lapse of time since the previous census, subject to disturbing factors such as epidemics or immigration. This, Lyell argued, would be analogous to the observed changes in the assemblages of species comprising the total faunas preserved in the various 'basins' of Tertiary strata.[30])

This was an effective analogy for Lyell to use, since regular decennial censuses had been introduced in Britain in 1801, and by the time he was writing there had been a succession of four such periodic censuses.[31]) The system had been introduced precisely in order to discover whether the population was changing in size, and if so, how and in what direction. The censuses were the empirical counterpart to the famous Malthusian debates on social policy in relation to population. Once again there is no special problem about Lyell's familiarity with these questions. Although he was too much a dedicated professional scientist to get involved in politics himself, he could hardly avoid being aware of political issues of such central importance.[32]) Furthermore — to mention just one example — one of his closest geological friends and allies, George Poulett Scrope, was also one of the most able political economists in the anti-Malthusian camp.

I suggest, therefore, that Lyell used the current Malthusian debate as a readily available source for an analogy that was of central importance in his geology: namely the demographic model of a human population in continual flux, as an analogue of the continual changes in the earth's fauna and flora through the epochs of geological time.[33])

The analogy with political economy
This brings me to the closely related analogy with political economy. I have suggested elsewhere how a casual metaphor in Scrope's geological work — the idea that geologists must make 'almost unlimited drafts upon antiquity' — shows that he was using a parallel between money and time as an analogy of great heuristic power; but there is no evidence yet that in this particular form the analogy operated outside Scrope's own mind. I also suggested, however, that in a more general way the statistical approach of political economy, as developed in the work of Scrope and his contemporaries, may have been profoundly influential on the geology of Lyell.[34]) On both sides of the Malthusian debates, political economists saw the whole economic system as one of antagonistic forces, operating through a complex mass of individual decisions and events. How far Lyell was familiar with theoretical political economy we do not know, though he was reading Adam Smith's classic *Wealth of Nations* as early as 1823.[35]) Yet in any case the political economists' image of a complex balanced system is strikingly parallel to Lyell's overall vision of the geological system.

It is probable, of course, that Lyell derived the general concept of balanced antagonistic forces in geology from the natural philosophy of Adam Smith's Edinburgh friend James Hutton, as mediated through John Playfair's *Illustrations of the Huttonian Theory of the Earth* (1802). Yet that orthodox interpretation of intellectual influences fails to acknowledge one subtle but profound contrast between the Huttonian tradition and Lyell's work. Like the 'conjectural history' of the Scottish Enlightenment, Hutton's system — and Playfair's re-interpretation of it — was one of 'conjectural' or idealised continents and oceans. It was almost as an afterthought that Hutton mentioned that the processes of erosion, sedimentation and elevation would not really be successive but simultaneous.[36]). Lyell, on the other hand, never left his readers in any doubt that every one of his more varied battery of 'causes now in operation' had been in operation simultaneously somewhere on earth, at every moment in the past. The categories of Lyell's analysis were never 'conjectural' continents and oceans, but *this* volcano, *that* river valley, an earthquake *here,* an ocean current *there.* In other words, in Lyell's view the dynamic equilibrium of the earth was the result of a summation of innumerable concrete local events. It is this that gives Lyell's system a statistical character that places it much nearer the political economy of his contemporaries like Scrope, than to the earlier natural philosophy of Hutton.

At the same time, however, a comparison between Lyell's work and Von Hoff's shows how the theoretical framework of balanced antagonistic forces — whether derived from the Huttonians or not — was as essential to Lyell's system as concrete local data on geological processes and their results. Alexander von Humboldt recommended Lyell to read Von Hoff's work, and Lyell learnt German specifically in order to do so.[37] Von Hoff's *Geschichte der durch Ueberlieferung nachgewiesen natürlichen Veränderungen der Erdoberfläche* was used extensively by Lyell; but when Scrope was about to review the *Principles of Geology,* Lyell told him:

> 'You should compliment him (Von Hoff) for the German plodding perseverance with which he filled two volumes with facts like statistics, but he helped me not to my scientific view of causes'.[38]

This was somewhat bluntly expressed, but nevertheless strictly accurate. Von Hoff listed his data under headings such as 'Vergrösserung der Meeresfläche' and 'Vergrösserung der Oberfläche des Landes', without distinguishing between the many *different* processes that had produced such extensions of the sea or land during historic times. Thus although Lyell could use Von Hoff's work profitably as a quarry for examples, its arrangement totally obscured the dynamic interplay of antagonistic forces which Lyell wished to expound.

It is significant that Lyell described Von Hoff's work as containing 'facts like statistics'. In the original sense of the word, 'statistics' were the empirical data on which it was hoped that the 'statist' or statesman could base rational policies. The fact-collecting model of 'statistics' would therefore have come naturally to a professional civil servant like Von Hoff. Yet it is important to note how far Lyell transcended Von

Hoff's work. Although he appreciated the value of such factual documentation in geology, Lyell saw the limitations of the empiricism of contemporary 'statistics', and he used Von Hoff's data in the service of a far more theoretical enterprise. I therefore suggest that what we can see in Lyell's work, in this respect, is a creative fusion between the abstract 'conjectural' model of a natural system in dynamic equilibrium, and the stochastic perspective that contemporary political economy derived from empirical 'statistics'.

Conclusion

In this paper I have suggested how a wide range of cultural resources was available to Lyell. This was simply a result of his own broad interests, which in turn were related to his favourable milieu at the centre of English metropolitan intellectual life. Indeed, since he was a good linguist and also in a position to travel extensively all over Europe, he was in personal contact with *savants* in Paris and other cultural centres, so that we can say that his milieu was at the centre of *European* intellectual life. It is therefore not surprising that he was familiar with contemporary studies of human affairs as diverse as historiography, linguistics, demography, and political economy. Lyell transposed these resources analogically into geology, and used them to help create his own rigorous conception of the proper method of geological research and his own distinctive vision of the past history of the earth. From historiography he gained an awareness of the deep causal connectedness of past, present and future, and the need for critical scrutiny of fragmentary evidence. From linguistics he gained an awareness that geology must be interpretative or hermeneutic throughout, and that the 'languages' of nature needed to be learnt. From demography he gained an awareness that gradual overall change could be the summation of piecemeal changes in innumerable discrete units. From political economy he gained an awareness that innumerable local events of a stochastic character could add up to a total system that was providentially in harmonious balance.

The cultural resources that Lyell used in his geology were equally available at the time to many other geologists of several nationalities: they too were generally gentlemen of liberal education and broad interests. Yet none of them produced a synthesising system of geology as comprehensive and coherent as Lyell's. I propose no explanation for this here, except to note that Lyell did step outside the institutional constraints that were imposed by the already well-established discipline of geology. He ignored the norms of theory-free empiricism that characterised the Geological Society of London; but equally he ignored the norms of esoteric technical publication that characterised the other great centre of geological research, in Paris. He openly erected a scheme of vast theoretical scope, and he aimed all his important work at a public that included but also extended beyond his fellow-specialists.[39])

A few years ago Kuhn commented that

> 'except in the rudimentary stages of the development of a field, the ambient intellectual milieu reacts on the theoretical structure of a science only to the extent that it can be made relevant to the concrete technical problems with which the practitioners of that field engage'.[40])

For the case-study I have discussed today, I think that this is true only in an indirect sense. Lyell may have thought that geology was in its 'rudimentary stages', but few of his contemporaries would have agreed, and also few modern historians of geology. On the contrary, when Lyell entered the field in the early 1820s, geology was already well established, both institutionally and cognitively. Thanks to the pioneer work of Hooykaas and other historians, it is now clear that Lyell was not such a radical innovator as some later geologists imagined (and as Lyell himself encouraged them to believe). At least in the formal sense, Lyell did *not* introduce some new 'actualistic method'; he did not formulate some cognitive panacea that was immediately relevant to the 'concrete technical problems' of contemporary geologists. Yet because certain aspects of Lyell's system had little influence on other geologists, we are now in danger of concluding that Lyell's achievement was no more than a public relations exercise, a popularisation of geology that almost accidentally helped forward some broad current of opinion labelled 'scientific naturalism' or 'evolutionary thought'.

This interpretation is surely at fault, because it assumes implicitly that a methodology such as actualism can only operate at one level. I suggest that the importance and influence of Lyell's work stemmed from the *confidence* with which he argued for the heuristic value of comparisons between present and past, and from the *persuasiveness* of the 'worked examples' that he accumulated. At this level his work was indeed relevant to 'concrete technical problems', yet we will not find evidence of his influence on other geologists if we search in their work merely for imitative features or open acknowledgements. At a deeper level, on the other hand, I believe we can see how Lyell's confidence in his approach gradually 'caught on' among his fellow-geologists of all nationalities.

I have not attempted to document this process here; instead, I have concentrated on the source of Lyell's own *self*-confidence. And it is here that the effect of an 'ambient intellectual milieu' on a developed science can be seen most clearly. Lyell reformulated the methods and aims of geology, partly by transposing a series of 'key' analogies from outside geology and indeed from outside natural science altogether. While I do not suggest for a moment that these analogies were his *only* source of inspiration, I do believe that the clear importance and success of these quite different fields of study gave Lyell a profound sense of confidence in his own enterprise. If this conclusion is valid, it implies that even a developed science can remain open to influences from its 'ambient intellectual milieu'; yet that influence may lie in the realm of intangibles such as overall 'confidence' in a certain approach, rather than in the more concrete realm of 'technical problems'.

In this paper I have used the example of Lyell's work for two reasons. I wanted to illustrate one way in which the abstract principle of actualism was given flesh and blood in the thinking and writing of one individual of international stature in the history of science. Secondly, however, I wanted to illustrate the working of one of the most important processes of creative thinking in science. Major innovations in natural science have generally come from individuals who have been thoroughly 'professional' with respect to their standards of work, but who at the same time have been prepared and able to transcend the narrow vision of most professionals, and to transpose key ideas and images from one area of thought to another. My example therefore has a lesson in it, I think, for all of us who are concerned with education in natural science.

REFERENCES

1) R. Hooykaas, *Natural law and divine miracle,* Leiden, 1959 (second impression, *The Principle of Uniformity,* Leiden, 1963).

2) W. Whewell, *History of the Inductive Sciences,* London, 1837, Book 18.

3) L.G. Wilson, *Charles Lyell, the years to 1841; the revolution in geology,* New Haven, 1972 (this biography contains valuable quotations from otherwise unpublished manuscripts); Hooykaas, *op. cit.,* and 'Catastrophism in Geology, its Scientific Character in relation to Actualism and Uniformitarianism', *Kon. Ned. Akad. Wet., Lett., Meded.* n.r., vol. 33 (1970), 271-316; W.F. Cannon, 'The Uniformitarian-Catastrophist debate', *Isis,* vol.51 (1960), pp. 38-55, and 'The problem of miracles in the 1830s', *Victorian Studies,* vol.4 (1960), pp. 4-32; M.J.S. Rudwick, 'A critique of uniformitarian geology', *Proc. Amer. philos. Soc.,* vol. 111 (1967), pp. 272-287, and 'Uniformity and progression: reflections on the structure of geological theory in the age of Lyell', in D.H.D. Roller (ed.), *Perspectives in the history of science and technology,* Norman (Okla.), 1971, pp. 209-227.

4) A selection of papers from this Symposium is published as the *Lyell Centenary Issue,* in *Brit. J. Hist.Sci.,* vol. 9, part 2 (1976).

5) C. Lyell, *Principles of Geology,* London, 1830-33, vol. 1, p.1 (I use the first edition of each volume throughout this article).

6) *Principles,* vol. 1, p.4. In a strikingly parallel passage Desmarest had earlier remarked that 'La théorie de la terre est à la géographie physique ce que le fable est à la histoire': *Encyclopédie méthodique, Géographie physique,* tome 1 (1975), Préface.

7) Quoted in Wilson, *Lyell,* p. 172, from a notebook dated about July 1827.

8) R. Porter: 'Charles Lyell and the principles of the history of geology', in *Lyell Centenary,* pp. 91-103. I now agree with Porter that Lyell was *not* claiming intellectual ancestry even among the Huttonians.

9) *Principles,* vol. 1, p.74.

10) B.G. Niebuhr, *The History of Rome,* trans. J.C. Hare and C. Thirlwall, vol.1, Cambridge, 1828 (original edition, *Römische Geschichte,* Berlin, 1811-12).

11) Anon. (T. Arnold): '*Römische Geschichte* von B.G. Niebuhr...,' *Quarterly Review,* vol. 32 (1825), pp. 67-92.

12) D. Forbes: *The Liberal Anglican idea of history,* Cambridge, 1952, ch.3. K. Dockhorn: *Der deutsche Historismus in England,* Göttingen, 1950. Lyell's membership of the Athenaeum in London (he joined the club in 1824 soon after its foundation: see Wilson, *Lyell,* p.35) was probably of particular importance in this respect.

13) G. Cuvier, *Recherches sur les ossemens fossiles,* 3e ed., Paris, 1825, vol. 1, 'Discours sur les révolutions du globe', p.1. Compare G. Buffon, *Époques de la Nature,* Paris, 1778, 'Premier discours'.

14) W. Buckland, *Reliquiae diluvianae,* London, 1823. See for example Lyell's reaction, quoted in Wilson, *Lyell,* p. 95.

15) Cuvier and Buckland are regarded as strongly actualist (in my opinion, correctly) in W.F. Cannon, 'Charles Lyell, radical actualism, and theory', *Lyell Centenary,* pp. 104-120. On the other hand, I consider that *most* of Buckland's research (and of most other geologists in the 1820s) was primarily *structural* rather than causal in its cognitive aims, so that the question of actualism rarely arose: see M.J.S. Rudwick, 'The emergence of a visual language for geological science, 1760-1840', *History of Science,* vol. 14 (1976), pp. 149-195.

16) Lyell's essay is printed (incompletely) in Wilson, *Lyell,* pp. 215-6.

17) Arnold, 'Niebuhr' (note 11), pp. 71-72. Porter (*op. cit.*) points out that Lyell con-

spicuously failed to apply Niebuhrian standards to his historical account of geology itself: here 'catastrophist' historiography suited his polemical purposes better!

18) Cuvier, *op. cit.* (note 13), my italics.

19) Mrs. Lyell, *Life, Letters and Journals of Sir Charles Lyell, Bart.,* London, 1881, vol. 1, p. 203.

20) *Ibid.,* p. 251.

21) *Principles,* vol. 3, pp. 461-2.

22) *Principles,* vol. 1, pp. 72-3.

23) *Principles,* vol. 3, pp. 33-4.

24) *Principles,* vol. 1, pp. 5-6.

25) See H. Aarsleff, *The study of language in England,* 1780-1860, Princeton, 1967, chs. 4, 5; J. Burrow, 'The uses of philology in Victorian England', in R. Robson (ed.), *Ideas and Institutions of Victorian Britain,* London, 1967, pp. 180-204.

26) Anon. (C. Lyell), 'Transactions of the Geological Society of London...', *Quarterly Review,* vol. 34 (1826), pp. 507-504: see p. 538.

27) G.B. Brocchi, *Conchiologia fossile subappenina,* Milano, 1814: see vol. 1, pp.227-8. Lyell later used Brocchi's historical information — with scant acknowledgement — in the *Principles:* see P.J. McCartney, 'Charles Lyell and G.B. Brocchi: a study in comparative historiography', in *Lyell Centenary,* pp. 175-189.

28) Wilson, *Lyell,* p. 215; *Principles,* vol. 3, pp. 32-33. In vol. 2, p.37, Lyell used the analogy to explain his concept of intra-specific variation.

29) It should go without saying that Lyell could have been searching for a *natural* (i.e. non-supernatural) mode of species-origin, without affecting his belief that these events must have been supervised providentially by an omniscient Creator. For a recent interpretation that stresses Lyell's negative reaction to Lamarck, see M. Bartholomew, 'Lyell and evolution: an account of Lyell's response to the prospect of an evolutionary ancestry for man', *Brit. J. Hist.Sci.,* vol. 6 (1973), pp. 261-303.

30) *Principles,* vol. 3, pp. 31-33. For the context of the analogy, see M.J.S. Rudwick, 'The strategy of Lyell's *Principles of Geology'*, *Isis,* vol. 61 (1970), pp. 5-33.

31) D.V. Glass, 'Some aspects of the development of demography', *J.roy.Soc. Arts,* vol. 104 (1956), pp. 854-869; M.J. Cullen, *The statistical movement in early Victorian Britain,* Hassocks (England), 1975: see pp. 10-13.

32) I use the word 'professional' in the sense that Lyell was trying to make himself financially independent through his geological writing and — for a short time — teaching. See J.B. Morrell, 'London institutions and Lyell's career, 1820-41', *Lyell Centenary,* pp. 132-146; M.J.S. Rudwick, 'Charles Lyell F.R.S. (1797-1875) and his London lectures on geology, 1832-33', *Notes. Rec. roy. Soc. Lond.,* vol. 29 (1975), pp. 231-263.

33) In view of the heavy ideological load that Darwin's later use of Malthus is sometimes made to bear, it is worth pointing out that Lyell's use of a demographic model was *neutral* with respect to any concept of struggle or disharmony in nature or society.

34) M.J.S. Rudwick, 'Poulett Scrope on the volcanoes of Auvergne: Lyellian time and political economy', *Brit.J.Hist.Sci.,* vol. 7 (1974), pp. 205-242: see pp. 236-242.

35) Wilson, *Lyell,* p. 115.

36) J. Hutton, *Theory of the Earth,* Edinburgh, 1795; see vol. I, p. 195-7. Significantly, Lyell was among the many readers of Hutton (and/or Playfair) who overlooked this disclaimer: see *Principles,* vol. 1, p. 473.

37) Wilson, *Lyell,* pp. 170, 267.

38) Mrs. Lyell, *Life,* vol. 1, pp. 268-9. Only the first two volumes of von Hoff's work (Gotha, 1822, 1824) were published before the *Principles.*

39) Of course he had mundane financial motives for writing for the general educated public: my point is, however, that this situation enabled him to be openly theoretical and

V

— not least — to employ the persuasive non-geological analogies I have discussed in this paper.

40) T.S. Kuhn, 'History and the history of science', *Daedalus,* vol. 100 (1971), pp. 271-304: see p. 280.

VI

CHARLES LYELL'S DREAM OF A STATISTICAL PALAEONTOLOGY

THE name of Charles Lyell (1797-1875) is well known to geologists and palaeontologists for at least two reasons: he is widely considered a pioneer of so-called 'uniformitarianism' in the earth sciences; and he is remembered as the originator of the terms Eocene, Miocene, and Pliocene, which even in the modern era of radiometric dating still dominate the descriptive stratigraphy of the Cainozoic (for historical re-evaluations of 'uniformitarianism', see Hooykaas 1959; Cannon 1960a; Rudwick 1971). This paper aims to show that Lyell's stratigraphical terms are 'conceptual fossils'; they are fragmentary relics of an ambitious theoretical project in stratigraphical palaeontology. This project faltered and failed almost as soon as it was launched, and the stratigraphical terms were 'metamorphosed' almost out of recognition. But Lyell's project is worth reconstructing none the less, because it serves to tie his practical work in applied palaeontology firmly into his broader programme for research on all aspects of the earth sciences.

Lyell himself had few personal followers and founded no distinct 'school' or research tradition. But it is difficult to over-estimate the influence of his compendious *Principles of geology* (1830-1833) and its later offshoot the *Elements of geology* (1838), which were widely translated and repeatedly updated in successive editions through nearly half a century. Lyell's persuasive interpretations of the accumulating empirical research of the mid nineteenth century were absorbed, selectively but pervasively, into the thinking of the first generations of professionalized geologists and palaeontologists; and his general approach has remained an essential element of the 'taken-for-granted' tacit knowledge of earth scientists to the present day.

THE PROBLEM OF THE TERTIARY FAUNAS

It is no accident that Lyell's permanent legacy to stratigraphical terminology should concern the Tertiary strata ('Tertiary' was used to cover all the 'Cainozoic' of modern geology except what were later interpreted as glacial and post-glacial deposits). This was not, however, because he sought to make an original 'contribution' by specializing in an 'under-developed' research field. Yet when Lyell first entered the community of active geologists centred on the Geological Society of London (Morrell 1976), Tertiary stratigraphy was in a certain sense under-developed. If the term 'paradigm' is used with proper caution in its historical sense (Kuhn 1962, *not* the palaeontological sense of Rudwick 1961, etc.) to describe a dominant style of research in a particular science at a certain period, then stratigraphy around 1830 was developing with great rapidity and success by following a paradigm that was primarily 'structural' in its cognitive goal (Rudwick 1976*b*) and increasingly palaeontological in its method. In other words, attention was focused on the discovery of the correct order of succession of formations, seen as a problem of three-dimensional structure; while 'characteristic fossils' were being used with increasing confidence as the most reliable (though not the only) criterion for the correlation of formations in different regions. The first aspect reached back to the highly fruitful research tradition of Werner and his followers (Ospovat 1969); the second aspect, while not altogether original to William Smith (1769-1839), certainly gained increasing emphasis from the manifest value of Smith's great geological map and its palaeontological illustrations (Smith 1815*a*, *b*; 1816-1819).

This paradigm was centred, however, on the strata with which Smith himself had had his greatest success: the so-called 'Secondary' formations (roughly equivalent to the Mesozoic and Upper Palaeozoic of modern geology). These could be divided relatively easily into formations of distinct and diverse character, many of them containing equally distinct and diverse fossils in some abundance. Below the Secondary strata, however, were the confusing 'Transition' strata and 'Primary' rocks (in modern terms, pre-Carboniferous strata and many metamorphic and igneous rocks), often highly disturbed and with few if any fossils. Above the Secondary strata were the Tertiary, which often seemed almost equally confusing, though for different reasons.

The Tertiary strata had indeed provided the paradigm of stratigraphy with an 'exemplar' that was at least as influential as Smith's, namely the study of the Paris region by Georges Cuvier (1769-1832) and Alexandre Brongniart (1770-1847). Their work showed the possibility of identifying a series of distinct formations over a wide area, using characteristic fossils as a tool of major importance; and their results were actually available in published form before Smith's (Cuvier and Brongniart 1808; 1811). But it soon seemed as if the Tertiary formations differed rather fundamentally from the Secondary. Tertiary strata seemed always to be confined to scattered 'basins'—the term was used in a quite literal sense, since the sediments were envisaged as having gradually filled pre-existing hollows in the underlying rocks. Cuvier and Brongniart described the Paris basin; Thomas Webster (1773-1844), the draughtsman at the Geological Society of London, soon afterwards described analogous Tertiary strata in the London and Hampshire basins (Webster 1814; 1816); and other basins were quickly added to the list during the 1820s. This apparent contrast between isolated Tertiary basins and widespread Secondary strata was heightened by the fact that the uppermost Secondary formation, namely the Chalk, was the most widespread and most distinctive formation of all.

A further difficulty derived from the very success of the French research. The most striking faunal distinctions among the strata of the Paris basin were given an interpretation that went far beyond the merely structural level of most stratigraphical work, reaching instead for a causal explanation. Using analogies with present-day faunas and floras, Cuvier and Brongniart interpreted the Parisian strata as an alternation of marine and non-marine formations; and a somewhat similar alternation was recognized by Webster in the Hampshire basin. This ecological dimension, coupled with the great difficulty of correlating the formations of one basin with those of another, led to a generally implicit but widespread belief that the Tertiary strata had accumulated under conditions that somehow differed radically from the Secondary.

This belief was probably reinforced by the analogous but even greater problems surrounding the most 'superficial' deposits of all. The unconsolidated, highly irregular, and often peculiar deposits that are now interpreted as glacial and post-glacial seemed naturally to suggest a major break in geological processes in the relatively recent past. Such an episode was sometimes labelled 'diluvial'; but when links with the biblical Deluge were expressed, this involved no mere literalistic interpretation (e.g. Buckland 1823), and in any case the phenomena themselves seemed to make some kind of drastic causal explanation inescapable (Rudwick 1970b; 1972, chapter 3).

At an early stage in his geological career, Lyell became dissatisfied with the prevailing tendency to assume the kinds of causal discontinuity that have just been summarized. It is difficult, however, to judge the relative importance, or mode of interaction, of theoretical and practical components in the development of his outlook. He may have been influenced by John Playfair's (1802) reinterpretation of the 'geological' aspect of the wide-ranging 'natural philosophy' of James Hutton. But since Lyell entered the field of geology in a period of self-consciously rapid development, he might well have found these authors merely 'old-fashioned' (see Porter 1976; 1977). He is likely to have been more impressed by seeing for himself the Tertiary strata around Paris. In 1823 he had as his guide Constant Prévost (1787–1853), who had already argued in public that the ecological explanation of the strata given by Cuvier and Brongniart could be brought even more closely into line with present-day analogies (Prévost 1823).

Lyell's first major scientific paper (Lyell 1826a) was devoted to showing that one characteristic Parisian lithology, a freshwater limestone, had a close modern analogue in the lakes near his Scottish family home. In his first essay on geology written for a more general audience (Lyell 1826b) Lyell extended this into the suggestion that Cuvier and other illustrious geologists had been premature in concluding that such analogues between present and past could not be found and used for explanation throughout geology. Lyell's confidence in this conclusion was strengthened by a persuasive case-study of the already classic area of Auvergne, published by his friend the political economist George Poulett Scrope (1797–1876) (Scrope 1827; Rudwick 1974). Lyell's review-essay on Scrope's work (Lyell 1827) shows clearly how he was adopting Scrope's picture of the very gradual development of the physical geography of central France, and trying to extend it to incorporate a picture of similar gradual change in its fauna and flora. Lyell explicitly used this example to illustrate how it was unnecessary to postulate sudden events of uncertain character in the past: all the observed phenomena of the Tertiary strata and their fossils could be explained without recourse to such events, simply by reference to processes observable at the present day.

At about this period, Lyell conceived the idea of writing a book that would reorientate geological interpretation along these lines. After many delays, and a change of plan that transformed it from a short elementary introduction into a massive three-volume work, the book was published as the *Principles of geology* (Lyell 1830–1833). Significantly, it was subtitled 'an attempt to explain the former changes of the earth's surface, by reference to causes now in operation'. Within this work, Lyell's reinterpretation of the Tertiary occupied a crucial position (Rudwick 1970*a*). In the first two volumes, Lyell surveyed the varied 'causes now in operation' that make for change in both inorganic and organic spheres, arguing that this repertoire of present-day processes was much more varied and more powerful than most other geologists had realized. In the final and culminating volume (1833), these processes were put to work in the causal interpretation of the stratigraphical record. Lyell illustrated this mode of interpretation chiefly by the example of the Tertiary. He chose the Tertiary because it was important for him to demonstrate the validity of his approach for the relatively recent epochs of earth-history. By doing so, he could both eliminate the supposed 'diluvial' break between the Tertiary and the present, and also go on to show that the contrast between the Tertiary and the Secondary was merely a difference of degree, not kind.

Within this over-all strategy, Lyell's concepts of Eocene, Miocene, and Pliocene played a crucial role, but a role that has often been misunderstood. To uncover Lyell's original intentions in coining these terms, it is necessary to look in more detail at previous interpretations of the Tertiary strata and at the development of Lyell's own ideas.

In analysing the formation of any novel scientific theory or project, it is important first to survey the range of other relevant theories or projects that were actually available as 'resources' that the principal figure under study could have utilized in his own construction. In the present case, at least four such existing pieces of work seem to have been relevant.

Firstly, the already classic description of the Parisian strata by Cuvier and Brongniart (1811) yielded a picture of the gradual accumulation of strata within a single basin under alternately marine and non-marine environmental conditions. Cuvier and Brongniart had given these ecological changes a 'catastrophist' interpretation, inferring that they had been caused by sudden changes in physical geography. Yet on Prévost's reinterpretation (1823) these variable conditions were to be expected, on present-day analogies, in any shallow gulf of the sea near the mouth of a large river. Not only was this the kind of explanation that Lyell intended to deploy more generally, but it also suggested that each Tertiary basin would have to be treated in the first instance as a separate entity. In other words, Lyell probably realized that it would be futile to search for exact correlations between the formations in different basins, if they owed their characteristics to essentially local factors.

A second major 'resource' for Lyell was the work of the Italian naturalist Giovanni Battista Brocchi (1772–1826), who had published a superb monograph on the molluscs of what he called the 'Subappenine' strata in north Italy, soon after the Parisian strata had been described (Brocchi 1814). It is probable that Lyell was familiar with this work at an early stage in his career: it was still perhaps the best monograph on any Tertiary fauna even ten years after publication. Its Italian was almost certainly no serious barrier to Lyell, since his father was a noted Italianist and he himself knew the language well enough in the 1820s to review a new edition of Danté (Wilson 1972, p. 187); and he also borrowed extensively from Brocchi, with scant acknowledgement, when later he was writing the historical introduction to the *Principles* (McCartney 1976). Brocchi prefaced his systematic work with a very long introduction, which included important interpretative comments on the relationship between his Italian fauna on the one hand and the Parisian and the present-day faunas on the other. Brocchi was a good enough naturalist to be aware of present-day regional faunal variations, and so he attributed the difference between the Subappenine fauna and the Parisian fauna to what we would term biogeographical factors. This was entirely reasonable, since there was no evidence to suggest any difference of age between the two basins. On the other hand, he was also aware that many of his Subappenine species had no living counterparts, and were probably extinct. However, he rejected Cuvier's 'catastrophist' explanation of extinction as inapplicable to marine molluscs—Cuvier had suggested that the extinct Tertiary land mammals had been annihilated by sudden marine incursions—and he found it a superfluous hypothesis anyway. In its place, Brocchi suggested an organismic explanation: species, like individuals, probably have a limited life-span, eventually losing their reproductive vigour and therefore dying out. On this model, extinction would be an essentially piecemeal process. This, he thought, might explain why his Subappenine fauna contained a mixture of extant and apparently extinct species. Furthermore, he pointed out that individuals of different biological groups have life-spans that vary from a few hours to a few centuries (e.g. insects and trees), and he thought that by analogy the average life-spans of species belonging to different groups might also vary widely. This could explain why the Tertiary strata contained many extant molluscan species, whereas their mammalian fossils were apparently all extinct.

Thirdly, Lyell was almost certainly familiar with the first full description of the basin in south-west France, published in 1825 by the young Parisian naturalist Barthélemy de Basterot (1800–1887). This memoir not only added another Tertiary basin to the growing list, but also included an important theoretical interpretation. Basterot derived his analysis explicitly from the new quantitative biogeography of the Swiss botanist August-Pyrame de Candolle (1778–1841). In his celebrated article on 'Géographie botanique', Candolle (1820) had used a wide range of published floras, from all parts of the world, to construct quantitative tables based on the numbers of genera and species common to various areas. These tables enabled him to distinguish twenty major biogeographical regions with distinct indigenous floras. He borrowed the then current meaning of the term 'statistics'—it denoted the collection of quantitative economic data for political use by the 'statist' or statesman—and he called his own work 'statistique végétale'. Basterot applied Candolle's botanical 'statistics' to his own palaeontological problem of the Tertiary molluscan fauna of the Bordeaux region. He had identified a total of 330 species in this fauna, and he analysed them quantitatively in two distinct ways, in relation to living species and in relation to other Tertiary faunas. Of the total number of species, only 45 were known to live in European seas and another 21 elsewhere in the world. Divided the other way, he noted that 91 species were also reported from Italy, 66 from the Paris basin, 24 from England (the Hampshire and London basins were not distinguished), and 18 from the Vienna basin; while 110 appeared to be peculiar to the Bordeaux basin. He suggested on the basis of these figures that the faunal similarity between any two basins might be roughly proportional to their geographical proximity.

Basterot's figures are not even completely consistent, and he did not attempt any more sophisticated analysis of them. His remarks are important, however, because they suggested at least the possibility of giving quantitative or 'statistical' precision to Brocchi's earlier inference that the faunal differences between the various Tertiary basins might reflect a pattern of Tertiary biogeographical factors. In other words, the diversity of the Tertiary faunas might be a function of space, not of time.

The fourth and last major 'resource' for Lyell's construction of his own interpretation of the Tertiary was the work of Scrope (1827), which has been mentioned already. Although Scrope was not concerned at all with the palaeontological aspect, he did describe the Tertiary strata of the Massif Central in enough detail for Lyell to realize that they were exclusively non-marine. More significantly, however, Scrope analysed the long history of the area, and showed that sporadic outbursts of volcanic activity had punctuated the gradual and continuous erosion of the Tertiary sediments. He argued that the lava-flows, isolated by subsequent erosion at various heights, formed 'a natural scale' or chronometer for estimating the relative *age* of the eruptions (Rudwick 1974). In other words, a quantitative measuring device, however approximate and uncalibrated, could be discovered within the geological phenomena themselves, and could convert the appearance of discontinuity into evidence of underlying continuity.

LYELL'S CONSTRUCTION OF A FAUNAL CHRONOMETER

Lyell's construction of his own distinctive theory for the Tertiary faunas, integrating the pre-existing 'resources' that have just been summarized, can be dated to the successive phases of his most important season of geological fieldwork. This was his long expedition through France and Italy in 1828-1829, partly in the company of his friend Roderick Murchison (1792-1871), who was not yet a rival. Lyell's development of his theory, which underlay his later terms Eocene, Miocene, and Pliocene, will be reconstructed here on the basis of the accessible records of his journey, namely his subsequently published letters from this period (K. Lyell 1881, pp. 182-251), some brief published extracts from his notebooks (quoted in Wilson 1972, pp. 187-261, and a few unpublished letters.

Lyell and Murchison first studied the areas that Scrope (1827) had described in the Massif Central. Lyell at least was completely convinced by Scrope's arguments for the very gradual erosion of valleys (Lyell and Murchison 1829a), and he must surely have recalled, whèn seeing the ancient lava-flows with his own eyes, how Scrope had used them as a quantitative 'natural scale' to measure geological time. Certainly he was impressed by the span of time that was implied by the hundreds of feet of thin-bedded Tertiary limestones, which were clearly the product of slow and tranquil deposition. All this confirmed his growing conviction, shared with Scrope, that most other geologists were seriously underestimating the sheer magnitude of geological time.

Lyell's interests, however, were more palaeontological than Scrope's, and he developed Scrope's pattern of interpretation in more biological directions. He checked that the Tertiary strata were indeed all freshwater in origin. Although strikingly similar to some of the Parisian limestones, they were also comparable to modern lake-marls, and contained similar freshwater molluscs and plants. On the other hand, the mammalian fauna of the Massif Central had evidently changed as much as it had in the Paris region, since local naturalists had discovered and described a fine fauna of extinct mammals (Croizet and Jobert 1826–1828; Deveze and Bouillet 1825–1827). But since there was no sign of any marine incursion into the area, Cuvier's explanation of the extinction of similar mammals in the Paris region was clearly invalid for the Massif Central. This may have confirmed Lyell's suspicion that some more gradual cause of extinction must be found. A further indication was that the mammalian fossils had not come from the freshwater limestone formation, but from a much younger—though still 'ancient'—river-gravel (Lyell and Murchison 1829a). The area thus contained a relatively recent mammal fauna that was full of extinct species, while the older Tertiary lake-deposits contained freshwater molluscs and plants much closer to those of the present day.

This evidence probably made Lyell recall Brocchi's suggestion that extinction might be caused by the intrinsic 'old age' of individual species, and that species in some groups (such as mammals) might have much shorter 'life-spans' than species in other groups (such as molluscs). Some such speculations along Brocchian lines are strongly suggested by the fact that while he was still in the Massif Central Lyell wrote a short essay in his notebook entitled 'On the laws which regulate the comparative longevity of species' (Wilson 1972, p. 215).

A few weeks later, Lyell and Murchison had left the Massif Central and reached Nice, where Lyell made use of the local knowledge of Giovanni Antonio Risso (1777–1845). He studied a thick Tertiary conglomerate formation and speculated on its causal origin; and he mentioned in a letter home that 'in the intervening laminated sands are numerous perfect shells, more than 200 in Risso's cabinet, 18 in a hundred of which are *living* Mediterranean species, whose habits are known' (K. Lyell 1881, p. 199). Risso himself, in his five-volume description of the natural history of the Nice area, had published long lists of molluscan species from the 'Formation Tertiaire' (Risso 1826, vol. 1, art. 4); but he had not distinguished the extant species among them, and the percentage expression was therefore probably Lyell's own gloss on Risso's work. It is in fact the first hint that Lyell was beginning to apply the quantitative or 'statistical' approach that Basterot had borrowed from Candolle. The context of Lyell's remark suggests, however, that he was using the numerical proportion purely as an ecological criterion, not as an indicator of geological age. The 'habits' of the extant species helped to show that the conglomerate formation had been deposited in conditions differing little from those probably still existing offshore.

After crossing the Appenines, Lyell and Murchison were able to study another important collection at Turin. Here Franco Andrea Bonelli (1784–1830) had already noted the faunal similarity between some of the local strata and Basterot's in south-west France (Wilson 1972, p. 221). Lyell recognized the Turin strata as similar to those he had been studying along the Mediterranean coast. In a letter from Milan shortly afterwards he referred back to 'the sub-Appenine beds from Montpellier to Savona, containing as they do nearly twenty per cent of decided living species of shells' (K. Lyell 1881, p. 201); and the context makes it clear that he was now using that percentage as an indication of *age*. This was because the strata near Turin were highly tilted; and Lyell already suspected from what he had seen in the Massif Central that there was a causal connection between tectonic disturbance and volcanic activity (Lyell and Murchison 1829*b*).

In pursuit of this hypothesis, Lyell left Murchison in northern Italy, and turned south towards Sicily to study active volcanoes, for in their vicinity he explicitly anticipated finding still more recent strata elevated above sea-level. This expectation was duly fulfilled (Wilson 1969; Rudwick 1969); and in the development of his hypothesis of elevation he made increasing use of the proportion of extant species in a fossil fauna as a guide to the relative age of the formation in which it was preserved. Brocchi had died only two years earlier, and so Lyell was denied the chance of a personal discussion with him; but Brocchi's Subappenine fauna, with its mixture of living and extinct species, was clearly playing a key role in Lyell's mind. Furthermore, unlike Basterot and even Brocchi himself, Lyell was now comparing different Tertiary areas primarily in terms of their relative ages, rather than in biogeographical terms. For example, after concluding his fieldwork in Sicily, he told Murchison how the strata he had studied there must be much younger than the Subappenine formation: 'I am come most unwillingly to this conclusion. But the numerous extinct species which characterise the Subapps. are wanting [missing] here, & living shells are present too plentifully, to admit a doubt that it is more related to our own epoch' (K. Lyell 1881, p. 233).

Up to this point, it seems that Lyell was using the proportion of living and extinct species in a purely empirical way and as a purely geological tool. But on his journey back through Italy a meeting with the botanist Domenico Viviani (1772–1840) at Genoa apparently turned his thoughts in a much more biological direction (K. Lyell 1881, p. 243). Lyell was so excited by what he called 'my new geologico-botanical theory' that he crossed the Alps in the depths of winter specially to talk about it with Candolle in Geneva. His report of his discussions with Viviani and Candolle merits quotation at some length.

I am now convinced that geology is destined to throw upon this curious branch of inquiry [i.e. biogeography], and to receive from it in return, much light, and by their mutual aid we shall very soon solve the grand problem, whether the various living species came into being gradually and singly at insulated spots or centres of creation, *or* in various places at once and at the same time. The latter cannot, I am already persuaded, be maintained. Viviani was puzzled to account for Sicily having so much less than its share of *peculiar* indigenous species; but this [is as it] should be, for *I* can show that three-fourths of this isle [i.e. Sicily] were covered by the sea down to a period when nine-tenths of the present species of shells and corals (and by inference of plants) were already in existence. Such an isle, like Monte Nuovo [the 'new' volcano formed in 1538 near Naples], has been obliged to borrow clothes from its neighbour, having scarcely had time to furnish any yet for its own nakedness. It has not yet seen out a tenth, perhaps not a twentieth part of a revolution in organic life. Give it the antiquity [i.e. as a *land* area] of the high granitic mountains of Corsica, and it will also boast its indigenous unique plants, unknown elsewhere either in the Mediterranean or other part of the globe [K. Lyell 1881, p. 246].

This important passage records the first rough outline in Lyell's mind of a wide-ranging *theory* of organic change, which would integrate evidence from biogeography, palaeontology, stratigraphy, and structural geology.

Like almost all naturalists at this period, Lyell believed that species were real entities, intra-specific variation being often considerable but always finite. There seemed to be good empirical grounds for rejecting Lamarck's earlier postulate of limitless variation.

The problem of accounting for the *origins* of these discrete units was therefore acute. Lyell probably shared the general belief that the production of new species must have been under divine providential control, in order to account for the precise adaptation of each species to the environment in which it was placed. Certainly he expressed this view, probably with complete sincerity, during his first public lectures on geology only three years later (Rudwick 1975; 1976*a*). But in the contemporary understanding of 'providence', such a view was quite compatible with a belief that God could have used some natural process or 'secondary cause' to achieve this end (Cannon 1960*b*). In the passage just quoted, Lyell seems to imply that new species would somehow originate spontaneously, if the appropriate ecological niches were vacant for long enough, as they might have been on an ancient and rather isolated island like Corsica. This process had not yet occurred on Sicily, he thought, because it had emerged from the sea-bed so recently, and it was so close to an existing land-mass that existing species had simply spread to it. The important point about Lyell's speculations, however, is that he clearly envisaged a process of *piecemeal* production of new species, in appropriate ecological situations that would tend to be scattered in both space and time. Also implicit in the quoted passage is Lyell's belief that the extinction of species was a process similarly piecemeal in character. As already mentioned, he had probably derived this idea from Brocchi and he may still have been using Brocchi's notion that the cause of extinction was the 'old age' of each species.

These two processes in conjunction produced an over-all pattern of continuous piecemeal change in the whole fauna and flora. With such a process of organic change, the specific composition of (say) the molluscan fauna, if followed from any given moment in past geological time, would gradually change until all the old species had been replaced by new ones. Lyell referred to such a cycle of change as a 'revolution'. For him this word carried no overtones of sudden violence; he used it in the older but still current sense of a complete cycle, such as the turning of a wheel or the circling of the Earth around the Sun. In the present context it meant a complete turnover in the specific composition of the fauna or flora (of some particular biological group).

This concept of slow cycles of organic change was given a quantitative dimension in Lyell's mind, by being geared to his previously empirical use of the proportion of extant species in various Tertiary molluscan faunas. This now became implicitly a *measure* of geological time. Thus in the passage quoted above, Lyell was measuring the age of Sicily as a land-mass by reference to the mere 5 or 10% of extinct molluscan species in its youngest marine strata; he was saying in effect that this represented an age of only 5 or 10% of one complete turnover or 'revolution', conceived as a major unit of geological time.

In the foregoing interpretation of Lyell's thought, I have made explicit a theoretical structure that was only hinted at in the available documentary record, and my interpretation has been guided by Lyell's own explicit statements of a slightly later date (particularly in the *Principles of geology*). Nevertheless, I believe that there is enough evidence to indicate that by the time he left Geneva early in 1829, Lyell had already constructed a theory of organic change that in principle provided the basis for a faunal chronometer for geological time.

From Brocchi he had drawn the idea of piecemeal extinction, and the notion that the molluscan species had intrinsically longer life-spans than, for example, the more spectacular mammalian species. He had then used Brocchi's emphasis on the mixture of extant and extinct molluscan species in the Subappenine fauna as the basis for a measure of geological age. He had seen the value of Basterot's quantitative comparisons of the faunas of different Tertiary basins; but he had reinterpreted Basterot's (and indeed Brocchi's) comments on these differences, seeing them not as a reflection of biogeographical factors, but primarily as a result of the different geological ages of the basins. Yet he had welcomed the biogeographical insights of Viviani and the great Candolle, integrating their stress on the spatial dimension with his own temporal emphasis, to produce a theory in which the piecemeal production and extinction of species was closely geared to the ever-changing local environment. Finally, he had seen the various Tertiary basins, like Scrope's sporadic lava-flows in Auvergne, as preserved 'moments' in a much longer history, for which a 'natural scale' or chronometer could be constructed.

Having sketched the theory just outlined, Lyell realized that its development required above all a sound knowledge of fossil and living molluscs. On his way back to London he therefore stayed in Paris to learn all he could from one of the best conchologists in Europe, Paul Gérard Deshayes (1797–1875). He was perhaps disappointed, yet also encouraged, to find that both Deshayes and another Parisian naturalist, Jules-Pierre Desnoyers (1800–1887), had independently reached somewhat similar conclusions about the Tertiary strata and their fossils: disappointed, because it might detract from the acclaim he hoped to receive for his work; encouraged, because the independent conclusions of others did at least confirm the validity of his own formulation. Although apparently Lyell did not know it at the time, yet another geologist, the young Heidelberg professor Heinrich Georg Bronn (1800–1862) had also been working on somewhat the same lines.

On the face of it, this looks like a striking case of what some historians and sociologists of science have analysed as 'simultaneous discovery' (e.g. Merton 1957; Kuhn 1959). There is indeed an element of simultaneity, but it is hardly surprising that several naturalists should have been working at the same time on the problem of the Tertiary strata and their fossils. The sheer accumulation of descriptive papers and monographs was making it increasingly evident that the Tertiary was not just a single major formation, but a highly complex series of formations, and this knowledge was readily accessible to the whole European geological community through an already well-developed system of scientific periodicals. Yet beyond this general concern to reduce the Tertiary to greater order and coherence, a detailed comparison of Lyell's work with that of other naturalists greatly reduces the element of coincidence.

Desnoyers had discovered a stratigraphical overlap that proved, by the ordinary principles of superposition, that the Tertiary strata of the Touraine region were younger than those in the Paris basin to the north. Desnoyers (1829) generalized this into a theory of the successive existence and filling of the various Tertiary basins, and used this theory to explain, for example, why the Crag of East Anglia was much closer faunally to the Touraine strata than to the London basin, although it was much further away geographically. For such a case, Desnoyers's view was clearly superior to Basterot's biogeographical explanation. Yet this palaeontological aspect played only a very minor role in Desnoyers's argument; and, furthermore, he attributed the formation of new basins to sudden tectonic events—the kind of interpretation that

Lyell was most concerned to eliminate from geology. So although Desnoyers added a footnote to his article in proof stage (1829, pp. 214–215), expressing his pleasure that Lyell had reached similar conclusions on the successive filling of the Tertiary basins, there was really only a small degree of overlap between their work. In the part of Desnoyers's memoir that he had completed before Lyell's stay in Paris, there is no sign of Lyell's more comprehensive vision of a complex pattern of continuous environmental and organic flux. Desnoyers was there concerned with the more conventional goal of dividing the Tertiary strata into successive *periods* of formation.

Deshayes shared this aim, complementing Desnoyers's stratigraphical fieldwork with a museum-based palaeontological study of Tertiary molluscan faunas. Before Lyell arrived in Paris, Deshayes had apparently already distinguished three groups of Tertiary formations on the basis of their fossils. Yet a detailed analysis of his published conclusions shows that he interpreted these groups of formations in conventional terms as the products of three successive and temporally contiguous periods of geological time. In his first brief announcement of his results, Deshayes (1831a) distinguished a first epoch with 3% extant species (e.g. Paris and London basins), a second epoch with 19% extant species (e.g. Basterot's Bordeaux strata and Desnoyers's Touraine strata), and a third epoch with 52% extant species (e.g. Brocchi's Subappenine strata and the English Crag). He referred to these as 'three great zoological epochs, completely distinct by the assemblage of species in each, and by the constant proportions between the number of living species and those that are lost' (Deshayes 1831a, p. 186). In fact Deshayes also mentioned a fourth group of strata with 96% extant species (e.g. the most recent of Risso's strata at Nice and Lyell's in Sicily), but he evidently thought these too recent to deserve the rank of 'epoch'. It is therefore clear that Deshayes was using his quantitative faunal analysis to define sharply *distinct* epochs: it was a useful tool for determining the temporal order of formations, in cases where the more conventional criteria of stratigraphical superposition and characteristic fossils happened to fail. In fact this is explicit in his fuller memoir (Deshayes 1831b), where he urged selection of characteristic fossils from the larger assemblages in which they occurred, and at various levels of specificity (e.g. '*Lucina divaricata*' for all the Tertiary strata, '*Cardium porulosum*' for the Parisian strata, and '*Cucullea cravatina*' for the lower *Calcaire grossier*). In other words, Deshayes was clearly concerned to use fossil assemblages simply to divide the Tertiary into discrete epochs; formations could then be assigned to the correct epoch by appropriate characteristic fossils, if the criterion of superposition was not available.

This project differed fundamentally from Lyell's in its cognitive goals and structure of thought. But the contrast has been masked by the fact that Lyell paid Deshayes, who badly needed financial support, to supply him with a complete 'statistical' analysis of the Tertiary molluscan faunas, for Lyell to use in the *Principles of geology*. This arrangement necessitated a certain compromise in Lyell's presentation of Deshayes's results.

Before analysing Lyell's use of Deshayes's work and the articulation of his own very different theoretical framework for Tertiary earth-history, a brief comment on Bronn's work is appropriate, although Lyell did not hear of it until later. Bronn toured northern Italy at about the same time as Lyell (though they did not meet), and used many of the same local informants. After his return to Heidelberg he published an elaborate quantitative analysis of the complete fossil record, with a more detailed analysis of the Tertiary faunas of the various European basins (Bronn 1831). Bronn's tables are an outstanding example of the 'statistical' approach in palaeontology at this time (remembering that 'statistics' meant simply a compilation of quantitative data). For each fauna he counted the numbers of genera and species belonging to various major groups, and converted these numbers into decimal fractions for easier comparison. For the Tertiary formations he calculated degrees of faunal affinity between the various basins; tabulated these affinities in terms of orders of similarity; and constructed an elaborate matrix to display them numerically. Yet only incidentally in the midst of all this 'Tabellenstatistik' (as its critics scornfully called it) did Bronn record the fractional proportions of extant species in the various basins. Only these figures—a minor feature of one of his seventeen separate tables (Bronn 1831, table 11)—are conceptually equivalent to the percentages that Lyell was to use in his later analysis of the Tertiary. Bronn did use the proportion of extant species to indicate that the strata near Turin were quite separate from the Subappenine strata (as Lyell had also concluded); but he did not develop this into a general criterion of geological age, and his published work shows no theoretical structure underlying his 'statistical' analysis. Like some more recent palaeontologists who have been keen to exploit statistics in the modern sense of the word, Bronn gives the impression of having had more figures than he knew how to handle, and of not having had any clear theory that he wished to test.

THE ARTICULATION OF LYELL'S FAUNAL CHRONOMETER

Lyell described his period of collaborative work with Deshayes in 1829 in the following terms: 'We planned together a grand scheme of cataloguing the tertiary shells of various European basins, that I might draw geological inferences therefrom.' The intended division of labour is here quite clear! Later in the same letter, Lyell made equally clear the ambitious scope of his project: 'My results will be an induction from nearly (perhaps more than) 3000 species in the tertiary formations alone, and I hope by other aid than Deshayes to carry it on through older strata also' (Mantell MSS., Lyell to Mantell, 24 Feb. 1829). In other words, Lyell hoped that quantitative faunal comparison—a 'statistical' palaeontology—would yield a *general* faunal chronometer for the whole of the fossil record, not just the Tertiary. Given Lyell's concept of successive turnovers or 'revolutions' in the specific composition of, say, the molluscan fauna, there was no reason in principle why the chronometer should be limited to the most recent complete cycle of organic change. If some earlier 'moment' of geological time were taken as the base-line for comparison, in place of the present, it would be possible to estimate the ages of still older strata by their degree of faunal approximation to that base-line. In this way the chronometer could in principle be extended backwards indefinitely through one faunal cycle after another. Such an ambitious scheme was never again explicitly stated by Lyell, except very briefly and much later (Burchfield 1975, p. 68). Yet my analysis of his intention is consistent with what Lyell did make explicit, and I believe it represents the full scope of Lyell's original 'dream' of what his faunal chronometer might achieve.

Lyell had to make this grandiose project concrete in the first instance, however, by dating the Tertiary strata alone. During a period of intensive work after his return to

England, hints in his letters reveal how far his conception of the Tertiary strati-graphical record differed from that of Deshayes and Desnoyers. For example, he told Murchison: 'My last tour persuaded me that altho' on enlarging our Geology in new countries the epochs will multiply, yet in any one country the groups [of strata] will be found to belong to much fewer geological epochs than our predecessors imagined' (Murchison MSS., Lyell to Murchison, 5 Oct. 1829). Here the word 'epoch', like the words 'statistics' and 'revolution', needs careful interpretation. In the early nineteenth century the older sense of 'epoch', meaning a *point* or moment in time, still co-existed alongside the secondary (and modern) meaning of a *period* of time. Lyell's use of the word in the passage just quoted was in a sense a blend of the two meanings. Of course he did not think that any group of strata had been deposited in a 'moment' of time, even in geological terms: on the contrary, he was centrally concerned to emphasize how slowly all sediments have accumulated. Yet the quotation does imply that he envisaged that the time taken for the deposition of any one formation was quite short in comparison with the totality of time represented by all the formations of the same general age in different regions. To put it another way, Deshayes and Desnoyers hoped that by faunal and stratigraphical comparison the Tertiary formations of different regions could eventually be linked together in an overlapping 'chain' that would provide a *continuous* record of sedimentation and marine life. Lyell, in contrast, believed that these formations represented mere isolated 'moments' (relatively speaking), separated from each other by vast spans of geological time of which no record had been preserved; and even future exploration would merely multiply the number of these 'moments' of preserved time without ever achieving a continuous record.

Lyell returned to Paris in 1830 to work with Deshayes again, but the publication of his interpretation of Tertiary earth-history was delayed by the unexpected magnitude of his task of describing and classifying the agents of change at present observable in the inorganic and organic realms. This work was published in the first two volumes of the *Principles of geology*, and only in 1833 did Lyell finally publish the last volume, in which the Tertiary was used as an 'exemplar' of his method of interpretation of the past 'by reference to causes now in operation'.

In his exposition of his new system for the Tertiary, Lyell used once more the Brocchian analogy between species and individuals, and wrote quite casually about the 'birth and death of species' (Lyell 1833, pp. 32, 33). He explained his theory that the extinction of species is a normal part of nature, and that it happens in a piecemeal manner at all times. In the previous volume he had in fact rejected Brocchi's 'old age' explanation of extinction and replaced it with a more 'modern' theory based on environmental change (Lyell 1832, pp. 128-130). But this scarcely reduced the value of Brocchi's general analogy, and Lyell used it with great effect in his exposition, adapting it into a *human* analogy based on the contemporary concern with population censuses (Lyell 1833, pp. 31-33; Rudwick 1977). Lyell likened the continual shifts in the areas of sedimentation (and hence of the possibility of preservation of the fauna) to the movements of itinerant 'commissioners' taking censuses of the population in different regions of a country. Each preserved fauna in a given Tertiary basin was thus like the 'statistical documents' that such officials might leave behind them, to record the state of the population in a given province at a certain time. On their next visit, the constituent individuals in the population would have changed, by deaths and births, in proportion to the time since the previous census. Likewise the change in the fauna between two successive formations would also be proportional to the unrecorded interval of time between them.

Obviously the analogy—like all analogies—had its limitations. Lyell was not postulating that the regularity of censuses had any analogue in the highly irregular fluctuations in the areas of Tertiary sedimentation. Conversely, he did want to use a quantitative comparison of faunas as a measure of time-differences not only between formations in one basin, but between formations in different basins. Nevertheless, the analogy was a helpful expository device for Lyell's readers, and it may also have been an important heuristic device at an earlier stage in Lyell's own mind. In any case, the population analogy served to make clear Lyell's distinctive concept of the various Tertiary formations as little more than momentary *samples* preserved from a vastly greater span of unrecorded time.

Lyell's introduction (1833, pp. 52-56) of the terms Eocene, Miocene, and Pliocene must be seen in the light of this conception of an extremely fragmentary geological record. Like other British scientists of this period, Lyell had applied to the polymathic William Whewell (1794-1866) for advice on his scientific nomenclature. Their letters reveal the barbarous Greek-based terms that might have entered the geologist's vocabulary—for example, 'Meiosynchronous', 'Meioneous', and 'Meiotautic'—before Whewell as an afterthought suggested 'Miocene', and of course 'Eocene' and 'Pliocene' too (quoted in Wilson 1972, pp. 305-307). Lyell also originally wanted an earliest 'Asynchronous' epoch, which in Whewell's hands became 'Acene'. Although the Acene was dropped from Lyell's published scheme, it is important for reconstructing his conception of the Tertiary epochs.

In earlier brief analyses of Lyell's work on the Tertiary, I interpreted Lyell's application of Deshayes's quantitative data in terms of a quite sophisticated conception of statistical survivorship rates (Rudwick 1970, pp. 24-26; 1972, fig. 4.5). This now seems likely to be anachronistic, since such statistics, in a form that Lyell might have recognized as relevant to his concerns, were not developed until the 1840s by the pioneer Belgian statistician Adolphe Quetelet (1796-1874). Lyell probably had a much simpler conception of the continual turnover of species during geological time, by which the percentage of extinct species in a Tertiary fauna would be directly proportional to its age.

It is probably consistent with Lyell's conception to depict his successive epochs in terms of a chronometer with a decimalized 'minute' hand (text-fig. 1). Suppose then that each 'hour' represents one complete faunal cycle or Lyellian 'revolution'. If the present (Lyell's 'modern' period) is represented by twelve noon, we can infer that Lyell conceived Deshayes's original estimates as follows: Deshayes most recent strata, which Lyell termed 'Newer Pliocene' (and later, 'Pleistocene'), were only 4% of the last 'hour' before 'noon'; his third epoch, Lyell's 'Older Pliocene', was just after the half-hour (52%); his second epoch, Lyell's 'Miocene', was only 19% through that last 'hour'; and his first epoch, Lyell's 'Eocene', was only 3% after 'eleven o'clock'. The provisional 'Acene' would then have represented some time *before* 'eleven o'clock', i.e. during the previous 'revolution', for it contained *no* extant species.

This analogy with a chronometer does not imply that Lyell thought that each of

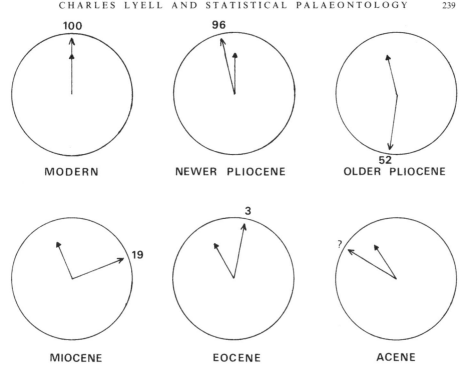

TEXT-FIG. 1. Diagrammatic interpretation of Lyell's original conception of the 'epochs' of Tertiary time, in terms of a faunal 'chronometer'.

Deshayes's epochs represented no more than 1% of the total duration of a 'revolution'; he was certainly aware that Deshayes's figures could not claim any such accuracy. Further collecting was continually enlarging the fauna known from each formation, and the figures that Deshayes finally submitted to Lyell for publication in the *Principles* differed slightly from his earlier estimates: Eocene $3\frac{1}{4}\%$, Miocene 18%, and (Older) Pliocene 49% (Lyell 1833, Appendix I, pp. 47–52). Yet even if Lyell realized the margin of error intrinsic to Deshayes's estimates, the suggested analogy with a chronometer does serve to emphasize that he did not think of his four periods (Deshayes's three, plus the Newer Pliocene) as periods that collectively recorded most of Tertiary time. He did, indeed, believe that the Newer Pliocene graded insensibly into the present. This was essential to his aim of breaking down the conceptual barrier between present and past, just as in Deshayes's view it disqualified the Newer Pliocene from being a real separate epoch. But between his four periods Lyell certainly believed that there were long spans of unrecorded time. He stated explicitly that the definition of these four periods was essentially a result of the accidents of preservation and collection, so that the periods were in a sense arbitrary; and he anticipated that intermediate periods would need to be defined in future (Lyell 1833, pp. 56–58). This makes it clear once more that to Lyell the known Tertiary faunas were no more than scattered samples from a far longer time-span of continuous change; they were *not*, as they were for Deshayes, a relatively complete record of a sequence of faunally distinct periods.

Lyell was always prudently cautious in public about suggesting any estimate in *years* for the magnitude of the geological time scale. Such reticence was prudent, not because Lyell had any reason to fear persecution or even ridicule by religious conservatives, but he knew that no estimate could be more than an enlightened guess, and that any such guess would be criticized by other geologists as a mere speculation running counter to the spirit of empirical science (the reception of Darwin's later rash guess is instructive: see Burchfield 1974).

In the *Principles*, the nearest that Lyell came to any calibration of geological processes in relation to human history was in his analysis of the volcano Etna. Here he estimated some 12 000 years for the accumulation of a small (unspecified) fraction of the volcanic cone, all of which post-dated the most recent Newer Pliocene strata of Sicily (Lyell 1833, pp. 97–101; Rudwick 1969). Privately, Lyell extended this estimate slightly in a way that hints at the kind of calibration that he envisaged for his faunal chronometer. In a letter to his sister, written while working in Paris with Deshayes in 1830, Lyell reported: 'This morning all my Etna shells were examined; out of sixty-three, only three species are not known to inhabit the Mediterranean; yet the whole volcano nearly is subsequent to them and rests on them. They lived on a moderate computation 100 000 years ago and after so many generations are unchanged in form' (K. Lyell 1881, p. 308). This and similar unpublished comments (see Tasch 1977) suggest that Lyell was estimating a period of the order of a million years for one complete faunal cycle or 'revolution', such as was represented by the whole of the Tertiary. This order of magnitude may seem unimpressive by modern standards, but it records a significant phase in the gradual stretching of the scientific imagination that enabled ever longer estimates to seem to geologists first conceivable, then plausible, and finally inescapable.

CONCLUSION

The first edition of Lyell's *Principles of geology* thus introduced the terms Eocene, Miocene, and Pliocene in a form that superficially resembles other stratigraphical terms of the nineteenth century. On the surface they seem to denote a sequence of contiguous periods of geological time, each with a distinct fauna. It might seem that the only difference between these Tertiary terms and the terms introduced for other parts of the stratigraphical record was that the Tertiary strata generally lacked easily identifiable 'characteristic fossils', so that they had to be distinguished by reference to the over-all character of the whole faunal assemblage. This was, roughly speaking, Deshayes's conception of the Eocene, Miocene, and Pliocene; and the later history of the terms shows that most other palaeontologists and geologists interpreted Lyell's terms in this way.

I have tried to show, however, that such an interpretation profoundly mistakes Lyell's own original intentions. In his early letters and notebooks, and even in his explanation of his terms in the first edition of the *Principles*, it is possible to detect a quite different and profoundly innovative conception. Lyell interpreted the Tertiary earth-history of Europe in terms of a continual flux of the physical and biotic environment, in which areas of sedimentation—and hence areas of preservation of the fauna and flora—had shifted irregularly in an almost aleatory manner. All the while the composition of the fauna and flora had been changing continuously but gradually, by the piecemeal introduction of new species (by *means* that remained obscure) and the equally piecemeal extinction of old ones. From this extremely complex network of interrelated inorganic and organic change, the fossil record hitherto examined had, as it were, 'caught on the wing' a few spatially scattered samples from four temporally isolated epochs. These were the Eocene, Miocene, Older Pliocene, and Newer Pliocene. Further collecting would doubtless add new samples, but the record could not be expected ever to approach completion. But such new samples could be fitted into a quantitative framework of geological time already provided in outline by the known Tertiary molluscan faunas: the slow but broadly uniform rate of faunal change provided a chronometer for the whole of Tertiary time, which constituted the last faunal cycle or 'revolution' in organic life.

Lyell's dream of a 'statistical' palaeontology that could provide a quantitative chronology for geology faded very quickly. The complete faunal discontinuity between the oldest Tertiary (Eocene) and the youngest 'Secondary' strata (Maastricht) perhaps discouraged him from even attempting to carry out his original project of extending his 'chronometer' backwards in time beyond the last faunal cycle. Even for the Tertiary faunas, his quantitative method found little understanding and much criticism from his contemporaries; and Lyell gradually diluted his original conception of the Tertiary epochs, stressing 'the per-centage test' of age less and less, until the epochs became almost indistinguishable from other more conventional stratigraphical units (e.g. Lyell 1838, pp. 286–289).

This fading of Lyell's original 'dream' cannot be traced here, but one major reason for it can be summarized briefly. The percentage test depended on the strict comparability of the specific units on which it was based. Although Lyell, like almost all other naturalists, believed in the ultimate reality of species, the distinction between taxonomic 'splitters' and 'lumpers' was already well known. Lyell therefore emphasized that the value of his percentage figures depended on the fact that they were based on identifications made by one single palaeontologist, Deshayes, who could be presumed to have used the same criteria of specific limits throughout (Lyell 1833, p. 51). Problems arose, however, as soon as other palaeontologists of a more 'splitting' disposition started calculating the percentage of extant molluscan species in the same Tertiary faunas (see Wilson 1972, chapter 14).

Lyell's quantitative analysis of the Tertiary molluscan faunas therefore failed as a result of the empirical difficulty of defining the limits of intra- and inter-specific variation. It failed at precisely the same period—the later 1830s—when Lyell's younger friend Charles Darwin (1809–1882) was privately struggling with a theory of trans-specific change that converted the taxonomists' empirical difficulty into a theoretical answer (see e.g. Limoges 1970; Gruber 1974). Darwin's evolutionary theory was based precisely on pursuing the taxonomists' problem to its limits, questioning the assumption of Lyell (and others) that there must be *some* intrinsic limit to variation. Thus the failure of Lyell's project for a quantitative faunal chronometer, and the success of Darwin's project for explaining the origin of new species, were both parts of a

242

much broader collective enterprise in which most naturalists in the 1820s and the 1830s were engaged.

Lyell's dream of a 'statistical' palaeontology is worth recalling because it records an important early attempt to construct a quantitative time-scale for the history of the earth, against which other events and processes, both inorganic and organic, could be plotted and measured. Although Lyell's attempt failed, his insistence on the vast span of geological time kept the problem alive. It was largely in response to his persistence that the project of constructing a geological 'chronometer' was taken up again in earnest in the second half of the nineteenth century, using a variety of mainly physical methods (Burchfield 1975). It was that tradition that led directly to the radiometric dating of the twentieth century, which has freed palaeontology from its subservience to stratigraphy and enabled it to 'come of age' as the temporal dimension of biological science.

Acknowledgement and note. I am greatly indebted to the Council of the Palaeontological Association for the invitation to give the Twentieth Annual Address. I have rewritten the lecture into a form more suitable for publication, while retaining the content essentially unchanged. At the specific request of some who heard the lecture, I have included fairly full cross-referencing to my other publications in this field, since most of them are in periodicals that will be less familiar to palaeontologists than they are to historians of science.

REFERENCES

BASTEROT, B. DE. 1825. Description géologique du bassin tertiaire du sud-ouest de la France. *Mém. Soc. Hist. nat. Paris*, **2**, 1–100.

BROCCHI, G. B. 1814. *Conchiologia fossile subappenina con osservationi geologiche sugli Appenini e sul suolo adiacente*. Milano.

BRONN, H. G. 1831. *Italiens Tertiär-Gebilde und deren organische Einschlüsse. Vier Abhandlungen*. Heidelberg.

BUCKLAND, W. 1823. *Reliquiae Diluvianae; or, Observations on the organic remains contained in caves, fissures, and diluvial gravel, and on other geological phenomena, attesting the action of an universal deluge.* 303 pp., Murray, London.

BURCHFIELD, J. D. 1974. Darwin and the dilemma of geological time. *Isis*, **65**, 300–321.

—— 1975. *Lord Kelvin and the age of the Earth*. xii + 260 pp., Macmillan, London, New York, etc.

CANDOLLE, A. P. DE. 1820. Géographie botanique. *Dict. Sci. nat.* **18**, 359–422.

CANNON, W. F. 1960a. The Uniformitarian-Catastrophist debate. *Isis*, **51**, 38–55.

—— 1960b. The problem of miracles in the 1830s. *Vict. Studies*, **4**, 5–32.

CROIZET, J. B. and JOBERT, A. C. G. 1826–1828. *Recherches sur les Ossemens fossiles du Département du Puy-de-Dôme*. Paris

CUVIER, G. and BRONGNIART, A. 1808. Essai sur la géographie minéralogique des environs de Paris. *J. Mines, Paris*, **23**, 421–458.

———— 1811. Essai sur la géographie minéralogique des environs de Paris. *Mém. Classe Sci. math. phys., Inst. imp. France*, an. 1810, (1), 1–278.

DESHAYES, P. 1831a. Tableau comparatif des espèces de coquilles vivantes avec les espèces de coquilles fossiles des terrains tertiaires de l'Europe, et des espèces de fossiles de ces terrains entr'eux. *Bull. Soc. géol. France*, **1**, 185–187.

—— 1831b. *Description de Coquilles charactéristiques des Terrains*. 264 pp., Levrault, Paris.

DESNOYERS, J. 1829. Observations sur un ensemble de dépôts marins plus récens que les terrains tertiaires du bassin de la Seine, et constituant une Formation géologique distincte; précédées d'un Aperçu de la non simultanéité des bassins tertiaires. *Annls. Sci. nat.* **16**, 171–214, 402–491.

DEVEZE DE CHABRIOL, J. S. and BOUILLET, J. B. 1825–1827. *Essai géologique et minéralogique sur les environs d'Issoire, département du Puy-de-Dôme, et principalement sur la Montagne de Boulade, avec la description et des figures lithographiées des ossements fossiles qui ont été recueillis*. Clermont-Ferrand.

GRUBER, H. E. 1974. *Darwin on Man. A psychological study of scientific creativity.* 495 pp., Wildwood, New York and London.

HOOYKAAS, R. 1959. *Natural law and divine miracle. A historical-critical study of the principle of uniformity in geology, biology and theology.* 237 pp., Brill, Leiden.

KUHN, T. S. 1959. Energy conservation as an example of simultaneous discovery. *In* CLAGETT, M. (ed.). *Critical problems in the history of science.* Madison (repr. 1969, pp. 321-356).

—— 1962. *The structure of scientific revolutions.* 172 pp., University of Chicago Press, Chicago.

LIMOGES, C. 1970. *La sélection naturelle. Étude sur la première constitution d'un concept (1837-1859.* 184 pp., P.U.F., Paris.

LYELL, C. 1826a. On a recent formation of freshwater limestone in Forfarshire, and on some recent deposits of freshwater marl. *Trans. geol. Soc. Lond.* Ser. 2, **2**, 72-96.

—— 1826b. [Review of] Transactions of the Geological Society of London. Vol. i, 2nd Series. London. 1824. *Quart. Rev.* **34**, 507-540.

—— 1827. [Review of] Memoir on the geology of central France, including the volcanic formations of Auvergne, the Velay and the Vivarais, with a volume of maps and plates, by G. P. Scrope, F.R.S., F.G.S. London, 1827. Ibid. **36**, 437-483.

—— 1830-1833. *Principles of geology, being an attempt to explain the former changes of the earth's surface, by reference to causes now in operation.* Vol. I, 511 pp. (1830); Vol. II, 330 pp. (1832); Vol. III, 398 pp. (1833). Murray, London.

—— 1838. *Elements of geology.* 543 pp., Murray, London.

—— and MURCHISON, R. I. 1829a. On the excavation of valleys, as illustrated by the volcanic rocks of central France. *Edinb. New phil. J.* **12**, 15-48.

—— —— 1829b. Sur les dépôts lacustres tertiaires du Cantal, et leurs rapports avec les roches primordiales et volcaniques. *Annls Sci. nat.* **18**, 173-214.

LYELL, K. (ed.). 1881. *Life letters and journals of Sir Charles Lyell, Bart author of 'Principles of geology'.* 2 vols. London.

McCARTNEY, P. J. 1976. Charles Lyell and G. B. Brocchi: a study in comparative historiography. *Br. J. Hist. Sci.* **9**, 177-189.

MANTELL, G. A. MSS. Mantell papers, in Alexander Turnbull Library, Wellington, New Zealand.

MERTON, R. K. 1957. Priorities in scientific discoveries. *Amer. sociol. Rev.* **22**, 635-659.

MORRELL, J. B. 1976. London institutions and Lyell's career: 1820-41. *Br. J. Hist. Sci.* **9**, 132-146.

MURCHISON, R. I. MSS. Murchison papers, in Geological Society, London.

OSPOVAT, A. M. 1969. Reflections on A. G. Werner's 'Kurze Klassifikation'. Pp. 242-256. *In* SCHNEER, C. J. (ed.). *Toward a history of geology: proceedings of the New Hampshire Inter-Disciplinary Conference on the history of geology September 7-12, 1967.* Cambridge, Mass. and London.

PLAYFAIR, J. 1802. *Illustrations of the Huttonian Theory of the Earth.* 528 pp., Creech, Edinburgh.

PORTER, R. 1976. Charles Lyell and the principles of the history of geology. *Br. J. Hist. Sci.* **9**, 91-103.

—— 1977. *The making of geology: earth science in Britain, 1660-1815.* 288 pp., C.U.P., London.

PRÉVOST, C. 1823. De l'importance de l'étude des corps organisés vivans pour la géologie positive, et description d'une nouvelle espèce de mollusque testacé du genre Melanopsis. *Mém. Soc. Hist. nat. Paris,* **1**, 259-268.

RISSO, G. A. 1826. *Histoire naturelle des principales productions de l'Europe méridionale et particulièrement de celles des environs de Nice et des Alpes maritimes.* Paris.

RUDWICK, M. J. S. 1961. The feeding mechanism of the Permian brachiopod *Prorichthofenia. Palaeontology,* **3**, 450-471.

—— 1969. Lyell on Etna, and the antiquity of the earth. Pp. 288-304. *In* SCHNEER, C. J. (ed.). *Toward a history of geology: proceedings of the New Hampshire Inter-Disciplinary Conference on the history of geology September 7-12, 1967.* Cambridge, Mass. and London.

—— 1970a. The strategy of Lyell's 'Principles of geology'. *Isis,* **61**, 4-33.

—— 1970b. The glacial theory. *Hist. Sci.* **8**, 136-157.

—— 1971. Uniformity and progression: reflections on the structure of geological theory in the age of Lyell. Pp. 209-227. *In* ROLLER, D. H. D. (ed.). *Perspectives in the history of science and technology.* Norman, Oklahoma.

—— 1972. *The meaning of fossils. Episodes in the history of palaeontology.* 287 pp., Macdonald, London and American Elsevier, New York.

VI

RUDWICK, M. J. S. 1974. Poulett Scrope on the volcanoes of Auvergne: Lyellian time and political economy. *Br. J. Hist. Sci.* **7**, 205–242.

—— 1975. Charles Lyell F.R.S. (1797–1875) and his London lectures on geology, 1832–33. *Notes Rec. R. Soc. Lond.* **29**, 231–263.

—— 1976a. Charles Lyell speaks in the lecture theatre. *Br. J. Hist. Sci.* **9**, 147–155.

—— 1976b. The emergence of a visual language for geological science, 1760–1840. *Hist. Sci.* **14**, 149–195.

—— 1977. Historical analogies in the early geological work of Charles Lyell. *Janus*, **64**, 89–107.

SCROPE, G. P. 1827. *Memoir on the geology of central France, including the volcanic formations of Auvergne, the Velay and the Vivarais.* 182 pp., Longman, London.

SMITH, W. 1815a. *A delineation of the strata of England and Wales with part of Scotland. . . .* London.

—— 1815b. *A memoir to the map and delineation of the strata of England and Wales, with part of Scotland.* 51 pp., Cary, London.

—— 1816–1819. *Strata identified by organized fossils, containing prints on colored paper of the most characteristic specimens in each stratum.* 32 pp., The author, London.

TASCH, P. 1977. Lyell's geochronological model: published year values for geological time. *Isis*, **68**, 440–442.

WEBSTER, T. 1814. On the freshwater formations in the Isle of Wight, with some observations on the strata over the Chalk in the south-east part of England. *Trans. geol. Soc. London*, Ser. 1, **2**, 161–254.

—— 1816. Additional observations on the strata of the island. *In* ENGLEFIELD, H. C. *A description of the principal picturesque beauties, antiquities and geological phenomena, of the Isle of Wight.* London.

WILSON, L. G. 1969. The intellectual background to Charles Lyell's *Principles of Geology*, 1830–1833. Pp. 426–443. *In* SCHNEER, C. J. (ed.). *Toward a history of geology: proceedings of the New Hampshire Inter-Disciplinary Conference on the history of geology September 7–12, 1967.* Cambridge, Mass. and London.

—— 1972. *Charles Lyell. The years to 1841. The revolution in geology.* 553 pp., Yale University Press, New Haven and London.

Caricature as a Source for the History of Science: De la Beche's Anti-Lyellian Sketches of 1831

THE VALUE OF CONTEMPORARY CARICATURES for the understanding of political and social history has long been recognized by general historians. The libelous lampoons of James Gillray, for example, can give not only amusement but also valuable insight into the affairs (in both senses) of the Prince Regent and into the way in which those affairs were viewed by the social class that bought the caricatures. Perhaps because of a concern to outgrow an anecdotal and antiquarian past, historians of science have been reluctant to give attention to the analogous resources that might help us to understand the *scientific* thought of other periods. There is much to suggest that these resources are substantial and still largely unused, at least within the period that saw the flowering of the political caricature. It might be felt, however, that caricatures relating to the history of science could hardly do more than illuminate the social circumstances in which the activity of science was carried out: for example, Gillray's entertainingly vulgar caricature of a Royal Institution lecture in 1802, showing Garnett and Davy demonstrating the properties of gases, is a valuable comment on the way in which science was seen by the fashionable classes at that period.[1] The purpose of this article, on the other hand, is to give an example of a caricature which throws light not only on the social context but also the substantive content of a scientific dispute of major importance. Furthermore, the discovery of the draft sketches which led up to this caricature enriches our understanding of its meaning and throws unexpected light on the mode of thought of the scientist who drew it.

Received July 1974: revised/accepted September 1974.

*Unit for History and Social Aspects of Science, Centrum Algemene Vorming, Vrije Universiteit, De Boelelaan 1083, Amsterdam-Buitenveldert, the Netherlands. I am grateful to Dr. Roy Porter for some valuable comments on a draft of this article.

[1] The caricature is reproduced in Draper Hill, *Fashionable Contrasts. Caricatures by James Gillray* (London: Phaidon, 1966), Plt. 93: "Scientific Researches!—New Discoveries in Pneumaticks!—or— an Experimental Lecture on the Powers of Air." See also Martin Sherwood, "Caricatures of Science," *New Scientist*, 1970, 47:382–384.

The publication of the *Principles of Geology* by Charles Lyell between 1830 and 1833 provoked a far-reaching discussion among British geologists about the fundamental methods and theories of their science.[2] To some historians Lyell's work has even seemed to mark the effective foundation of geology as a genuine science, or at least to introduce a radically new "paradigm" of explanation,[3] whereas other historians have stressed the continuity of method and approach between Lyell and his opponents, while emphasizing the specific points on which they differed.[4] In either case, however, there is agreement that the publication of the *Principles* stimulated a period of active self-reflection by geologists about the foundations of their science.

The Lyellian debate needs to be seen in terms that encompass more than a simple conflict between enlightened Lyellians and scientifically conservative geologists motivated by theological concerns. The traditional historiography of "conflict" between science and religion in this period fails to account adequately for some of the most important figures in the opposition to Lyell's geological system.[5] Among such opponents was Henry Thomas De la Beche (1796–1855), Lyell's almost exact contemporary.[6] De la Beche was a prominent member of the small circle of active geologists who formed the core of the Geological Society of London and a distinctive subgroup within the Royal Society. Like most other members of that circle, he was a gentleman with sufficient private means—derived from a Jamaican sugar plantation he had inherited—to indulge his taste for geology as he wished. But by 1832 his income from this source had been so reduced that he was forced to apply to the government for a grant to continue the geological fieldwork he had already begun in southwest England. It was this work that led eventually to the foundation of the Geological Survey and to De la Beche's appointment as the Survey's first director-general—

[2] Charles Lyell, *Principles of Geology, Being an Attempt to Explain the Former Changes of the Earth's Surface, by Reference to Causes Now in Operation* (London, 1830–1833).

[3] Leonard G. Wilson, *Charles Lyell. The Years to 1841: The Revolution in Geology* (New Haven/London: Yale University Press, 1972); Thomas S. Kuhn, *The Structure of Scientific Revolutions* (Chicago: University of Chicago Press, 1962); p. 10.

[4] W. F. Cannon, "The Uniformitarian–Catastrophist Debate," *Isis*, 1960, *51*:38–55; M. J. S. Rudwick, "Uniformity and Progression: Reflections on the Structure of Geological Theory in the Age of Lyell," *Perspectives in the History of Science and Technology*, ed. Duane H. D. Roller (Norman, Oklahoma: University of Oklahoma Press, 1971), pp. 209–227. R. Hooykaas, "Catastrophism in Geology, Its Scientific Character in Relation to Actualism and Uniformitarianism," *Koninklijke Nederlandse Akademie van Wetenschappen, afdeling Letterkunde, Med.* (n.r.), 1970, *33*:271–316; reprinted in Claude C. Albritton, ed., *Philosophy of Geohistory 1785–1970* (Stroudsburg, Pa.: Dowden, Hutchinson & Ross, 1975), pp. 310–356.

[5] A classic source is Andrew D. White, *A History of the Warfare of Science with Theology in Christendom* (New York: D. Appleton, 1901, reprinted 1955), Ch. 5, "From Genesis to Geology." The more recent standard account, though far more sophisticated historically, still sees the debate in essentially the same terms, as a gradual retreat of a "providentialist" view of nature, backed by natural theology, before the advance of objective science: Charles Coulston Gillispie, *Genesis and Geology. A Study in the Relations of Scientific Thought, Natural Theology, and Social Opinion in Great Britain, 1790–1850* (Cambridge, Mass.: Harvard University Press, 1951).

[6] Almost alone among the group of outstanding English geologists of his generation, De la Beche did not receive the standard Victorian "Life and Letters" treatment after his death. For a brief account, see V. A. Eyles' entry on him in the *Dictionary of Scientific Biography*, Vol. IV (New York: Scribner's, 1971), pp. 9–11. He was not French, and his name is correctly spelled with a capital D (as in his own autograph signature) and should be indexed under that letter.

the first English field geologist to be in permanent full-time employment by the state.[7]

There are hints in De la Beche's correspondence, and in Lyell's, that there was some personal antipathy between them. This may have aggravated their methodological and substantive disagreements, but the latter cannot be dismissed simply as the products of that antipathy. Nor is there any sign in De la Beche's published or unpublished work that his criticism of Lyell's geology was motivated by any religious desire to maintain, for example, the reality of geological "catastrophes" in the past.[8] On the contrary, in early-nineteenth-century Britain De la Beche is almost a paradigm of the self-consciously "professional" scientist, concerned above all for the autonomy of the science from extra-scientific concerns and for its scientific and philosophical respectability. If the geology of this period is to be interpreted at all in ideological terms (and I do not deny the possible validity of such interpretations), then Lyell should be grouped *with* some of his catastrophist opponents, as those who were aware of the wider implications of the science for human self-understanding, while De la Beche should be placed against them as a prototypical modern scientist who was more concerned with the progress of geology as an autonomous source of objective and useful knowledge. Indeed, Lyell himself seems to have held such a view of De la Beche's work, admiring its professional thoroughness and reliability but finding it heavy reading. He told his fiancée, for example, "If you are not frightened by De la Beche, I think you are in a fair way to be a geologist; though it is in the field only that a person can really get to like the stiff part of it."[9]

De la Beche's criticisms of Lyell's geology were primarily criticisms of his methodology and of his specific theoretical conclusions. This is clearly evident in all his published work after 1830, particularly in his important and—by historians—underrated volume of *Researches in Theoretical Geology*.[10] It is confirmed, I suggest, by the series of anti-Lyellian sketches which form the subject of this article.

De la Beche was a talented amateur artist who used his skills primarily to develop the range and efficacy of the visual "language" of geology—maps,

---

[7] F. J. North, "H. T. De la Beche: Geologist and Business Man," *Nature*, 1939, *143*:254–255; "The Ordinance Geological Survey: Its First Memoir," *ibid.*, 1052–1053; "Geology's Debt to Henry Thomas De la Beche," *Endeavour*, 1944, *3*:15–19.

[8] Gillispie, *Genesis and Geology*, pp. 139–140, notes correctly that De la Beche was not a "scriptural catastrophist" and that his work was "entirely without Mosaic allusions or overtones," but finds it in consequence "rather dry" and devotes only half a page to him out of 228. It could be argued that De la Beche's lack of "scriptural" concerns made him unimportant for Gillispie's chosen theme; but my point is that this theme should not be taken as adequate *by itself* for the interpretation of British geology at this period, within which the importance of De la Beche merits an allotment of far more than 0.2 per cent of the space available.

[9] Charles Lyell to Mary Horner, Sept. 24, 1831: Mrs. [Katherine M.] Lyell, ed., *Life, Letters and Journals of Sir Charles Lyell, Bart.* London, 1881), Vol. I, p. 341. Lyell was probably referring to De la Beche's recently published elementary work *A Geological Manual* (London, 1831).

[10] H. T. De la Beche, *Researches in Theoretical Geology* (London, 1834); see also his *Manual* for an earlier implicit critique of Lyell's work.

sections, geological landscape views, and so on.[11] But he also embellished his scientific correspondence with lively caricatures that comment on current geological debates; some of these he then drew on stone and distributed as lithographs to his friends and colleagues.

One of the more entertaining of these lithographs, which has already been reproduced in a modern work, is entitled "Awful Changes" (Fig. 1). It shows a "Professor Ichthyosaurus" lecturing to an audience of Jurassic reptiles and demonstrating a human skull.[12] Hitherto this has seemed to be a straightforward humorous comment on the most popular and entertaining geological lecturer of the period, William Buckland, whose Oxford lectures were well known for their use of visual aids in the form of spectacular fossil specimens.[13] The caricature has appeared to be a fantasy of the tables turned on Buckland, with an ichthyosaur analyzing human functional morphology as effectively and entertainingly as Buckland in his lectures analyzed that of the Jurassic reptiles and other fossils. The interpretation I shall offer here will suggest that it embodies a much more serious scientific meaning, and that the ichthyosauran lecturer was not Buckland but Lyell.

The evidence for this interpretation is found in a geological notebook compiled by De la Beche around 1830–1831 while he was doing fieldwork in southwest England.[14] The front of the notebook is filled with geological notes, probably written after each day's work, and illustrated with fine sketches of significant rock outcrops, sections of strata, and so on. At the back of the notebook are many pages of more informal sketches and caricatures. Among these is a sequence of ten sketches, the last of which is unmistakably a rough draft for the drawing that he lithographed. I shall argue from an analysis of their captions and iconography—if the term is not too pretentious for caricatures— that they embody an important critique of Lyell's *Principles*.

In any case there is circumstantial evidence that De la Beche drew the preliminary sketches at least within about a year of the publication of the first volume of the *Principles*. In De la Beche's notebook these sketches are flanked on one side by a sequence of three political caricatures, entitled "John

---

[11] See particularly his *Sections and Views, Illustrative of Geological Phaenomena* (London, 1830), entirely illustrated with his own lithographed drawings.

[12] Fig. 1 is reproduced from a copy in a scrapbook which belonged to Roderick Murchison and was probably sent to him by De la Beche (Archives of the Institute of Geological Sciences, London, GSM1/558, p. 3). The lithograph is also reproduced (much reduced in size) in Richard J. Chorley, Anthony J. Dunn, and Robert P. Beckinsale, *The History of the Study of Landforms* (London: Methuen, 1964), Vol. I, p. 104, and as the frontispiece to Francis T. Buckland, *Curiosities of Natural History* (London, 1859).

[13] See the engraving reproduced in Chorley et al., *Landforms*, Vol. I, p. 103. The existence of an enlarged reproduction of the caricature (now in the Bodleian Library in Oxford), evidently intended as a visual aid for Buckland's own lectures, makes this interpretation improbable, and lends further support to my contention that the caricature was a visual comment on *Lyell's* theories, which would have been useful for Buckland's public critique of that work.

[14] Archives of the Institute of Geological Sciences, London (GSM 1/123). I am indebted to the librarian of the Institute for access to the archives and for permission to publish some sketches from the notebook in this article, and to Mr. John Thackray for much helpful assistance.

[Bull] and the Grandees." This series shows a group of aristocrats dressed as clowns or jesters, insisting on their right to live off the hard-won earnings of the ordinary Englishman but being thrown into disarray in the final scene by the inexorable progress of "The Knowledge Locomotive Engine" hauling a wagon loaded with globes and other scientific apparatus (Fig. 2). The series surely reflects the contemporary agitation over the Reform Bill (as well as being perhaps a comment on the Society for the Diffusion of Useful Knowledge); and the final scene may have been prompted by the untimely end of the unfortunate William Huskisson at the opening of the Manchester and Liverpool Railway in 1830.

Flanking the geological sketches on the other side are two dated caricatures. One is entitled "Receive the bread of life" and shows a publication of the Religious Tract Society being offered to an apparently undernourished young woman, while others kneel in the background bearing other tracts in their upturned hands as if receiving the Sacrament. With its appended note "Sketch from real life, Totness, Sept. 1831," this would seem to be a sharp comment on one of the establishment's most approved responses to the acute social problems of the time.[15]

The other dated sketch shows a geologist, probably De la Beche himself, in a sitting room, with his hammer and collecting bag on the table behind him, looking disconsolately at the pouring rain outside the window and evidently fretting at his inability to get out into the field. The caption refers sardonically to the only scientific opportunity the situation offered: "Opportunity of studying the effects of rain on glass. Devon. Oct. 1831." Since a similar figure appears in the first of the geological sketches, it is possible that they were drawn on this occasion in October 1831; that is, that De la Beche used his enforced confinement indoors to reflect on the most important theoretical challenge to his own geological synthesis and to express his thoughts in a characteristic visual form. This date would be consistent with my inference later in the article that the "Professor Ichthyosaurus" of the final caricature should be identified as Lyell; for in April 1831 Lyell had been appointed Professor of Geology at the newly established King's College in London—the first teaching position in geology outside Oxford and Cambridge.[16]

There is one serious problem, however, about dating the final lithograph to 1831; this is that some (but not all) copies bear the date 1830 after De la Beche's signature. But this is probably a mistake which De la Beche made in all innocence when redrawing the caricature for a second edition, believing

---

[15] Such sketches of social and political comment are worth stressing for the light they throw on De la Beche's broader attitudes. His Reformist position is not invalidated by the fact that he was the proprietor of a slave-owning plantation. He was evidently uneasy about the slavery question, yet could not feel disinterested. A year's residence in Jamaica had impressed upon him the need for reform in the treatment of slave labor, but he favored legislative reform rather than immediate abolition. His attempt to describe the "facts" of slave labor objectively forms an instructive parallel to his Baconian attitude to the precedence of facts over theories in geology. H. T. De la Beche, *Notes on the Present Condition of the Negroes in Jamaica* (London, 1825).

[16] Martin Rudwick, "Charles Lyell F.R.S. (1797–1875) and His London Lectures on Geology, 1832–33," *Notes and Records of the Royal Society, London,* 1975, 29:231–263.

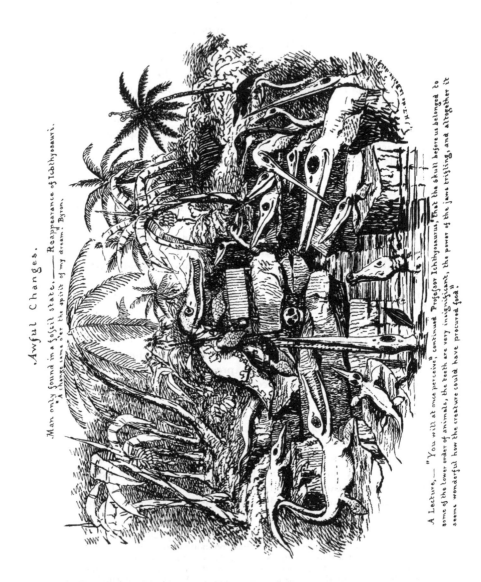

Figure 1. "Awful Changes." "Man only found in a fossil state,—Reappearance of Ichthyosauri. 'A change came o'er the spirit of my dream.' Byron." "A Lecture,—'You will at once perceive,' continued Professor Ichthyosaurus, 'that the skull before us belonged to some of the lower order of animals, the teeth are very insignificant, the power of the jaws trifling, and altogether it seems wonderful how the creature could have procured food.' "

that he had first drawn it in the year that the first volume of the *Principles* was published.[17] On the other hand, apparently he did draw *another* anti-Lyellian caricature at the earlier date, for on September 15, 1830, Buckland wrote to him as follows:

> Many thanks and much praise to you for your Caricature of Actual Causes & the Huttonian Theory rediviva. The book is written in a very seductive Style & will no doubt make many Converts. I wish Conybeare wd take it in hand & be stirred up to a Reply. I shall probably be going to see him in about 10 Days & shall stimulate Him if I can to resume his Pen on Geology.[18]

The Lyellian reference in this passage is unmistakable. The first volume of the *Principles* had been published only a few weeks earlier, and William Conybeare was indeed prevailed upon to write a long critique of the work—one of the first to be published and one of the most important.[19] But Buckland's description of this earlier caricature does not at all fit the lithograph "Awful Changes," and I am therefore inclined to prefer for the latter a dating in 1831. In any case, however, the appearance of only *one* volume in the geological sketches suggests that they date at the latest from before the publication of the second volume of the *Principles* in February 1832.

I turn now to the analysis of the main series of sketches. Although they are superficially diverse, they can be understood as a connected sequence linked by common themes and images. I suggest that De la Beche started the series intending to devise an anti-*Principles* caricature that he could circulate among other geologists, and that he experimented with several related themes before finding the one that he finally used.

The first theme to come to De la Beche's mind was a comment on the methodology of Lyell's work. The first three sketches all reflect Lyell's own emphasis on the need to find the proper way of *seeing* the phenomena of geology. Lyell had emphasized that geological theorizing had been distorted by a failure to correct for our human viewpoint as subaerial terrestrial beings whose existence was of fleeting duration compared with the history of the earth.[20] To De la Beche, as to other geologists, much of Lyell's argument seemed aprioristic. The Geological Society's president, Adam Sedgwick, was

---

[17] There are copies of both versions in the Institute of Geological Sciences in London. A close comparison of them shows that the later version must have been traced with some care from the earlier, but that the shading was redrawn; in other words the later version was not simply a second impression from the original stone. This is also indicated by the inversion of two words in the subtitle. I am inclined to think that the undated version reproduced here was the original, and that its success with his friends led De la Beche to redraw it again later—adding an incorrectly remembered date—in order to have more copies to distribute.

[18] William Buckland to De la Beche, Sept. 15, 1830 (minimal punctuation added), De la Beche papers, Department of Geology, National Museum of Wales, Cardiff. I am indebted to Dr. Douglas Bassett for access to these MSS. I have not been able to trace the letter to which this was a reply.

[19] W. D. Conybeare, "An Examination of Those Phaenomena of Geology, which seem to Bear Most Directly on Theoretical Speculations," *Phil. Mag.*, 1830–1831, 8:359–362, 401–406; 9:19–23, 111–117, 188–197, 258–270. The first installment was published in the Nov. 1830 number of the journal.

[20] *Principles*, Vol. I, pp. 81–83.

Figure 2. "John [Bull] & Grandees, No. 3." "Yew up, ya scum, out of the way there." "The Knowledge Locomotive Engine."

Take a view, my dear Sir, through these glasses, and you will see that the whole face of nature is as blue as indigo.—

Figure 3. "Take a view, my dear Sir, through these glasses, and you will see that the whole face of nature is as blue as indigo."

not alone in feeling that Lyell had borrowed from his barrister's training too much of the "language of an advocate," and had distorted facts to suit his own theory.[21]

De la Beche's first sketch (Fig. 3) shows an elegantly dressed figure wearing a barrister's wig—inferentially Lyell himself. He is significantly without a hammer or any other symbol of the working geologist. Instead he is holding a pair of spectacles, like those worn in fact by Lyell, who had bad eyesight. He is standing on ground marked "Theory," which forms a spectacular viewpoint looking out over a broad valley to a range of mountains. He is offering his tinted spectacles to a "genuine" geologist—perhaps De la Beche himself—who

[21] Adam Sedgwick, "Anniversary Address of the President," *Proceedings of the Geological Society of London*, 1831, *1*:281–316, at p. 303.

holds a hammer in his hand and has a collecting bag over his shoulder. Lyell is saying to the geologist, "Take a view, my dear Sir, through these glasses, and you will see that the whole face of nature is as blue as indigo." In other words De la Beche felt that Lyell's perception of the geological scene was systematically colored by the theoretical presuppositions inherent in his viewpoint (the double meanings were surely intentional).[22] The choice of indigo possibly symbolizes Lyell's argument that even the loftiest mountain ranges (such as that in the background) have been elevated at some time from beneath the sea, and that even such apparently "primitive" rocks as granite and gneiss, often found in the cores of mountain regions, were nothing but the metamorphosed remains of ordinary marine sedimentary rocks.[23] Alternatively, or in addition, the blue of indigo may represent the conventional cartographic color for rivers and water generally and may refer to Lyell's "Huttonian" claim that even the deepest valleys (such as that in the middle distance) have been excavated entirely by the slow erosive action of the rivers now flowing in them.

But while the figure of Lyell claims to be offering with one hand an aid to better perception, which might be interpreted as his methodological *principles* (the title *Principles of Geology* was not idly chosen), in fact he possesses an unacknowledged *theory*, which he is holding with his other hand but concealing behind his back—a volume marked "Theory of the Earth." This was the label, deliberately reminiscent of James Hutton's work, that many of Lyell's critics applied to the *Principles* in order to stress not only its all-explanatory theoretical ambitions, which seemed out of date in the Baconian atmosphere of the 1830s, but also the neo-Huttonian steady-state content of Lyell's system. This becomes explicit later in the series of sketches.

The second sketch (Fig. 4) pursues the perceptual theme with an implicit accusation that Lyell's way of seeing geological nature involves not merely the systematic interpretative coloring of phenomena, but also a deliberate blindness to certain aspects of them. Lyell (my identification of the figure is still inferential) now appears in the guise of a doctor, though he is still wearing his barrister's wig. He is applying an eye cover to a decrepit hammer-bearing geologist and reassuring him: "There, there, my dear Sir, don't you see far far clearer than before,—the whole force of vision being properly directed there is not that danger of stumbling that there was previous to the application of the plaster." And on the following page is the additional comment: "By blocking one eye, the whole power of vision is directed through the other—the confusion caused by two is avoided and all is clear." De la Beche surely did not think that the ordinary geologist was in fact decrepit—that

---

[22] For a later use of the same metaphor of colored spectacles in the same Lyellian context, see M. J. S. Rudwick, "A Critique of Uniformitarian Geology: A Letter from W. D. Conybeare to Charles Lyell, 1841," *Proceedings of the American Philosophical Society*, 1967, *111*:272–287, at p. 281.

[23] Blue not only suggests the sea, but had become a standard conventional color on geological maps for depicting limestone formations. Granite, by contrast, was almost always depicted by a bright red tint.

Figure 4. *"How to see clearly." "There, there, my dear Sir, don't you see far far clearer than before,—the whole force of vision being properly directed there is not that danger of stumbling that there was previous to the application of the plaster."*

is, in need of Lyell's theoretical ministrations—for in the previous sketch he had been depicted as in perfect health, and indeed much better equipped than Lyell for the practical rigors of fieldwork. Rather, it reflects De la Beche's belief that the geologist would be better off without this "medical" attention.

With its title "How to see clearly," this caricature probably represents De la Beche's criticism of Lyell's refusal to accept the observable phenomena of the past as a valid source of theoretical inference in themselves, but only as interpreted in the light of the phenomena of the present. De la Beche and others among Lyell's critics had no quarrel with actualistic reasoning (i.e., from present to past) as a prescriptive methodology—on the contrary, they used it much of the time themselves—but they felt that in certain cases the observable products of past geological events indicated that "actual causes" (i.e., processes operating in the present) were not *wholly* adequate for causal explanation. Thus for example the observed deformations of strata in many mountain regions (such as that depicted in the first sketch) seemed to De la Beche to be inexplicable in terms of very gradual and relatively gentle forces, no matter how long the periods of time during which those forces were inferred to have operated.[24]

The third sketch (Fig. 5) refers again to the impaired vision induced by Lyell's conception of geology but alludes more directly to its persuasive power within the geological community. The central figure—again inferentially Lyell—is here leading a crowd of others, and all without exception are wearing colored spectacles. Even an owl, the very symbol of wisdom, who contemplates the scene from a perch in the background, is similarly disadvantaged. Furthermore, the other figures are dressed in period costume and seem to be intended to suggest Guy Fawkes and his fellow conspirators, or some such historical episode. Verging on the libelous, De la Beche entitled this cartoon, "One fool makes many, or green glasses the go [i.e., the fashion]." In other words, he is suggesting that the superficial attractiveness of Lyell's work is such that everyone is in danger of having their geological vision systematically distorted in the same way, and perhaps that there is even a conspiracy to promote Lyell's viewpoint.[25]

The figure I am identifying in Figure 5 as Lyell—he is now wearing a chef's cap over his barrister's wig—is bearing a large bowl of steaming food, perhaps toward the convivial table of the Geological Society Club,[26] and it is the delightful aroma of what he has cooked which is luring the others to follow him. The ingredients of his cookery are too small to be easily identifiable, but probably represent components of his theory. For example, at the back of the bowl is a miniature scythe-bearing Father Time (an identification that might seem precarious were it not for the undoubted reappearance of this figure in a later sketch), and to the right is a tiny representation of one of the four winds in their traditional cartographic form—perhaps a reference to the major explanatory work done in Lyell's system by such agents as winds and ocean currents.

It is probable that De la Beche fully intended the notion of *cookery* to be understood in its pejorative metaphorical sense; that is, that Lyell had to some extent "cooked" his results. De la Beche may also have known Gillray's famous cartoon in which the political world was represented as a pudding, a product of cookery, to be carved up between Pitt and Napoleon.[27] But there is some

[24] See his later arguments on this point in his *Researches*, Ch. 6; see also Rudwick, "Uniformity and Progression."

[25] The letter already quoted (n. 18) shows that Buckland shared this view.

[26] The independent dining club formed in 1824, at which the élite of the society could (and still do) dine together informally on the days of the meetings.

[27] Reproduced in Hill, *Fashionable Contrasts*, Plt. 39: "The Plumb-pudding in danger;—or—State Epicures taking un Petit Souper."

Figure 5. *"One fool makes many, or green glasses the go."*

evidence that Lyell had laid himself open to the application of the metaphor of cookery in a much more specific sense.

Earlier, in 1830, he had described his "grand new theory of climate," which played a pivotal role in the strategy of the *Principles,* as a "receipt," a recipe in the cookery sense. This was in a private letter to Gideon Mantell.[28] But Mantell may have passed the notion on to others, or Lyell himself may have boasted quite openly in these terms about his forthcoming work. In either case it is at least possible that Lyell's *own* description of his work in terms of cooking had become generally known in the Geological Society circle.

The cooking "receipt" that Lyell had offered Mantell (and perhaps others) was "for growing tree ferns at the pole, or if it suits me, pines at the equator"; in other words, for explaining the production of any kind of climate at any latitude, with the appropriate organisms, simply as a result of alterations of the configuration of land and sea. If this boast had become generally known, the emphasis on growing might perhaps account for De la Beche's choice of green for the colored spectacles in this sketch.

De la Beche's next sketch was to refer unambiguously to Lyell's steady-state theoretical synthesis, and at first sight this seems to form a major break from the first three, more methodological drawings. But if, as I have suggested, De la Beche was already thinking of Lyell's climatic theory when he drew the third sketch, the continuity of thought would be complete. For the function of Lyell's climatic theory in the strategy of the *Principles* was to undermine the validity of the current "directionalist" interpretation of the climatic history of the earth—in other words, to argue that the fossil record did not necessarily imply that the earth's surface had gradually become cooler and climatically more diverse. This was an important part of Lyell's broader aim of replacing the directionalist synthesis as a whole with a steady-state interpretation of every aspect of geology.[29] The third drawing thus forms a natural transition from methodological to substantive criticism.

The fourth sketch (Fig. 6) refers clearly to Lyell's steady-state tectonic hypothesis of oscillating crustal blocks in dynamic equilibrium (a hypothesis later developed by the young Charles Darwin, notably in his work on coral reefs). It shows "Europe and Africa" lying in one pan of a chemical balance and "America" in the other, the two being almost exactly balanced. Lyell argued that such continental blocks of the earth's crust were gradually rising or falling, but almost "insensibly" to human observers.[30] As if to emphasize the fanciful foundations of this hypothesis of balanced crustal blocks, De la Beche attached a small winged fairy to each pan of the balance. But perhaps these also

[28] Lyell to Gideon Mantell, Feb. 15, 1830, Mantell papers, Alexander Turnbull Library, Wellington, New Zealand. I am indebted to the Manuscripts Librarian for the loan of a microfilm of the Lyell–Mantell letters, which enabled me to check the accuracy of the transcription in *Life, Letters and Journal,* Vol. I, p. 262.

[29] Rudwick, "Uniformity and Progression"; "The Strategy of Lyell's *Principles of Geology,"* Isis, 1970, 61:4–33.

[30] E.g., *Principles,* Vol. I, pp. 472–479 (the conclusion of the volume). The sketch shows some misunderstanding of Lyell's hypothesis, which supposed continental blocks undergoing erosion to be oscillating with temporarily *oceanic* blocks receiving the extra weight of sedimentation.

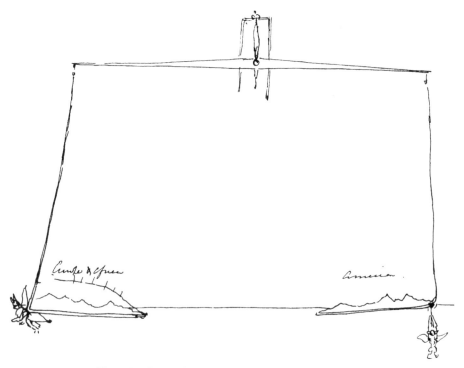

Figure 6. "Europe & Africa" being balanced by "America."

represented the "dusky melancholy sprite" (Alexander Pope's Umbriel) that Lyell himself imagined as the being whose subterranean viewpoint could serve to correct the human perspective on geological phenomena.[31] If so, this image would be another transitional feature, linking De la Beche's thought back to the preceding drawings. In any case, one fairy seems to be lending its insubstantial lift to Europe and Africa, while the other is apparently lending its equally insubstantial weight to America. Only such fanciful disturbances to the alleged balance across the Atlantic, De la Beche seems to be saying, could produce Lyell's system of crustal blocks slowly oscillating in steady-state equilibrium.

For the fifth sketch (Fig. 7), De la Beche redrew the continents suspended from a balance, and the accompanying text has the form of a see-sawing nursery rhyme:

> Here we go up, up, up,
> Here we go down, down, down.

The balance is now held by the figure of Father Time, seated on clouds, complete with his traditional scythe and hour glass. But he is viewing his

[31] *Ibid.*, p. 82.

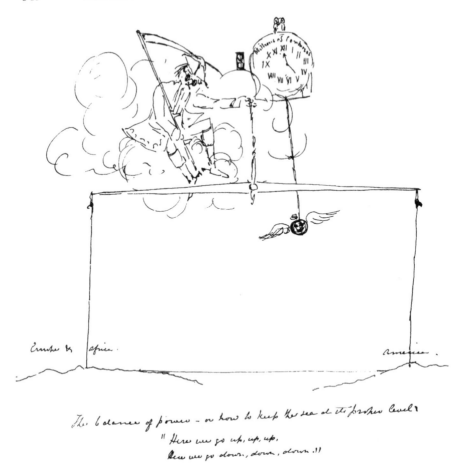

Figure 7. "The balance of power—or how to keep the sea at its proper level: 'Here we go up, up, up,/Here we go down, down, down.'"

balance through colored spectacles, like those worn by everyone in the third sketch. He also has a more nineteenth-century piece of equipment, a large pendulum clock, on which perches an owl (carried over from the third sketch). But the clock is driven, not by anything as mundane as the weight which would normally be suspended below such a clock, but by an animated pendulum with a pair of angelic wings (perhaps transmuted from the fairies' wings of the previous sketch). In other words, time itself is being measured—or even driven—by forces far more fanciful than those in any ordinary clock. This reference to the specific character of *Lyellian* time is made more explicit by the fact that the face of the clock is marked not in hours but in "Millions of Centuries." Like other geologists, De la Beche felt that Lyell tended to use a vastly extended scale of time as an almost material force, the mere invocation of which was held to be adequate to explain even the most catastrophic phenomena in geology. In this sketch the particular reference is again to Lyell's explanation of continental elevation and subsidence. The title, "The balance of power—or how to keep the sea at its proper level," uses the political phrase to refer to Lyell's insistence that changes in relative sea level are to be explained in terms of endogenous movements of the earth's crust rather than in the more traditional terms of eustatic changes in the level of the sea itself.

Father Time reappears in the sixth sketch (Fig. 8), but he is now accompanied by an iconographically more original figure representing Space. Approaching these demiurgic beings is an ordinary mortal—again inferentially Lyell— apparently wearing his barrister's gown as well as his wig. He is thrusting toward them a book—inferentially the *Principles*—and crying, "Behold my book, Sirs, Time & Space." This is probably De la Beche's somewhat cynical comment on the vast explanatory ambitions of the *Principles,* and it seems to anticipate the grandiloquent passage, comparing time in geology with space in astronomy, with which Lyell concluded his third and culminating volume three years later.[32]

De la Beche and other critics of Lyell were not of course averse to a geological time scale of millions of years, so long as it was genuinely required for the explanation of phenomena. What they objected to was Lyell's ad hoc use of time to explain away features that did not fit his overall theory of the earth. Prominent among such features was the "diluvium," the irregularly distributed and peculiar deposits (in modern terms, of glacial origin) that were generally attributed to an unusual diluvial episode or episodes in the geologically recent past.[33] The diluvium lay above (and was therefore more recent than) the huge thickness of "regular strata"; but in turn it was overlain by the still more recent "alluvium," which was manifestly the product of ordinary "actual causes." The peculiarity of the diluvium thus seemed to belie Lyell's claim that there had been an unbroken continuity of ordinary causation since the most remote periods of the earth's history. To overcome this objection, Lyell characteristically used a vast time scale to iron out the apparently violent origin of the diluvium into a much longer period of normal causes. He regarded it as a natural connecting link between an alluvial period—also vastly extended in duration— and the ordinary Tertiary strata. Most other geologists, however, even if they did not share Buckland's earlier belief that there had been only a single diluvial episode and that this could be equated with the biblical flood, still felt that to eliminate the peculiar character of the diluvium altogether was an implausible and ad hoc maneuver.[34]

De la Beche's seventh sketch (Fig. 9) reflects this feeling. "Alluvium," which most geologists at the time (like their successors today) considered as the product

[32] *Ibid.,* Vol. III (1833), pp. 384–385.

[33] Chorley *et al., Landforms;* Gordon L. Davies, *The Earth in Decay. A History of British Geomorphology 1578–1878* (London: Macdonald, 1969), Chs. 7, 8; M. J. S. Rudwick, "The Glacial Theory," *History of Science,* 1970, 8:136–157.

[34] Gillispie, *Genesis and Geology;* cf. Leroy E. Page, "Diluvialism and Its Critics in Great Britain in the Early Nineteenth Century," *Toward a History of Geology,* ed. Cecil J. Schneer (Cambridge, Mass.: M.I.T. Press, 1969), pp. 257–271.

Figure 8. "Behold my book, Sirs, Time & Space."

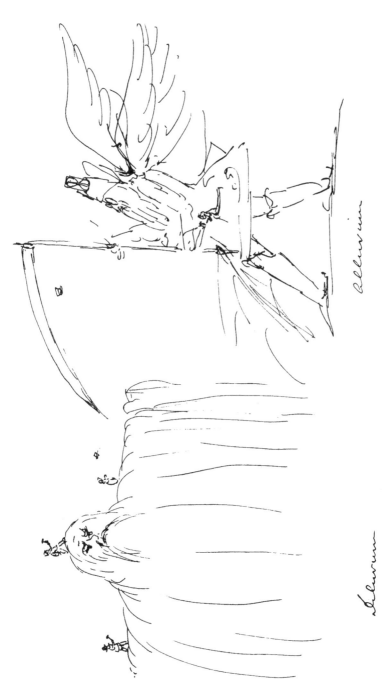

Figure 9. "Diluvium" and "Alluvium."

of merely the few thousand years of post-diluvial (i.e., post-glacial) history, is personified as Father Time himself, whose clock, in the previous sketch, had been calibrated in "Millions of Centuries." In accordance with this fanciful concept, Father Time has now sprouted a pair of angelic wings—perhaps those previously attached to the pendulum of his clock.

If the alluvium was of such unimaginable antiquity, what conception could be attached to the age of the diluvium, let alone the proportionately far greater time represented by the still older "regular strata"? De la Beche made this point by drawing alongside "Alluvium" a figure of "Diluvium," personified as the "Ancient of Days" Himself—the figure is perhaps based on the Creation scene in William Blake's *Book of Job*—over whom tiny hammer-bearing geologists are clambering irreverently.[35]

Perhaps the thought of all those geologists puzzling over the diluvium provided the connecting link to De la Beche's eighth sketch (Fig. 10), which takes up the social theme again by depicting an argument at a formal scientific meeting, probably of the Geological Society, which was the institutional setting for the main Lyellian debate. The characters are however transmuted into animal form. A long-snouted ichthyosaur is addressing the chair, saying, "Mr. President crocodile, Allow me to . . ." The president, who has a lion[?] and an elephant as secretaries by his side, apparently interrupts by saying, to a different figure, "Allow me to state, Mr. Plesiosaurus, that Mr. Longirostrus Ichthyosaurus's remarks were confined—." The long-necked plesiosaur replies, "I bow Mr. President."

Both the iconography and the meaning of the dialogue are obscure. I am inclined to think that the scene may be based on De la Beche's recollection of an earlier encounter between Lyell and his opponents, at which the diluvial question was central to the debate. Even before the first volume of the *Principles* had been published, Lyell's Playfairian arguments for the very gradual erosion of valleys by the rivers flowing in them—a test case for his whole approach—had been countered by Conybeare, who termed the debate one of "Fluvialists" versus "Diluvialists"; and another of Lyell's critics had summarized the Lyellian position as "Give us time, and we will work wonders."[36] If De la Beche was indeed recollecting this or a similar exchange at the Geological Society, at least some of the iconography would become intelligible.

Conybeare is an obvious candidate for the plesiosaur in the sketch, since the reconstruction of that reptile had been one of his most distinguished pieces of research.[37] The plesiosaur's apparent concession of defeat would then

---

[35] William Blake, *Illustrations of the Book of Job* (London, 1826); see the engraving entitled "When the morning Stars sang together & all the Sons of God shouted for joy" (Job 38:6—the conclusion of the passage in which God asks Job, "Where wast thou when I laid the foundations of the earth?"). The border of the engraving is decorated with six small vignettes of the days of creation, which of course many geologists at this period regarded as a symbolic summary of the long history revealed by their science. The figure of the Creator is strikingly similar in pose to De la Beche's sketch. That De la Beche should have known Blake's work is not implausible in view of his strong artistic interests.

[36] Reported in Lyell to Mantell, Apr. 1829, *Life, Letters and Journals*, Vol. I, p. 252.

[37] W. D. Conybeare and H. T. De la Beche, "Notice of a Discovery of a New Fossil Animal,

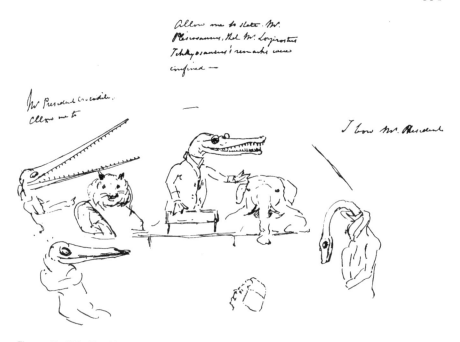

Figure 10. "Mr. President crocodile, Allow me to." "Allow me to state, Mr. Plesiosaurus, that Mr. Longirostrus Ichthyosaurus's remarks were confined—." "I bow Mr. President."

represent Lyell's belief that he had come off best in that round of the debate.[38] This suggests Lyell himself as the ichthyosaur, though I do not know what prompted that characterization, unless he had a reputation for snapping ferociously at his opponents. During the relevant period the president was Sedgwick and later Roderick Murchison, one of whom would thus be the crocodile, and the secretaries were (at different times) William Broderip, Murchison, and Edward Turner; but again I do not know why any of them should have been characterized by the animals shown. De la Beche might have depicted himself as the only human figure, the onlooker in the foreground.

Obviously this interpretation is highly conjectural. There are other possibilities. A few years later De la Beche caricatured the prominence of his own nose;[39] this might be taken to suggest that in the present sketch he was depicting

Forming a Link between the Ichthyosaurus and the Crocodile; Together with General Remarks on the Osteology of the Ichthyosaurus," *Transactions of the Geological Society of London*, 1821, 5:558–594. De la Beche had collaborated with Conybeare in this work by providing the stratigraphical context of the discovery.

[38] Lyell to Mantell, June 7, 1829, *Life, Letters and Journals*, Vol. I, p. 253.

[39] De la Beche to Sedgwick, Dec. 11, 1834 (University Library, Cambridge, Sedgwick papers, Add. 7652). This too was a caricature with substantive meaning; see Martin J. S. Rudwick, "The Devonian System 1834–1840. A Study in Scientific Controversy," *Actes du XIIe Congrès international d'Histoire des Sciences* (Paris: A. Blanchard, 1971), Vol. VII, pp. 39–43.

himself as the ichthyosaur, to whom the president gives the otherwise superfluous epithet of "Longirostrus." But then his opponent would either be Conybeare—which is clearly improbable—or if the plesiosaur was Lyell, then Lyell would be shown in an attitude of defeat—which is equally improbable. I am therefore inclined to prefer the first interpretation I have suggested, without claiming that that it is more than a conjecture.

But whatever persons are represented by the various animals, they are curiously mixed in geological age. In the terminology of the time, the ichthyosaur and plesiosaur were Secondary, the crocodile, lion, and elephant Tertiary, and the human being Recent. Thus the scene cannot be attributed imaginatively to any single geological epoch. This confusion of epochs is unlikely to be accidental, for most geologists felt strongly that fossils were above all a record of the history of progressively higher forms of life on earth. The mixture would be understandable, however, if it were De la Beche's comment on Lyell's idiosyncratic denial that successive epochs had been characterized by any such distinctive sequence of faunas.

In any case it is obvious that the sketch as a whole is incomplete, and perhaps the most plausible conclusion is that De la Beche became dissatisfied with it and abandoned it before he had drawn enough, and written in enough of the dialogue, to enable us fully to understand the scene. A temporary mental block is suggested also by the fact that on the page facing this sketch De la Beche drew what is little more than a doodle (Fig. 11). Yet even this is worth analysis. Here a book appears again, and I infer once more that this was the *Principles*. But now it has left Lyell's hand and is being borne freely upwards, wafted by a pair of angel's wings and buoyed up by a balloon—at this date perhaps still a hot air balloon! This may be a comment on the lack of solid, empirical, down-to-earth foundations in Lyell's work; its proper place, in De la Beche's view, is up in the realms of fantasy.

After this pause (assuming he did draw the sketches in the order in which they appear), De la Beche made a second attempt to characterize Lyellian theory in terms of animal representation (Fig. 12). This time he was more successful: the last sketch is unmistakably the precursor of the lithograph of

Figure 11.

Figure 12. "Return of Ichthyosauri &c." " 'Principles &c.' " "You will at once preceive that the [skull] before us belonged [to] some of the lower order of animals, the teeth are very insignificant, the power of the jaws trifling, and altogether it seems wonderful how the creature could have procured food."

"Awful Changes," and the sketch itself was left unfinished, evidently because he completed the drawing on stone.

Up to this point, as I have admitted repeatedly, my identification of Lyell and Lyellian theory in the sketches has been based on inference from what we know from other sources about Lyell's work and De la Beche's attitude to it. But in this last sketch the theme is at last stated unambiguously: for after the title "Return of Ichthyosauri &c" De la Beche wrote "Principles &c." Taken together with the internal consistency of the iconography and the external evidence (from the rest of the notebook) that the sketches were drawn some time after the first volume of the *Principles* was published, I consider that there is a strong case for believing that the whole series was indeed an anti-Lyellian polemic.

The reference to the *Principles* in the caption of the last sketch was not repeated on the lithographed versions which De la Beche distributed. But the discovery of this vital clue forces us to reassess the possible meaning of the final caricature. I have already mentioned that it has seemed hitherto to be a simple humorous comment on Buckland's lectures, with the tables turned and the Jurassic reptiles in command. But the discovery of its context as the culmination of a series of sketches that seem to comment on Lyell's steady-state system of geology forces us to reconsider the identification of the professorial lecturer.

If I am right in suggesting that Lyell had been a central figure in all the earlier sketches, it would be improbable that Buckland should take over that position in the final version. It is surely more likely that Lyell himself was the lecturer. Certainly he had an appropriate position by 1831 (if the later date for the sketches is adopted), Professor of Geology at King's College in London. In other words, the "Mr. Longirostrus Ichthyosaurus" at the Geological Society (in the previous sketch) could have been transmuted into the "Professor Ichthyosaurus" at King's College in the final caricature—both figures representing Lyell himself. If on the other hand the caricature dates from 1830, before Lyell's appointment, the lecturer could still represent Lyell, as a fanciful successor (or usurper?) of Buckland at Oxford. De la Beche's choice of a human skull for the lecturer's object of attention, and the caricature's provisional title "*Return of Ichthyosauri*," now force us to take the *future* reference of the drawing more seriously, as a comment on Lyell's work.

An important section of the first volume of the *Principles* was devoted to an attempt to undermine the validity not only of the evidence for directional change in the inorganic environment during the history of the earth, but also that for a similarly directional or progressionist history of life.[40] In other words, as mentioned earlier, Lyell extended his steady-state theory even to the fossil record. To most geologists this seemed to be particularly implausible. It involved a complete rejection of the clear trend of palaeontological discovery, which seemed to be uncovering a broadly progressive history from the simpler or lower organisms, through the successive appearance of more complex or higher forms of life, to the geologically recent creation of man himself. It is well known that Lyell circumvented this evidence by attributing it to differential preservation, and that he asserted that even the higher forms of life (e.g., mammals) had existed at the remotest known periods.

In this Lyellian setting the text of De la Beche's ichthyosauran lecturer takes on new meaning, as a comment not on Buckland's methods of functional analysis of fossils (which were indeed followed by Lyell), but rather on Lyell's denial of any trend from lower to higher in the history of organic life. In

---

[40] *Principles*, Vol. I, Chs. 6–8 and Ch. 9 respectively.

the final version of the caricature (copied with only trivial changes from the last sketch), this text reads as follows:

> "You will at once perceive," continued Professor Ichthyosaurus, "that the [human] skull before us belonged to some of the lower order of animals, the teeth are very insignificant, the power of the jaws trifling, and altogether it seems wonderful how the creature could have procured food."

In other words, in Lyell's interpretation a human being is no "higher" than a reptile; indeed, its functional anatomy might even make it seem "lower." This refers to Lyell's argument, based on a rigid dualism between the physical and the moral realms, that the recent appearance of man was no evidence of organic progress, on the grounds that man's superiority was exclusively moral, and that physically he was no higher than any other mammal.[41]

But Lyell's idiosyncratic interpretation of the fossil record had deeper implications that linked it indissolubly to the rest of his steady-state system. Not only had there been no overall progression in organic life in the past; there would be none in the future either. Lyell tried to guard against being interpreted as postulating a strictly cyclical theory of indefinite repetition on the Greek model, but it is clear that many of his contemporaries did regard his theory as being for all practical purposes one of cyclical repetition.[42] Indeed he laid himself open to this interpretation. For example, having told Mantell his "receipt" for producing organisms of diverse climatic habits more or less to order at any latitude, he explicitly applied this to the future as well as the past. Mantell had discovered the fossil terrestrial reptile *Iguanodon* in Sussex, and Lyell told him that "All these changes are to happen in future again, and iguanodons and their congeners must as assuredly live again in the latitude of Cuckfield as they have done so." Likewise in the published *Principles* Lyell wrote: "Then might those genera of animals return, of which the memorials are preserved in the ancient rocks of our continents. The huge iguanodon might reappear in the woods, and the ichthyosaur in the sea, while the pterodactyle might flit again through umbrageous groves of tree-ferns."[43] Lyell may not have believed that organisms would ever reappear as precisely *identical* species after their earlier embodiments had become extinct. But his concept of new species being "created" in ecologically and adaptively appropriate space/time locations, combined with his geological concept of a continual flux at the earth's surface which was repeatedly producing virtually identical environments, led him inexorably to a view that *very similar* species must have been produced repeatedly in the past and would continue to be produced indefinitely into the future.

---

[41] *Ibid.*, pp. 162–164.

[42] See, e.g., his reply to Whewell on this point in 1837: *Life, Letters and Journals*, Vol. II, pp. 2–7.

[43] See n. 28 and *Principles*, Vol. I, p. 123. Gideon Mantell, "Notice on the Iguanodon, a Newly Discovered Fossil Reptile, from the Sandstone of Tilgate Forest, in Sussex," *Philosophical Transactions of the Royal Society*, 1825, pp. 179–186.

The future reference of De la Beche's final caricature therefore needs to be taken as a serious comment on this aspect of Lyell's system. The title of the drawing, "Awful Changes," refers to a repeated line in Byron's *The Dream* (1816): the dreamer moves forward in time from one visionary episode to the next, introducing each successive vision with the words "A change came o'er the spirit of my dream." In such a manner, De la Beche implies ironically, Lyell the visionary dreamer might conceive such a scene as he has drawn, from the far distant *future* history of the earth.[44] The subtitle of the caricature, "Man found only in a fossil state—Reappearance of Ichthyosauri," refers even more clearly to this supposed future state of the earth.

The centrality of Lyell's steady-state theories as a target for De la Beche's criticism will be no surprise to those Lyell scholars who take seriously the many critical published reviews and unpublished comments on the *Principles*. I have argued earlier in *Isis* that the whole strategy of the three-volume work was dominated by this steady-state component; and these caricatures give further support to the view that it seemed equally important to Lyell's critics.[45] More recently it has been suggested that the steady-state component was extrinsic to Lyell's original research program: that his reason for adopting a theory so far out of line with current geological thinking lay in his fear of the evolutionary implications of progressionist theory for the place of man in nature, and that he adopted a steady-state theory as a defense against these implications only after reading Lamarck in 1827.[46] This interpretation is not, however, incompatible with my own; on the contrary, the urgency of Lyell's putative fear of Lamarck's theory may be reflected in the thoroughness with which he developed his geology into a watertight steady-state system that would serve to defend the uniqueness of man. In any case, there is no doubt that the steady-state component of Lyell's uniformitarianism was a conspicuous target for criticism by almost all of his contemporaries, even by those who, like his friend George Poulett Scrope, most fully approved of his emphasis on the adequacy of actual causes within a vastly extended time scale.[47]

I therefore conclude that De la Beche's caricature of "Awful Changes," which

---

[44] De la Beche's irony may extend further, applying the final doom of Byron's "Wanderer" ("To end . . . in misery") to Lyell's overambitious quest for the final secrets of the earth:

". . . with the stars
And the quick Spirit of the Universe
He held his dialogues; and they did teach
To him the magic of their mysteries;
To him the book of Night was open'd wide,
And voices from the deep abyss reveal'd
A marvel and a secret—Be it so." (*The Dream*, viii).

[45] See nn. 4 and 29.

[46] Michael Bartholomew, "Lyell and Evolution: An Account of Lyell's Response to the Prospect of an Evolutionary Ancestry for Man," *British Journal for the History of Science*, 1973, 6:261–303.

[47] [G. Poulett Scrope], "Art. IV. [Review of] *Principles of Geology* By Charles Lyell, F.R.S. 2 vols. Lond. 1830," *Quarterly Review*, 1830, 43:411–469; Martin J. S. Rudwick, "Poulett Scrope on the Volcanoes of Auvergne: Lyellian Time and Political Economy," *Br. J. Hist. Sci.*, 1974, 7:205–242.

is superficially so trivial in meaning, has a much more weighty significance in the light of the series of sketches of which it was the final product. It serves to emphasize that Lyell's geology by no means swept the English geological community off its feet (despite the fears expressed in De la Beche's third sketch); that criticism of it was by no means confined to scientific or theological conservatives or motivated by their prejudices; and that the steady-state component of Lyell's uniformitarianism seemed so implausible that the persuasive impact of his use of actualistic reasoning must also have been dulled.

In the light of this example, I suggest that historians of science who encounter caricatures by other scientific figures of the past, whether published at the time or not, would do well to study them carefully for their possible substantive meaning, before dismissing them light-heartedly (or with academic embarrassment) as materials unworthy of serious historical attention.

De la Beche's use of the medium of caricature illuminates not only the substantive content of a scientific argument, but also—perhaps less controversially—its social context. It emphasizes the close parallel between the social worlds of geology and politics at this period: De la Beche's drawings, like Gillray's, reflect the rivalry, polarization, and partisanship of contemporary debate. His sketchbook shows that he himself felt no restrictive boundary between the scientific and the political as fields of caricature, even though the avowedly nonpolitical norms of the Geological Society may have inhibited him from circulating his political drawings as freely as his geological sketches.

Granted, however, the social circumstances of the caricature I have analyzed, the way in which De la Beche conceived his final design still remains a separate question. I suggest that his sequence of preliminary drawings deserves attention for what it reveals about the mode of thinking of a man who in his publications appears as a prototypical sober, rational scientist. If my interpretation of the sketches is even approximately correct, it must be clear that the sequence displays all the dreamlike qualities of genuine free-associative thinking. It shows the characteristic fluid transmutation of images and ideas from one sketch to the next (e.g., spectacles, angel or fairy wings) and the uninhibited use of metaphors, puns, and double meanings (e.g., seeing, coloring, cooking, viewpoints, balance of power, hot air).

But De la Beche was not one person when he drew such scientifically meaningful caricatures and a different person when he went out into the Devonshire countryside to study geology. He was a single human being. If we are to understand more fully the nature of scientific thinking, we shall surely need to take more seriously the mental levels on which such free-associative thinking occurs. But if we are also to avoid writing spurious "psychohistory," it is equally clear that we shall need to tie down such analyses as closely as possible to interpretable evidence. I do not claim that my interpretation of these sketches is likely to be correct in every detail. But I do believe that, like any other suggested reading of documentary material, it is at least testable and improvable by clear criteria based on its coherence with other sources of evidence.

VIII

CHARLES LYELL, F.R.S. (1797–1875) AND HIS
LONDON LECTURES ON GEOLOGY, 1832–33

IN his survey of *The Organisation of Science in England*, Donald Cardwell describes how the foundation of University College London in 1826, closely followed by that of King's College in 1828, reflects the polarization between Radical-Dissent and Anglican-Tory groups among those concerned with the promotion of middle-class higher education in the metropolis (1). He points out, however, that while the teaching staff (but not the students) at King's were required to be Anglicans, and the courses included religious instruction, the syllabus was no less liberal than at University College, and included substantial provision for the teaching of science. Among the eight professorships proposed by the chartered committee in April 1830 (in addition to three Chairs with guaranteed salaries), there were four in clearly scientific fields. By February 1831 James Rennie had been appointed for natural history and zoology, Henry Moseley for natural and experimental philosophy, and J. F. Daniell for chemistry (2). At about the same time Charles Lyell (F.R.S., 1826) made enquiries about the possibility that geology might be added to the list of scientific posts, and in April 1831 he was appointed Professor of Geology (3). King's College opened for teaching in October 1831, and Lyell gave courses of lectures on geology the following summer (1832) and again in 1833. Concurrently with the latter, he also gave a course at the Royal Institution. But while the first of his courses at King's was still in progress, Lyell began to consider resigning from his Chair, and he duly did so in October 1833.

There are several features of historical importance about Lyell's brief tenure of the first British teaching post outside Oxford and Cambridge to be devoted specifically to geology. On a traditional interpretation of his sympathies, it might be expected that he would have applied to the relatively 'free-thinking' University College rather than to the ecclesiastically orthodox King's College. Moreover, it has long been known that one member of the Council at King's queried the theological implications of Lyell's geology before acceding to his appointment. It has therefore been inferred that Lyell's apparently premature resignation was connected with the unpalatable heterodoxy of his views, as expressed in his lectures and his publications. Thus for example the centenary historian of King's inferred that Lyell's assumptions about the vast time-scale of the Earth's history were unacceptable to the authorities at the College, and that Lyell's true reason for resigning was that he wished discreetly to avoid a Galileo-like confrontation with the College Council (4). Likewise, Lyell's most recent biographer considers that Lyell was less than honest in stating his true beliefs when obtaining the post, and accuses Lyell's clerical critic of 'tortuous hypocrisy' (5).

In this article I shall suggest that such interpretations derive from pre-suppositions of dubious validity about Lyell's religious position and social attitudes. They are part of a broader interpretation in which Lyell's activities have been seen primarily in terms of a 'conflict' of science and religion in early 19th-century Britain. That 'conflict' thesis is derived in turn from the crusading attitude of late 19th-century writers who saw themselves as the prophets of a fully secularized society under the guiding light of Science (6). The thesis is historically suspect, not only because it projects back into the earlier part of the century a polarization that belongs to much later decades. It also conceives the relation between the scientist and his society in one-dimensional terms, as a confrontation between the hypostatized entities of Science and Religion, instead of seeing both as complex social activities in which the same individuals participated. Furthermore, it ignores all questions relating to the material substructure underlying the activity of science; and it completely evades the question, which has become so acute in our own day, of the social responsibility of the scientist for the effects of popularizing his scientific ideas. Once the presuppositions of this outdated historiography are abandoned, however, much that has seemed puzzling about Lyell's activities, and particularly his behaviour over his tenure of the Chair at King's, becomes relatively straightforward and consistent. Lyell can be exonerated from deviousness and his critic from hypocrisy. More importantly, the episode then becomes a valuable example of the self-conscious professionalization of science in early 19th-century Britain (7).

LYELL'S APPOINTMENT

When Lyell first began to make an indirect approach to King's he was already well known among the London intelligentsia as a geologist of great promise. The first volume of his *Principles of Geology* had been published a few months earlier, and its attractive style and persuasive argument had won for it a wider readership than most other similar scientific treatises would have had (8). Among the smaller circle of those actively engaged in geological research his work was already highly respected, even by those who disagreed with some of his theoretical conclusions. On academic grounds alone, he was obviously a strong candidate for any teaching post in London that might be available.

It is possible that University College did consider him for a post in geology, or that he himself approached the College. He certainly knew the Warden, Leonard Horner, a fellow-geologist who was soon to become his father-in-law. But with the chemist Edward Turner and the botanist John Lindley already on the staff, and capable of teaching geology quite adequately between them, there need be no surprise that University College did not in fact offer him a post (9). Conversely, there is much to suggest that his sympathies were closer to King's than tradition has allowed. He had first become known in intellectual circles through his series of essay-reviews for the *Quarterly Review* in the late 1820s; and while he privately expressed some wry amusement at the thought of contributing to the Tory periodical, it is clear that he was well aware of the opportunity it gave him to reach the influential audience of the *Quarterly* with a more enlightened view of science (10).

He also had a more concrete reason for applying to King's. He had been called to the Bar in 1822 and had practised briefly on the western circuit (11) but his legal qualification failed to provide him with any regular income, and it is clear from his correspondence that his financial problems gave him continuing anxiety. He hoped that the sale of the *Principles*, like the fees for his articles in the *Quarterly*, would help to provide him with enough income to devote himself fully to geology, but he could not be certain of this in advance (12). The same uncertainty characterized the Chair at King's, since the lectures would be optional, and the income derived from the post would therefore depend on the number of fee-paying students and members of the public that he was able to attract. On the other hand he realised that he could give the lectures with relatively little extra work, by condensing the material he was assembling for the *Principles* and by illustrating the lectures with large-scale drawings and geological sections (13). In this way he hoped he could derive some extra income from lecturing, without serious delay to the larger work and the more substantial rewards he hoped that would bring. A further consideration may have been the encouragement of William Whewell's favourable review of the first volume of the *Principles* in the *British Critic*, which led him to expect support from intellectual Anglican circles (14).

His method of applying for a Chair at King's supports this inference. He first approached Adam Sedgwick, the Woodwardian Professor of Geology at Cambridge and, like Whewell, a Fellow of Trinity College. A year earlier, in the first of his Anniversary Addresses as President of the Geological Society of London, Sedgwick had praised Lyell's early work on the geology of central France, and had explicitly approved Lyell's inferences about the vast time-scale involved. In his second Address, delivered just at the time that Lyell was thinking of applying to King's, Sedgwick welcomed the first volume of the *Principles*, praising it warmly in many respects—including again Lyell's assumption of a vast time-scale—but at the same time criticizing trenchantly some of its fundamental methodological presuppositions and theoretical conclusions (15). Whewell's 'catastrophist' label has since been applied to Sedgwick, but the supposed polarization between this view and Lyell's 'uniformitarianism' by no means precluded a warm personal relation between the two men. Nor did it prevent Sedgwick from giving Lyell his full professional support by agreeing to recommend him to Charles Blomfield, who as Bishop of London held a key position on the Council at King's.

There need be no surprise that Blomfield, together with William Howley, the Archbishop of Canterbury (whom Lyell also canvassed), and other members of the Council, approved of Lyell's suggestion. Geology, as a relatively new and increasingly popular science, was an attractive candidate for addition to the list of scientific Chairs at the College; and since the post would be paid for out of the fees of those attending the lectures, it would cost the College virtually nothing. Furthermore, the opportunity of acquiring as Professor one of the most promising younger geologists in the country was not to be missed, for it would add needed prestige to the new institution.

Their only possible reservation, of course, derived from the sensitive position of geology—unlike most other areas of science at the time—as bearing upon the traditional interpretation of scripture. But Blomfield and his committee were no blinkered literalists. Their attitude to Lyell's geology, which as non-geologists they admitted they found novel and even 'startling', was simple and consistent. As Lyell himself told Gideon Mantell, the Council were satisfied 'that whether the facts [i.e. of his geology] were true or not, or my conclusions logical or otherwise, there was no reason to infer that I had made my theory from any hostile feeling towards revelation' (16). They were not trying to foreclose scientific conclusions by ecclesiastical edict; but they were properly concerned that those appointed to teaching positions should not misuse the authority of their situation by deliberately attacking the foundations of the institution. In other words, the issue was that of the public responsibility of the man of science.

Once the issue is seen as one of social responsibility, rather than an attempted interference with the autonomy of science, the attitude of the one member of the Council who had reservations about Lyell becomes intelligible. Edward Copleston, the Bishop of Llandaff and Dean of St Paul's, had been Provost of Oriel College while Lyell was an undergraduate at Oxford, and had been a prominent member of the reformist Noetic group there. His theological views were liberal, and he deplored both literalism in biblical interpretation and (later) the anti-intellectual tendencies of the Tractarians (17). It is worth noting that he was instrumental in persuading King's to appoint his friend N. W. Senior, the distinguished Oxford economist, to the Chair of political economy, a subject which—as Lyell himself commented—was much more sensitive to conservative opinion than geology (18). Lyell may have had little time for 'our least of all great men, our Oxford Copleston' (19), perhaps because Copleston was also a friend of William Buckland, the leader of the 'Oxford school' of geology and Lyell's chief target for geological criticism. But Copleston was no bigot, and on the matter of Lyell's appointment, he merely wanted rather firmer reassurance that Lyell would not use his position irresponsibly. He accepted Lyell's scientific qualifications, but enquired through J. G. Lockhart, the editor of the *Quarterly*, about Lyell's views on the two points where he felt that the concerns of revealed religion impinged on the findings of geology. The questions at issue were (1) the creation of Man at a (geologically) recent period, and (b) the historicity of a universal Flood at a somewhat later date.

Lyell replied that the still unpublished part of his *Principles* would explicitly criticize the hypothesis (i.e. Lamarck's) that dispensed with divine action in the creation of each species; and that he knew no evidence to contradict the traditional period for the creation of Man (20). On both these points the sincerity of Lyell's reply is amply confirmed by the explicit arguments of the second volume of the *Principles* (21). On the other question, about the Flood, Lyell admitted unambiguously to Copleston that, like many other geologists, he did not believe that 'a deluge has passed over *the whole earth* within the last 4000 years', but he agreed that the area then occupied by Man might perhaps have been inundated, for all that geology could tell.

Copleston's response was to declare himself fully satisfied about Lyell's views on the creation of Man, while expressing continuing concern about the possible effects of Lyell's views about the Flood on his potential audience at King's (22). 'Truth will ultimately prevail', he told Lyell, 'and I by no means wish to check free enquiry'. The rest of his letter confirms the sincerity of that remark. He accepted that with regard to the Flood the progress of science might involve an adjustment of 'the language of Scripture as understood and interpreted by the Church', but he felt a pastoral responsibility to ensure that a change 'which perhaps would put in jeopardy the faith of many' should not be forced on Lyell's audience prematurely, i.e. before it was necessitated by firmly established scientific research. (As we shall see, Lyell's work did persuade him later to concede the need for this change.) In other words, Copleston was aware how easily a scientist may use the authority of his position to draw conclusions of general human significance, which are not strictly inferrable from the science itself.

An awareness of this danger has become praiseworthy and even commonplace when applied to the modern debates on, for example, race and intelligence. It is surely only the 'conflict' assumptions inherited from the last century that prevent some modern historians from seeing that the same principle was involved in an earlier episode such as this. Cautious ecclesiastics may not be the most attractive personalities in any age, and Copleston may have been over-zealous—like many before him and since—in seeking to protect a faith which he assumed to be fragile and vulnerable among 'weaker brethren' than himself. But such judgments should not prevent us from seeing that according to his own conception of his responsibilities he was making reasonable enquiries about Lyell's attitude. He was not trying to censor the scientific contents of Lyell's lectures. He was concerned only that Lyell's geology should 'not be proposed in a form offensive to religious minds'; or in other words that Lyell should not misuse the authority of his teaching situation.

This exchange of letters, as already mentioned, has been censured by Lyell's biographer as one of regrettable insincerity capped by one of a 'most tortuous hypocrisy' (23). Lyell's prejudices and apparent deviousness become intelligible, however, once a simplistic 'conflict' historiography is abandoned. There is nothing to suggest that in wishing to 'free the science from Moses' (24) Lyell intended to undermine other people's religious beliefs. On the contrary, his intention is expressed precisely by that quoted phrase: to liberate geology from the impediment which he felt that traditional scriptural interpretation had imposed, or in other words to make geology an autonomous science. Privately, he may indeed have regarded the punitive aspect of the doctrine associated with the Flood as morally distasteful: for example, he commented sarcastically that Copleston 'concedes that the drowning of all men ought to satisfy all reasonable folk' (25). But whatever his private religious beliefs may have been at this time—and the question is too large to tackle here—there can be no doubt about his public attitude to the beliefs of others. He felt that a respect for the religious sensibilities of others was a proper extension of ordinary good manners into a realm of deeper significance: one might privately consider those beliefs unfounded or even idolatrous, but that was no excuse for undermining them.

Thus, for example, Lyell likened Toulmin's aggressive assertion of the un-created eternity of the world—a view possibly derived from James Hutton's geological system—to the equally insensitive behaviour of the British soldiers who had 'dug up the idols of the Burmese temples in the late campaign, and sent them home as trophies' (26). This example indicates that Lyell's scruples were not confined to the beliefs of his own society: it was not a matter of expediency but of principle. There is thus no evidence to suggest that he would have chosen to use his lectures at King's to propagate deliberately anti-Christian arguments, even if—which is questionable—he held such views himself.

Lyell's friend George Poulett Scrope—his geological mentor in central France and already the author of reformist pamphlets on economic and social questions—was being unduly pessimistic in thinking that if Lyell's geology were to be 'taken into the bosom of the Church' by his appointment at King's, it would be more surprising than the passing of the Reform Bill (27). Lyell himself was more realistic, appreciating that 'If we don't irritate, . . . we shall carry all with us'—as he had told Scrope a few months earlier, explaining how he wished the *Principles* to be reviewed in the *Quarterly*. He added on that occasion that then 'the bishops and enlightened saints will join us in despising both the ancient and modern physico-theologians' (28). By the former he meant the leaders of Anglican thought and such intellectuals as Whewell and Sedgwick; by the latter, such 'diluvialist' geologists as Buckland and the scientifically worthless 'scriptural geologists'.

Yet once again the fluidity of the 'party-lines' involved in the debate is indicated by the fact that it was Buckland's friend and champion William Conybeare who volunteered to write to Copleston, to urge him to withdraw his reservations about Lyell (29). But there is no paradox, once it is accepted that the debate was primarily about the scientific and social respectability of geology, rather than a conflict between geology and Genesis. For Conybeare, like Sedgwick, recognized Lyell's exceptional abilities as a geologist, even though they both disagreed with him on important points of geological method and theory; and they knew that his lectures at King's would do nothing but good to the cause they all had at heart, namely the promotion of geology as an established branch of science.

In the event, Copleston withdrew his opposition to Lyell's appointment, perhaps as a result of Conybeare's intervention, or in view of the Council's otherwise unanimous approval. Lyell was formally appointed to the Chair on 15 April 1831, and at their next meeting the Council voted him £100 for the purchase of specimens to form the nucleus of a College collection and to illustrate his lectures (30). Lyell himself, however, evidently regarded the lecture course as a chore to be kept as short as possible. He reported that some of his friends thought his post 'a foolish fancy', and perhaps his own confidence was undermined by this; but Whewell encouraged him by stressing the value of the lectures to Lyell himself, as a means of ordering his own thoughts on geology (31). Later in the year, William Somerville (Mary's husband) likewise encouraged him indirectly, by urging him to write a popular introduction to geology that would be more digestible than Henry De la Beche's recently published *Geological Manual*; and Lyell then saw that he could use the lectures as a means of drafting the chapters for such a book (32). In the event, this work was repeatedly delayed by the need to revise the *Principles* for successive editions, and Lyell's *Elements of Geology* did not appear until 1838. Nevertheless, this idea of writing a simple popular book on geology—an idea which dates back to his earlier thoughts about the work which became the *Principles*— indicates his continuing concern with spreading an 'enlightened' view of the science as widely as possible. If he was reluctant to devote time to the lectures, it was not because they would need to be elementary in content, but because they would reach only a small audience, and because the financial rewards would also be slight. As he told his fiancée, 'If in England a man really wants to have the satisfaction of influencing public opinion in his favourite science and getting fame at the same time, the way is certainly to give Murray [his publisher] a readable book rather than K[ing's] C[ollege] a popular course of lectures' (33).

In view of the suggestion that Lyell's later resignation from King's was related to the content of his lectures or of the *Principles*, it is important to note that he was considering resigning several months *before* he even began to lecture. Before the end of 1831 he was talking about his 'decided wish to get free of the professorial chair', and he added explicitly that he would resign more readily 'if I could secure a handsome profit from my work', i.e. from the *Principles* and other future publications. His ambivalence towards the College was not related to any antipathy to its Anglican foundation and curriculum; on the contrary he thought highly of it and recommended his aunt to send a young cousin there. He simply doubted whether the work that the lectures would require would be sufficiently repaid in terms of money, time and prestige. 'Do not think that my views in regard to science are taking a money-making, mercantile turn', he told his fiancée candidly. 'What I want is, to secure the power of commanding *time* to advance my knowledge and fame' (34). His attitude was not quite as venal or careerist as it may seem, for the widening of his knowledge of geology did indeed require substantial material support. As he himself realized as acutely as any geologist of his generation, the progress of the science required above all else the experience of travel, of seeing a wide range of phenomena with one's own eyes; and that needed money. In this respect he must surely have envied his more affluent friend and colleague Roderick Murchison, with whom he had shared part of his important Continental tour in 1828–1829. Doubtless Lyell was as much concerned with the advancement of his own fame as with that of geology; but when he said 'I am determined to make science a profession' (35) it is not only charitable but also justifiable to allow the ambiguity of the phrase, and to concede that he wanted financial independence through science not only for himself but also ultimately for others too.

Lyell's doubts about keeping the Chair at King's were particularly disappointing to Conybeare, who had canvassed hard on his behalf with Copleston, and he urged Lyell to think again. Likewise William Fitton and Babbage, although no friends of the Anglican establishment at King's, also urged him not to resign, at least until he had discovered whether lecturing did in fact interrupt his other work, and particularly if the College were content for him to give only a short course. So Lyell went ahead with his preparations for a course in the summer of 1832, getting the topographical artist George Scharf to paint him enlargements of some of his own and Scrope's drawings to use as visual aids in the lecture room (36).

It is significant that Lyell himself, before his course began, had mentioned to a correspondent that his geological views had been known to, and implicitly approved by, the episcopal divines on the Council at King's when appointing him to his Chair; and he explicitly associated himself with Buckland, Conybeare and Sedgwick, on the specific question of the vast time-scale of geology in comparison with human history:

I consider such an inference to be as fully borne out by the proofs, as almost any conclusion in physical science which does not depend on evidence capable of strict mathematical demonstration. Although the divines of my acquaintance who are professionally geologists 'The Revd. Dr. Buckland of Oxfd. & Revd. A. Sedgwick of Cambridge' may differ perhaps as to the quantity of time which geoll. changes must have occupied I may say that they & the Revd. W. Conybeare well known by his geologl. works all agree in ascribing a much higher antiquity to the Earth (Profr. Sedgwick even incalculably higher) than the date of the creation of man. I have advanced this opinion again & again in my 'Principles of Geoly —' after the publication of which I was selected by the Archbp. of Canterbury & Bp. of

London to fill the chair at King's College, my work having been duly consid [?] by several divines on the council of that College. I regret to hear that any conscientious persons shd. have taken alarm at the effect which such discussions may have in regard to Christianity which I feel assured will no more be injured by the advance of Geoly. than by the establishment of the earth's motion, a doctrine once thought so subversive of the bible, & which like many truths in geoly. is still considered by many irreconcilable with the strict letter of particular passages in the sacred writings (36a).

This statement must surely weaken still further the suggestion that Lyell's later resignation was related to the Council's disagreement with his geology or their fears of its implications for biblical interpretation.

LYELL'S INAUGURAL LECTURE

Lyell gave his first lecture at King's on 1 May 1832. The month could hardly have been less auspicious for a course which he hoped would attract many of the London intelligentsia, for it coincided with the height of the crisis over the Reform Bill. But his lectures contained no hint of the turbulent political situation. Lyell's totally unpolitical attitude was explicitly related to his belief in the political powerlessness of the nascent scientific profession. 'As for public affairs', he had told his fiancée a few months earlier, 'I have long left off troubling myself about them, as knowing that one engaged in scientific pursuits has as little to do with them in point of influencing their careers, as with the government of the hurricanes or earthly motion; and if one becomes annoyed, there is an end of steady work'. A necessary condition of useful scientific work and of a happy life, he believed, was 'to keep out of the excitement of politics' (37). In this non-political careerist attitude towards science, as in so much else, Lyell was a true forerunner of the modern scientific profession.

Lyell began his first lecture with 'a few observations on the present peculiar situation of the science of geology' (38). It was 'anomalous' in that its rapid progress made any theoretical generalizations seem hazardous:

> One of our poets, alluding to the incessant fluctuations of our language after the time of Chaucer, complains that,
>> We write on sand, the language grows,
>> And like the tide our work o'erflows;
> and the Geologist might well indulge the same feeling in regard to the want of stability in the subject matter of his own science during the period of its most rapid progress. . . . He might well think that his theories would only

be written upon the sands and that the tide of new discovery would soon sweep away and obliterate [them] for ever.

In this situation Lyell did not set himself apart from or in opposition to the avowedly 'Baconian' *public* policy of the Geological Society. 'This school of English geologists', he said, 'had no disinclination to indulge in speculative views and to discuss theoretical questions in their own circle'; but they were properly reluctant to publish general theories in a popular form, when they knew how the growing empirical foundations of the science were continually changing the plausibility of such theories. Lyell argued that lectures were a particularly appropriate medium at such a time, in that someone personally involved in research could give an up-to-date outline of a theoretical synthesis without the commitment to permanence that a publication would imply.

It could be argued, of course, that Lyell was in fact engaged in just such a theoretical publication on a massive scale, in the *Principles*. But although that work had first been planned as popular *Conversations on Geology*, it had changed under the pressure of Lyell's own theoretical development into a work on a far higher technical and theoretical level. At that level, however, its theoretical content was relatively concealed from all but the geologically well-informed reader, by the sheer bulk of detail. His lectures at King's, on the other hand, and the *Elements of Geology* which eventually sprang from them, gave him an opportunity to spell out his theoretical system in a more open and popular form.

After his preliminary remarks, Lyell began the substance of his lecture in a characteristic way by outlining the general structure of the Earth's crust (the only part accessible to direct observation) by reference to the 'sample' area of north-west Europe (the region best known to geologists and most familiar to his audience), illustrating this with maps and sections. But it is clear from his notes that, lacking experience of lecturing, he did not get as far with this outline as he had planned.

His audience at this lecture was small—only about 80, he reported. This was a serious disappointment, because he counted on the initial lecture, which was free and open to all, to gain him fee-paying subscribers for the rest of the series. The weather was bad, and it was still the Easter Vacation for the Inns of Court, to which he looked for support from his fellow-lawyers (39). Yet those who heard him were enthusiastic, and the following day, at the Geological Society's meeting, Lyell was pressed to allow women to attend (40). Since this was the issue which was the overt cause of his later resignation, it is important to note that Lyell himself was at first opposed to the request. His reason is clear: the admission of women to the lecture room would be 'unacademical' (41). This view is consistent with his great concern for the academic prestige of the College, and hence of the Chair that he held and of the lectures that he gave. Much as he criticized the curricular conservatism of the ancient universities, he clearly wanted to see King's emulate them in the outward forms of academic respectability (42).

On the other hand, he told his fiancée, 'I could have a class of 600 or 700 if I admitted ladies' (43)—that is, a *fee-paying* class. His estimate may have been over-optimistic, but his evaluation of the situation was surely not far wrong. Geology enjoyed particular popularity among upper- and middle-class women at this period, because much could be read and understood without training in the more 'rigorous' branches of science, and much could be seen and appreciated at first-hand within the compass of excursions around any place of residence in the country or any holiday at the sea-side. Several prominent geologists—including, after his marriage, Lyell himself—regularly took their wives on even arduous geological expeditions in Britain and on the Continent; and several of these wives—for example Mary Buckland and Charlotte Murchison—aided their husbands' work by being accomplished at sketching geological views. Other women amassed important fossil collections and observed local geological details which were of great value to the small number of 'full-time' geologists: Etheldred Benett of Warminster and Mary Anning of Lyme Regis are examples.

Caught in this dilemma between his concern for academic respectability and his desire for subscribers, Lyell seems to have acquiesced in the pressure that his friends put on the authorities at King's; and William Otter, the Principal, evidently agreed to allow women to attend, pending a definitive ruling by the Council. In view of this, Lyell decided to give a second introductory lecture later the same week, for the benefit of what he hoped would be an enlarged audience. Though it did not reach his earlier estimate, nearly 300 attended this lecture.

He began by recapitulating his first lecture briefly, and then completed his outline of the nature of the Earth's crust by describing not only the stratified sedimentary rocks and the volcanic rocks, but also the more problematical 'Primary' rocks such as granite. These were a problem, particularly at this level of popularization, because he could not demonstrate their mode of origin by analogy with processes observable in the present. He argued against the conventional view that they were truly 'primary' in the sense that they had been formed in the 'nascent state' of the Earth. Instead, he asserted (as later in the third volume of the *Principles*) that 'primary' rocks were of many different ages, being formed either by injection of fluid material at great depths or by 'metamorphic' changes of pre-existing rocks. The theoretical significance of this was that it left the preserved geological record of the history of the Earth without any beginning.

This Huttonian inference had already, in the *Principles*, laid him open to the charge of advocating the uncreated eternity of the world; and he therefore used this (second) introductory lecture as an opportunity to defend himself publicly against this charge. A long concluding section of the lecture was written out in full, which in the MSS. of the lectures is a sure sign that he had thought out the passage with particular care. He defended Hutton—and by implication himself —by following Playfair in making the 'eternalism' merely epistemological:

> It may be perfectly true that there may be a boundary to the material Universe and yet the most powerful telescope that man may ever be able to invent may only serve to disclose to us the myriads of new worlds. Why then should we not be prepared to expect from analogy that as the Author of Nature has not permitted man by the aid of his feeble powers to scan the limits of the universe as regards space, so also he may have hidden from us the limits of past time as regards the history of our planet. . . . There is no termination to the view of that space which is filled with manifestations of Creative power; why then, after tracing back the earth's history to the remotest epochs, should we anticipate with confidence that we shall ever discover signs of the beginning of the time that has been filled with acts of the same creative power.

The theological form in which this characteristic analogy with astronomy is cast was no mere rhetoric. Lyell defended his Huttonian vision of an Earth that was in a balanced steady state, by the explicit argument that this enhanced the orderliness of the system and hence its expression of divine design.

Yet having denied the validity of one conventional demonstration of divine power (i.e. in the creation of the Earth), albeit in order to enhance another, Lyell then went on to affirm that geology *could* demonstrate the creative beginnings of each and every organic species, and particularly of Man. Geology alone had shown conclusively that organic species had come into existence at definite points in the geological time-scale, and above all that Man himself was of recent origin (44). Lyell used this conclusion to make an unambiguous assertion that each such species-origin had required almost unimaginable creative wisdom:

> It implies in the first place an intimate knowledge of all the properties and laws of inorganic matter—it implies a fore-knowledge of the results of

future combinations of physical causes, it requires us to suppose that the maker possessed an exact knowledge of the attributes of all the other animals and plants with which the newly created species was destined to co-exist. It would in short be endless to attempt to enumerate the multitude of present and future conditions which it would be necessary to know and provide for, before it would be possible to determine whether a new species endowed with certain physical powers and instincts would be enabled to endure even for a few centuries. If we may presume to hazard a conjecture on a subject involved in so much mystery, we might perhaps incline to the opinion that the creation, not merely of the reasoning powers of man, but even the calling into being of one of the inferior animals endowed with various instinctive faculties, implies a higher exertion of creative power than the mere formation of an uninhabited planet.

Claiming that in this respect geology was even superior to astronomy, Lyell asserted that no other science afforded such diverse or extensive evidence of design, and he quoted with approval from Whewell's review of the *Principles* to this effect. As for evidential theology, he maintained—quoting Sedgwick with approval—that it was simply inexpedient in the fluid state of the science to argue particular correlations between geology and scripture, since any particular suggestion might be invalidated by the progress of the science, and its rejection might then be taken—wrongly—to imply a weakening of the basis of religion itself. Lyell concluded the lecture by quoting from Blomfield's sermon at the opening of the College, to the effect that truth from any source could not undermine but only enhance an appreciation of 'the glory of the Creator' (45).

This lecture was evidently a great success. Friends such as Babbage and Fitton, who would have castigated Lyell for any compromise towards 'scriptural geology', were as pleased as Otter; and another clerical member of the audience urged him to publish the lecture (46). In his concluding passage, Lyell had certainly lost no opportunity to quote from writers who would be approved by the authorities at the College; but there is no good reason to suppose that he was less than sincere in the opinions he expressed in this 'high flight' of oratory. If, for example, he had quoted the Bishop of London merely to create a favourable impression with the Council, he would surely not have referred to it privately (writing to his fiancée) as 'a truly noble and eloquent passage', or paraphrased Blomfield with approval as saying 'that Truth must always add to our admiration of the works of the creator [and] that one need never fear the result of free enquiry' (47). It is surely an important historiographical rule that one should assume sincerity unless there is strong evidence against it. I suggest that in this lecture Lyell was not 'arguing for the independence of science from theology' (48), except in the limited sense that he felt it was inexpedient and undesirable for geology to be continually looking out for correlations with scripture. At the much deeper level articulated by natural theology, we must surely accept, as sincere and genuine, Lyell's emphatic stress on the value of geology for the validation of a world suffused with divine design.

LECTURES AT KING'S COLLEGE, 1832

After this promising start to the series, Lyell must have been disappointed in the number of subscribers who paid the fee of a guinea and a half to attend the subsequent lectures. He gained fewer than seventy subscribers, and of these, only two were students of the College (49). He was encouraged to be told by John Lubbock, then Vice-President of the Royal Society, that he was 'doing good to science by these lectures', but he remained fearful that he might 'lose the substance by catching at the shadow—the solid reputation in many countries which my work [i.e. the *Principles*] may earn, for the exciting ready-money profit and applause of a lecture-room'. Once again his thoughts turned to resignation, and there is no reason to doubt his truthfulness when he told his fiancée that he would give Otter 'all my reasons—being all connected with my pursuing my original researches without disturbance'. Fitton at first agreed to take his place for the following year, but later withdrew, thinking that Lyell's reason would seem inadequate to the outside world and that it might reflect embarrassingly on the College (50).

After his two introductory lectures, Lyell got down to the ordinary business of the twice-weekly lectures, which was to illustrate his conception of the principles of geological interpretation by means of a few examples worked in detail. Lyell used his lectures to illustrate, on a more accessible level than in his printed work, the application of his 'principles' to the interpretation of the fossil record. To a great extent the substance of the lectures is parallel to, or at least selected from, his treatment of these topics in the third and culminating volume of the *Principles*, which at this time he was still writing (51). Thus in the third and fourth lectures he explained the principles by which the sequence of strata could be interpreted in chronological terms, particularly if the localized distribution of stratal formations and faunal provinces was understood rightly— i.e. by analogy with the present.

In the fifth lecture he began to apply these principles to the major example of the European Tertiary strata. He allowed full credit to Georges Cuvier,

Alexandre Brongniart and younger French geologists, to Brocchi in Italy and to Thomas Webster in England, for their elucidation of the sequence of strata and their awareness that the Tertiary strata were not of the same age in the different regions. But the importance of his own work in Calabria and Sicily in 1828–1829 was it showed that between even the most recent of these strata and the present there were 'Monuments of an intervening period' and that further 'intercalations of new periods' were to be expected as the area of exploration was extended. This argument implied that the apparent discontinuities between the different formations, and their fossils, were due simply to imperfect knowledge. This led him to begin his sixth lecture with his only criticism of the 'catastrophist' geologists. As in the *Principles*, he contrasted his own geology with theirs on the *methodological* level:

> I consider all reference to those irregular and extraordinary causes which are supposed to belong either to a pristine state of the planet or to recur only after distant intervals for the sake of producing geological phenomena, as inventions not simply without value but as extremely mischievous in the present state of the science.

Lyell asserted that it was *premature* to resort to 'revolutions, catastrophes, deluges, periods of repose, refrigeration, annihilations, [and] paroxysmal elevations'. This *a priori* 'assumption of discordance' between present and past was contrasted with his own approach. Such 'irregular causes', he said, 'we avoid because we are assured that we have not yet exhausted the resources which the study of present causes' could supply. In other words, it was premature and counter-heuristic to invoke causal agents unknown in the present, when so little was yet known of the possible explanatory efficacy of agents that could be *seen* in operation. As in the *Principles*, Lyell was criticizing the catastrophists on methodological grounds; he was not accusing them of invoking supernatural causation for geological processes. The argument was not about the relation of geology to religion, but about the proper method by which geology could be pursued most fruitfully.

Once again, as I have already stressed, the party-labels which have been applied to such geological controversies should not be taken to imply personal antagonisms. Babbage, in typically aggressive style, said of this lecture that Lyell 'tore Buckland's theory to tatters before his face'. But Lyell told his fiancée that Buckland had been 'more good-humoured' since the lecture, and reported that the Principal himself had praised him for his handling of his differences from Buckland (52).

On the same day, the Council resolved not to admit women to the College's lecture courses in future; but there is no evidence from Lyell's copious lecture notes that he had said anything offensive to the Council which might have made them use such a devious manoeuvre to eject him from his Chair. Their decision was in any case not put into effect until the following year; and it seems likely to have been simply a formal policy decision, taken at their first meeting since the issue had arisen at the beginning of Lyell's course (53). The Council's reason for their decision was probably the same as that which had made Lyell himself reluctant to give his consent to the admission of women: namely, the question of the academic respectability of the College. It is significant that the Council at King's was not the only body that doubted at this time whether the presence of women was compatible with serious intellectual work: the same issue was a matter of controversy at the British Association.

In his sixth lecture, Lyell had argued that his own 'method of philosophizing', when applied by others in the past to such problems as the origin of fossils and of basalt, had shown its superiority by results: 'this plan has put geologists into the right road, the other has led to contradictory systems'. He now had to show its efficacy by applying it to the more complex problem of the interpretation of Tertiary Earth-history. In his seventh lecture he argued that areas of deposition of sediment were now, and must always have been, localized in extent. Hence a 'change of species [is] everywhere in progress but [the] fossilization of animals and plants [is] partial [i.e. localized]'. There are no evolutionary overtones to this statement; as in the *Principles*, Lyell was referring to a continual overall change in faunas and floras by the piecemeal creation and extinction of individual species. His point was that, like the periodic visit of census 'commissioners', the natural recording of this continual change was bound to be discontinuous, owing to the 'commemorating process visiting and revisiting different tracts in succession'. Every time a region was exposed to erosion, it involved the partial destruction of this record, 'like the burning of documents' in human history.

If fossils were to be used as the records of the passage of geological time, as censuses plotted the populational changes of human history, then the most effective fossils had to be chosen for this purpose. In his eighth lecture Lyell began by referring to the spectacular work of Georges Cuvier with fossil mammals; but this led him into his only major digression (after the introductory lectures) from the straightforward business of geology.

Cuvier had died less than two weeks before. In terms of the traditional polarization between catastrophists and uniformitarians, it might be expected that Lyell's reference to Cuvier's work would have been critical or at least perfunctory. In fact, his lecture notes—and a long section written out in full—show that it was lengthy and generous. Lyell correctly distinguished between Cuvier's geology, relatively little of which had been based on first-hand field-work, and his palaeontology:

> I may observe that altho' many of his geological speculations may require to be modified or rejected, yet in those departments of Natural History which bear upon our science, especially fossil osteology, when the facts came under his own observation, nothing could exceed the caution with which he drew his inferences, and I may venture to say that there are few authors who will have less to retract.

Significantly, however, what Lyell stressed most about Cuvier's work was the material and organizational substructure of his research activities. Cuvier had commonly been criticized for involving himself in politics to the detriment of his science; but Lyell defended him—with evident fellow-feeling—by pointing out that Cuvier had been unable to gain much income from his research: 'his writings produced but small emolument, being by their nature too expensive to obtain many purchasers'. Consequently, to the lasting loss of science, 'his circumstances rendered [it] almost necessary to engage in some public duties of a more lucrative kind, but which men of inferior and more common talent might have performed with equal ability'.

> You will ask then in what manner this great naturalist was enabled to complete so many works of laborious research. . . . In the first place he had a power of turning his thoughts in succession to a variety of subjects and of directing his whole mind exclusively to them for an allotted period. But he also pursued his studies with a degree of method most rare to be remarkable in men of equal genius.

Lyell described—from his own visit—Cuvier's famous study with its various desks devoted to different research projects. But he particularly emphasized the value of Cuvier's method of using many research assistants, although this had led the envious to ask 'what part if any of Cuvier's work he had ever written himself except the preface'. Lyell defended Cuvier's method by arguing that it was inherently appropriate for those areas of science 'in which there must necessarily be much compilation and where the union of many observers may be advantageously directed to one point'. Lyell clearly approved of the hierarchical nature of this teamwork:

> His [Cuvier's] was the master spirit that superintended the whole; directing even the details, systematizing the results worked out by others, and

pointing to the general views deducible from them. He justly appreciated the value of the high power of his mind and would not waste his time on that part of a work which, to use Sir W. Scott's expression, may be done by steam.

Lyell's approval of this aspect of 'the French system' of organizing science led him naturally to some comments on the current debate on the alleged decline of science in England. It was not so much a question of decline, he said, as that 'we are not where we should be. We ought to lead when we follow. No country has such resources, none [is] so dependent on [the] application of higher departments of science to the arts'. The French excelled, he argued, because they had adopted a 'subdivision of labour' in science, and because science was a 'profession' to which men were 'devoted for life'.

Have they greater resources? No. Numbers? No. Sale of publications? So much the reverse [that] they often sell as many copies in England as at home. Talent and zeal? No, [their] disinterested devotion [is] far less than here.

Lyell referred to Babbage's and John Herschel's comments on the state of the physical sciences in England, and then suggested that their criticisms were equally applicable to geology. He contrasted the 'system' that was visible in the organization of science in France with the unplanned swings of fashion in England—'a rush to Chemistry then to Geology'. So, finally, he asked, 'what is the cause' of the disappointing state of science in England? He answered (in note form):

In France though the numbers are few they are systematically organized. It is not left to chance—sciences subdivided—each given professionally to some branch—not wealth but honor—and a competency. Compare French to small regular standing army—English to desultory multitudinous host of irregulars.

Lyell's long digression is of great interest as showing his concern at this time with the problem of the organization and planning of science, its professional status and its financial basis. His remarks, while generally characteristic of the 'declinist' party among English men of science, have particular importance in the present context in that they confirm an interpretation that regards Lyell's attitude towards his own activities, both of authorship and of his lectures at King's, as governed primarily by 'professional' considerations.

After his long digression on Cuvier and the state of English science, Lyell returned to the task of exposition by arguing that molluscs were the most suitable fossils for geological dating, particularly if uniformity of taxonomic treatment were ensured by their being identified by a single naturalist—as Deshayes in Paris was doing for Lyell's own collections. In the following lecture, Lyell presented and defined—for the first time in public—the periods of Tertiary time which this research had distinguished: the Recent, Newer Pliocene, Older Pliocene, Miocene and Eocene, in order of increasing disparity from present faunas and inferred increasing age. He presented a 'Synoptical Table' with the intention of showing that similar marine and freshwater sediments and volcanic rocks were known from each period, proving that conditions had been broadly similar throughout the Tertiary. As usual, he made this abstract conclusion more concrete by presenting a worked example, choosing the classic French region of Auvergne, which he and Murchison had travelled through (following Scrope's footsteps) four years earlier. This region, he noted, provided a 'perfect restoration of [the] Eocene period, [its] lakes, seas, volcanos, land quadrupeds, reptiles, testacea, plants, insects, Geography'. The great extinct volcanoes of the Massif Central had continued in activity through several periods, just as Etna had its roots in the Pliocene and might even outlast the human race.

In his tenth lecture, Lyell improved his 'Table' by the addition of 'sub-terranean' (i.e. intrusive igneous) rocks for each period. He argued again that the apparent 'distinctiveness of [the] periods indicates our imperfect information' rather than any real discontinuities, and that the piecemeal faunal change that was so imperfectly recorded had been 'like fluctuations of a population'. As later in the final volume of the *Principles*, he then began a detailed interpretative survey of Tertiary history in Europe, employing a retrospective order of description in order to penetrate from conditions approximating to the present towards those less familiar. He concentrated on his own research in Sicily, because this had shown most clearly the evanescent dividing line between the present and the past. He devoted the whole of his eleventh lecture to a descrip-tion of the thick but geologically recent Sicilian strata, stressing the 'proofs of [their] gradual accumulation'. In the following lecture he argued that the huge cone of Etna was still younger, although extremely ancient in human terms; he guessed that it might represent some 60 or 70 thousand years, though this was only a small fraction of the (relatively recent) Newer Pliocene. As in his pub-lished work, he argued that 'confined notions in regard to time have cramped the freedom' of the geologist to interpret his phenomena correctly (54). His geological opponents were not averse in principle to invoking a vast time-scale, though they may have lacked Lyell's imaginative insight into its explanatory potential. All that Lyell felt bound to oppose at this point was the catastrophists' assertion that violent 'diluvial waves' had passed over low-lying land areas at a geologically recent time. Lyell described the well-preserved cones of loose volcanic débris around the flanks of Etna, showed that they had been erupted at intervals from pre-historic times into the present, and concluded that they proved that 'no wave has passed over the forest zone of Etna'. This of course implied that no *general* deluge could have affected the area.

From his usual report to his fiancée, it is clear that Lyell was pleased with this lecture and its distinguished audience. What is particularly noteworthy is that he said that *Copleston* had been convinced by his argument (55). This is not surprising, however, for Copleston was no obscurantist, but a fair-minded if conservative scholar who could recognize a well-reasoned argument when he heard one.

For his thirteenth and final lecture, Lyell argued that the eruption of the vast mass of material forming Etna had been almost negligible compared to the simultaneous injection of material at great depth, which had caused the gradual elevation of central Sicily. The conclusion was that the physical geography of the region had changed drastically even within the lifetimes of species still in existence. It was 'striking to the imagination to contemplate these wonderful changes since living species predominated'. 'Yet why', Lyell asked, should it seem so surprising, if not 'because of [the] dogma' that physical and organic changes were occasional and sudden rather than continuous and gradual. His own interpretation showed that the adaptation of species to their habitats was more subtle than had been supposed; species had the power of migrating and changing their distribution continuously, colonizing areas of land that had not even existed when they themselves were first created. 'Thus new light [is] thrown on adaptation of instincts and migrating habits and attributes of species to incessant changes of surface'. On this somewhat Darwinian (but not evo-lutionary) note, Lyell ended his lecture course, concluding simply with a brief recapitulation.

LECTURES AT KING'S COLLEGE, 1833

Lyell was pleased with the popular success of his lectures and the approval of other men of science whose opinion he valued, but he evidently judged the material rewards (about £86) to have been small in comparison with the work that the lectures had involved. Before the course ended, he reported to Babbage that the Council were regretting their 'hasty resolution' to debar women from the lecture courses (56). Presumably they had begun to realize that the presence of women was not in fact lowering the academic tone of the College and that the success of Lyell's lectures was bringing credit to its reputation.

That summer Lyell went out to Bonn, married Mary Horner and took that long-suffering lady on a geological honeymoon. On returning to London he continued to work mainly on the final volume of the *Principles*. But the lectures for 1833 were not forgotten, and in the New Year he was arranging further visual aids for them. He told Babbage that, since King's had debarred the large potential female audience from attending his lectures there, the Royal Institution had invited him to give a course which would not be under that restriction (57). Accordingly, at the end of April he began to give two courses concurrently. If the issue of the admission of women had been merely a pretext for removing a lecturer with unacceptably heterodox notions, we might expect that Lyell would have used his new opportunity to explain with greater freedom his true views on the relation of geology to religious belief. In fact, his lecture notes show no contrast whatever in this respect between his second course at King's and his new course at the Royal Institution.

Lyell began his new course at King's with a historical retrospect of the science:

> These early speculators were entirely unconscious of the extent of their own ignorance and how much they had yet to learn respecting the operations of the existing agents of change . . . and hence they indulged their fancy in a boundless field of visionary speculation which brought discredit upon the very name of Geology and persuaded many that it would never rise to the rank of a science.

But he also criticized the extreme empiricist position that had been adopted by the 'practical men' of geology in reaction against this earlier theorizing. Their attempted distinction between an observational and factual 'geognosy' and a theoretical 'geology' had proved unworkable, and had led them to abandon the search for the *causes* of geological phenomena in favour of purely factual description.

> But they could not . . . disconnect the ordinary language which they made use of from theoretical views, and hence there was a manifest inconsistency between their professions and their practice, [and] they added . . . an additional source of prejudice—a determination to have no theory.

Lyell was thus having to disengage himself from two contrasting traditions, the over-speculative and the over-empirical, and to establish the validity of his own theoretical outlook. As usual, he illustrated his approach with a concrete example, using again the Auvergne and stressing the vast time-scale over which the successive volcanic eruptions had occurred.

After this free introductory lecture, the Council's decision had its anticipated effect. Lyell found himself with a class of only fifteen subscribers, and consequently he felt justified in cutting the number of lectures from twelve to eight. Since once again only two of his subscribers were students of the College, the Council's argument that the presence of women might have distracted their academic attention seemed to Lyell somewhat unrealistic. He told John Fleming, the presbyterian minister and amateur geologist, 'I regret that the bishops cut short my career at King's College' (58); but the context makes it clear that he was not accusing the episcopally-dominated Council of any ecclesiastical interference with his freedom. He simply regretted that their over-cautious concern for the College's academic respectability had led them into a decision that made his lectures financially unprofitable.

In his second and third lectures Lyell introduced the general principles for establishing the chronology of the Earth's history, stressing the value of fossil evidence, since the 'same species [are] not repeated at different eras but [the] same rocks have been'. He then outlined the 'principal groups of strata in ascending series', using his own new 'subdivisions of [the] Tertiary epoch'. He explained his method of dating the Tertiary strata by the proportion of extant molluscan species, concluding in characteristic style that 'in a word, shells are the favourite medals which Nature has selected to record the history of the former changes of the globe and its inhabitants'. This led him to explain, more clearly than anywhere in his published work, how the value of fossil shells derived from the greater average longevity of molluscan species compared to, say, mammalian species:

> It appears to result . . . that those laws by which the successive extinction of species is brought about operate more powerfully on the mammalia, that they [i.e. mammals] have less means of resisting the destructive causes by which the extermination of particular species is effected.

Lyell thus attributed the difference to a contrast in environmental tolerance. He illustrated it by the example of Cuvier's Eocene *Palaeotherium*, a genus which was radically different from any living mammal, yet had been contemporaneous with the still-living and widely-distributed molluscan species *Lucina divaricata*. Finally, as in the *Principles*, he denied that the fossil record gave any evidence of the 'progressive development' of life from simple beginnings, saying that in favour of this theory 'the only argument must be derived from the comparatively modern introduction of man'.

Lyell used the subsequent lectures to work out in detail an example of his method of integrated geological interpretation. He chose a familiar and accessible area—south-east England—and reconstructed its history since 'Secondary' times. He argued that the broad anticlinal dome of the Sussex Weald had been uplifted not suddenly but very gradually over a long period, that it had been eroded concurrently by the sea (the scarps of the North and South Downs being former sea cliffs), and that some of the material eroded had been deposited during the same period to form the Tertiary strata of the London and Hampshire 'basins' to the north and south of the Weald. This was an excellent small-scale example of his global concept of 'antagonist powers' (in this case, erosion and deposition) working to maintain an overall steady-state balance at the Earth's surface.

In his last two lectures this model of Earth-history was taken to its foundations, with an outline of his theory that the so-called 'Primary' rocks were not all ancient in date, and would be better termed 'Hypogene'—he used the term here for the first time—to indicate their origin at great depths in the Earth's crust. Lyell apparently ended his course at this point, without any general summary or statement of its implications.

It is possible to detect in the form of Lyell's lecture notes a certain perfunctory or even casual attitude towards their preparation. In view of the very small and unprofitable class that the Council's policy had produced, his attitude is understandable. He formally resigned from the Chair in October 1833. The following January, John Phillips, the Keeper of the York Museum and a nephew of the veteran 'practical geologist' William Smith, was appointed to succeed Lyell at King's, but held the post only in a part-time capacity and continued to live at York (59). Perhaps this also indicates that, like the other Chairs at King's, the position continued to be relatively unattractive, for reasons more related to its financial rewards and prestige than to any ideological constraints.

LECTURES AT THE ROYAL INSTITUTION, 1833

Meanwhile, Lyell had been lecturing to far larger numbers at the Royal Institution. There the attendance at the first lecture was 238, and only declined gradually (as happens to the best lecturers) to 149 at the end of the course of seven weekly lectures. Nearly half his audience each time was composed of members of the Institution, who paid no subscription, but Lyell collected thirty-six new subscribers (the remainder having already paid for other courses). Of these, nineteen were women, including several ladies of title, and also the wife and daughters of the Principal of King's, who would surely not have sent them there if he had had any doubts about the religious implications of Lyell's views (60).

Nevertheless, the course was a poor substitute for King's in financial terms. There the fee for his course had been set at a guinea and a half, most of which accrued to Lyell himself; whereas at the Royal Institution he received only ten shillings out of each subscription of a guinea, making the lectures much less profitable *per capita*. He received £65 for the course (61), compared with £86 the previous year at King's.

At the Royal Institution Lyell's level of exposition was rather more elementary than at King's, and the passages which he wrote out in full show that he was developing a more relaxed and informal style than previously. As at King's he made frequent use of visual aids in the form of large-scale geological sections and other wall charts, and specimens on the lecture table. Thus in his first lecture he used an enlargement of Conybeare's recent section across Europe (62) to show how the successive formations of strata were 'arranged in chronological series like volumes on [the] history of different nations'. He then explained how strata are formed by deposition of material from suspension; he cited observations on the Ganges and the Orinoco, and apparently performed an experiment in the lecture room to show the gradual settling of emery powder in a tank of water. By such processes, he concluded, even the vast thicknesses of strata to be observed in some mountain ranges had been accumulated gradually beneath the sea.

In the second lecture, Lyell supported this contention by stressing the importance of 'attending to the minuter points of analogy' between processes observable in the present and their geological results in the past. As an example, he analysed in detail the origin of the ripple-marked sandstones of the Wealden strata in Sussex—he evidently had a large slab of the rock in the lecture room.

You will perceive on this slab that there are the marks of animals which have been burrowing beneath the surface of the sand, very analogous to those which are sometimes seen on our shores.

I am inclined to suppose the former existence of an ancient beach at Horsham from this circumstance. I have seen in the quarries at Stammerham near Horsham the clay underneath the slabs of furrowed or rippled sandstone traversed by small cracks or cavities, like those which you observe in clay or argillaceous mud when dried by the sun at the bottom of ponds, or on a great scale in the mud of our estuaries, where many square miles are laid dry at low water. If a stratum of sand was strewed over such dried and cracked clay, it would of course enter into the cracks; and afterwards, when consolidated into a slab of sandstone, you would find the underside presenting in relief a cast of those cracks in the clay, into which it would be formed as in a mould.

I accordingly observed on the upper part of these slabs the ripple marks and in the lower casts of these rents or small fissures in the clay.

. . . I infer that the drying up and cracking of the clay took place during an interval of low water, or between the ebb and flow of the tide.

He then mentioned similar ripple-marked sandstones of many different geological ages, and described in detail how he had watched wind-blown ripples forming and re-forming on coastal dunes.

Now it is highly interesting to consider . . . that such minute circumstances as these ridges and furrows in sand are repeated through rocks of every age, & you will find in all of these the same waving of the ridges & furrows, the same general parallelism & occasional branching off of a furrow or a ridge into two parts. You will also find that the average magnitude of these furrows accords exceedingly well with the ordinary size of those now seen on the sea shore.

I have been the more particular in calling your attention pointedly to these minute analogies, because it has been too much the fashion in geology to adopt a different course & to assume that some extraordinary & irregular causes have produced the results which we witness in the structure of the earth. If, instead of speculating on the works of mysterious agents during the original formation of the earth, geologists had been employed for the last half century or more in instituting a close comparison between effects as minute as these ripple marks on the sea beach with the internal structure and arrangement of the strata in mountain regions, we should I am persuaded ere this have arrived at the solution of nearly all the most difficult problems in the science.

I have quoted from this example at some length, because it illustrates particularly clearly the nature of Lyell's method and style. Such close actualistic comparisons between present and past would have been acceptable to all Lyell's geological colleagues, whether 'catastrophist' or not. The difference was that Lyell was much more convinced of the power of such comparisons to explain not only the minute details but also the major features of geology. Furthermore, Lyell used a persuasive method of presentation, by focusing the argument on concrete examples. While this method is used throughout the *Principles*, this more informal example may serve to show more clearly how much the success of Lyell's enterprise depended on his skill at working out his method of inter-pretation in detail with reasonably familiar material. If there is a sense in which Lyell's geology embodied a new 'paradigm' for the science, it must be sought not so much in his large-scale theorizing—much of which was rejected both at the time and later—nor even his rhetorical advocacy of the elusive principle of uniformity, but rather in his presentation of a series of persuasive 'exemplars' which convinced others of the efficacy of actualistic explanations in geology (63).

Lyell's demonstration that the Horsham sandstones had been formed at sea-level, although hundreds of feet of marine strata had been deposited subsequently on top of them, proved that the Earth's crust in that region must have subsided and since been elevated again to its present position. 'By what machinery has this mighty change been brought about', Lyell asked, '[and] are there any powers now acting in nature capable of reproducing similar effects?' In other words, Lyell now had to convince his audience that even such major effects could be explained in terms of ordinary processes still acting.

Other geologists tended to argue, from the more disrupted and contorted appearance of many older strata, that the forces of elevation must have become progressively weaker, and the Earth's surface more tranquil, in course of time. Lyell countered this argument in his third leature, using an analogy drawn, as so often, from the interpretation of human history:

> If in a particular country which has been frequently exposed to the ravages of war we find evident signs in the different cities, towns and fortresses of the havoc which has been committed by hostile armies, but we find that the newer towns present less signs of ruin & dilapidation than the most ancient, we should not infer that the wars of remote ages had been more calamitous & destructive or that the artillery of modern times had been less effective, but merely that the more ancient forts & cities had been shattered & thrown into ruins again & again before the foundation of the more modern cities.

In the same way, Lyell asserted, there was no reason to infer that the forces of elevation had ever been more—or less—powerful than at present. Their modern manifestation, he argued, was to be seen in earthquakes; and he described as an example the elevatory effects of the great 1822 earthquake in Chile (which Charles Darwin was investigating at first-hand at just this time). In his fourth lecture he described the 'parallel roads of Coquimbo' (also in Chile, and also studied by Darwin) as a series of elevated sea-beaches, and suggested that the similar 'parallel roads of Glen Roy' in Scotland were likewise evidence of large-scale gradual elevation (64). He also discussed the relation of such earthquakes to the activity of volcanoes (which he saw as the surface effects of the same processes), and described the extinct volcanoes of Auvergne. The following week he analysed the volcanic phenomena of Campania as being the most suitable for the comparison between present and past, describing the history of

Vesuvius and showing how the so-called Temple of Serapis at Pozzuoli proved that there had been movements of both elevation and subsidence since Classical times. To make this point, he used an enlargement of the famous picture of the 'temple' columns, which he had chosen as the frontispiece of the first volume of the *Principles*.

In his sixth lecture, he used an analysis of the activity and structure of Vesuvius, Etna and other volcanoes to argue against the 'elevation crater' theory that had been introduced by the veteran Prussian geologist Leopold von Buch. It should be noted that this theory was of strictly technical import, and although it postulated a sudden violent process it was not 'tainted' with the apologetic overtones of, for example, Buckland's diluvial theory. Lyell rejected it quite explicitly on the usual methodological grounds. 'We see clearly that the ordinary action of Volcanos would produce & is causing such cones as Vesuvius, Somma & Etna. . . . Why then should we introduce [an] unknown agency?' This was to employ a 'false mode of philosophising', 'founded on no analogy' with the observable present. In his final lecture Lyell described the intrusive 'dikes' of volcanic lava that could be seen in the interior of Etna, and the granite veins of different ages in other regions. Such proofs of the forcible intrusion of liquid material from the depths of the earth indicated the adequacy of these processes as the true cause of elevation.

Lyell's notes for his lectures at the Royal Institution, like those for King's, contain no general summing up or conclusion. But what is most striking about this series is the complete absence of any polemical material on the relation of geology to scripture. Buckland and his fellow-'catastrophists' were criticized—though not by name—*only* on the strictly methodological grounds that they tended to invoke unknown agencies prematurely, without examining sufficiently closely the possible adequacy of present processes for causal explanation. Furthermore, Lyell made no attack on the more popular but scientifically negligible 'scriptural geologists'.

Of course, we have only his notes as evidence of what he said, but it is clear from the form of these notes that whenever he reached a topic of importance, where he wanted to express himself clearly and unmistakably, he went to great trouble to write out the passage in full, even drafting it several times. I believe, therefore, that the notes as they are preserved are a reliable indication of the content of his lectures, and that the absence of the polemical material mentioned is reliable evidence that he never touched on such subjects. I therefore conclude that Lyell did *not* use the freedom that was available to him at the Royal Institution to expatiate on subjects which—it might be suggested—he felt inhibited from mentioning in the Anglican atmosphere of King's.

On the contrary, his ignoring the 'ancient and modern physico-theologians' was surely an indication that in this respect he felt the professional status of geology was now becoming more secure, and needed no such polemical defence. He was in no way unusual in this conclusion. Of all specialist scientific groups in London, geology was perhaps the most self-assured and confident of its own future, and Lyell was merely adopting an attitude which he shared with all his friends and colleagues at the Geological Society, including those who were also divines.

CONCLUSION

In conclusion, I suggest that Lyell's three lecture courses are important for the light they shed on Lyell's own conception of his science at the time he was writing the *Principles*. It is striking that his lectures are drawn principally from the material he was preparing for his third volume. For the 1832 lectures, this emphasis might be attributed to the fact that the material of the third volume was still unpublished and therefore had novelty value. But for the 1833 courses this explanation no longer holds, and anyway it is unlikely that even in 1832 Lyell would have assumed that his audience was familiar with the 800 pages of the earlier volumes. I suggest instead that the material that Lyell chose to use for his lectures (in both years) reflects his own view of the essential business of geology, namely the reconstruction of the geological past. The earlier volumes of the *Principles* were of course important to him, but only as means to an end. In his own phrase, the close study of processes operating in the present provided the 'alphabet and grammar of geology' but this was only a prelude to the business of deciphering the language in which Nature had recorded the past history of the Earth (65).

Finally, I conclude that Lyell's attitude towards his Chair at King's College becomes intelligible and consistent once it is seen in the light of his overriding concern with the status of his science. He applied for the post because he believed it would enhance his own personal prestige and that of his science, and also bring him at least some financial reward. He remained ambivalent towards the task because he felt that the time and trouble it required was incommensurate with the reward, and that it was inevitably a distraction from the more important work of completing his *Principles of Geology*. When the College Council resolved not to allow women to attend lectures in future, Lyell decided to resign as soon as he decently could, because he correctly judged that the Council's resolution would drastically reduce his potential fee-paying audience. Accordingly he accepted an invitation to lecture at the Royal Institution, where this restriction would not apply, although *per capita* the lectures there were much less lucrative. Once he had completed the publication of the *Principles*, however, and its sales assured him of a steady income from successively updated editions, he could afford to drop all lecturing activities and concentrate on research and writing. I suggest that there is no evidence that the decision of the Council at King's to exclude women was connected in any way with disapproval with the content of Lyell's lectures. Like Lyell himself—until he realized the financial implications—they were merely concerned with the academic respectability of their institution.

NOTES

(1) D. S. L. Cardwell, *The Organisation of Science in England A Retrospect* (London, 1957, reprinted 1972), p. 47. For a contemporary expression of this polarization, see the caricature reproduced in F. J. C. Hearnshaw, *The centenary history of King's College London 1828–1928.* (London, 1929), opp. p. 42, showing the supporters of the two institutions on opposite ends of a see-saw.

(2) Hearnshaw, *op. cit.* (1), p. 86.

(3) Leonard G. Wilson, *Charles Lyell. The Years to 1841: The Revolution in Geology* (New Haven and London, 1972), p. 308; Mrs Lyell (ed.), *Life, Letters and Journals of Sir Charles Lyell, Bart.* (London, 1881), Vol. I, p. 315.

(4) Hearnshaw, *op. cit.* (1), p. 109; Cardwell, *op. cit.* (1), p. 48.

(5) Wilson, *op. cit.* (3), pp. 309–313.

(6) J. W. Draper, *History of the Conflict between Religion and Science* (New York, 1875, reprinted 1970); A. D. White, *A History of the Warfare of Science with Theology in Christendom* (New York, 1895, reprinted 1955).
    A more recent standard work on the period, although far more sophisticated historically, still portrays the debate in essentially the same positivist terms as the gradual advance of a truly scientific geology over ground surrendered by natural theology: Charles Coulston Gillispie, *Genesis and Geology. A study in the relations of scientific thought, natural theology, and social opinion in Great Britain, 1790–1850* (Cambridge, Mass. 1951).

(7) For an important recent analysis of this process, see J. B. Morrell, 'Individualism and the structure of British science in 1830,' *Hist. Stud. Phys. Sci.*, **3**, 183–204 (1971).

(8) Charles Lyell, *Principles of Geology, being an attempt to explain the former changes of the earth's surface, by reference to causes now in operation*, 3 vols. (London, 1830, 1832, 1833; reprinted New York, 1970).

(9) Hearnshaw, *op. cit.* (1), p. 91; Cardwell, *op. cit.* (1), p. 46. Both Turner and Lindley were Fellows of the Geological Society, and Turner was one of its Secretaries at this time.

(10) Lyell, *op. cit.* (3), 1, p. 164–5. Lyell's articles are listed in Wilson *op. cit.* (3), p.529–530 (except that the Wellesley Index has established that the earliest one mentioned there was by Copleston, not Lyell). Although the articles were anonymous, their authorship was generally known among the circle of 'reviewers.'

(11) Wilson *op. cit.* (3), p. 112, 137.

(12) Lyell, *op. cit.* (3), Vol. I, 234.

(13) *ibid.*, Vol. I, p. 315.

(14) Wilson, *op cit.* (3), p. 308; [William Whewell], '[Review of] Principles of Geology . . . By Charles Lyell . . . Vol. I . . Murray 1830,' *British Critic.* **9**, 180–206 (1831).

(15) A. Sedgwick, 'Anniversary Address[es] of the President,' *Proc. Geol. Soc. Lond.* **1**, 187–212, 281–316 (1830–1).

(16) Lyell, *op. cit.* (3), Vol. I, p. 317.

(17) W. Tuckwell, *Pre-Tractarian Oxford. A Reminiscence of the Oriel* 'Noetics.' (London, 1909), pp. 17–50. I am grateful to Dr M.J. S. Hodge for this reference.

(18) Lyell, *op. cit.* (3), Vol. I, p. 322.

(19) Quoted in Wilson, *op. cit.* (3), p. 308.

(20) *ibid.*, p. 309–310.

(21) Lyell, *op. cit.* (8); see the analysis in Martin J. S. Rudwick, 'The Strategy of Lyell's *Principles of Geology*,' *Isis*, **61**, 4–33 (1970).

(22) Wilson, *op. cit.* (3), p. 310–2.

(23) *ibid.* p. 312.

(24) Lyell, *op. cit.* (3), Vol. I, p. 268.

(25) Quoted in Wilson, *op. cit.* (3), p. 312–3.

(26) Lyell, *op. cit.* (3), Vol. I, 172–3. I am indebted to the Manuscripts Librarian at the Turnbull Library, Wellington, New Zealand, for supplying me with a microfilm of the Lyell-Mantell correspondence, from which Toulmin's name (omitted from the published version) has been added. On Toulmin, see G. L. Davies, 'George Hoggart Toulmin and the Huttonian theory of the Earth', *Bull. Geol. Soc. Am.*, **78**, 121–4 (1967), and R. Hooykaas, 'Geological Uniformitarianism and Evolution,' *Archs. int. Hist. Sci.*, **19**, 3–19 (1966).

(27) Quoted in Wilson, *op. cit.* (3), p. 309. On Scrope's influence on Lyell, see Martin J. S. Rudwick, 'Poulett Scrope on the volcanoes of Auvergne: Lyellian time and political economy,' *Bri. J. Hist. Sci.*, **7**, 205–242 (1974).

(28) Lyell, *op. cit.* (3), Vol. I, p.271.

(29) *ibid.*, Vol. I, p. 316.

(30) King's College London, Council minute book 'B'. Lyell had spent most of his allowance by the beginning of his second lecture course: Lyell to H. Smith (the then Secretary), 16 April 1833. I am indebted to the present Secretary of the College and to Mr H. A. Harvey for access to the archives.

(31) Lyell, *op cit.* (3), Vol. I, p. 329.

(32) *ibid.*, Vol. I, p. 354.

(33) Quoted in Wilson, *op. cit.* (3), p. 340.

(34) Lyell, *op. cit.* (3), Vol. I, pp. 357–360.

(35) *ibid.*, p. 376.

(36) *ibid.* pp. 358–362, 365.

(36a) Lyell to Edgar (Professor of Divinity at Belfast), 15 February 1832. I am indebted to the Historical Society of Pennsylvania for permission to publish this extract, and particularly to Mr J. B. Morrell of the University of Bradford for informing me of the existence of this important letter and for many other valuable comments on the present article.

(37) Leyell, *op cit.* (3), Vol. I, pp. 352–3, 367.

(38) Edinburgh University Library, Lyell MSS. 8, 'Lectures on geology, King's College London, Royal Institution 1833 etc.' Lyell's voluminous lecture notes, which have

not apparently been utilised by historians hitherto, are of course vital for the understanding of this episode.

I am indebted to Mr C. P. Findlayson for access to these MSS. and for much helpful assistance, and to the Librarian for permission to publish some extracts here. I have deposited a handlist with the MSS., outlining my reconstruction of the notes and fragments of text that relate to each of the lectures in the three lecture couses concerned. For this reason, I shall not give detailed citations of the MSS. for the quotations which follow.

(39) Wilson, *op. cit.* (3), p. 353.

(40) Lyell, *op. cit.* (3), Vol. 1, p. 381.

(41) Quoted in Wilson, *op. cit.* (3), p. 353.

(42) See his remarks on the academic dress and 'gentlemanly tone' of the College: Lyell, *op. cit.* (3), Vol. 1, pp. 352, 382.

(43) Quoted in Wilson, *op. cit.* (3), p. 353.

(44) On Lyell's attitude to the origin of Man, see Michael Bartholomew, 'Lyell and evolution: an account of Lyell's response to the prospect of an evolutionary ancestry for Man,' *Br. J. Hist. Sci.*, **6**, 261–303 (1973).

(45) Charles James [Blomfield], Lord Bishop of London, *The duty of combining religious instruction with intellectual culture. A sermon preached in the chapel of King's College London at the opening of the Institution on the 8th of October 1831* (London, 1831), pp. 9–10.

(46) Lyell, *op. cit.* (3), Vol. 1, p. 382.

(47) Quoted in Wilson, *op. cit.* (3), p. 354.

(48) *ibid.*, p. 355.

(49) Lyell, *op. cit.* (3), Vol. 1, p. 397; Wilson, *op. cit.*, (3), pp. 356–8; Lyell to H. Smith, 8 December 1831 (King's College archives). Unfortunately the subscription lists do not seem to be extant.

(50) Lyell, *op. cit.* (3), Vol. 1, pp. 383–4; Wilson, *op. cit.* (3), pp. 360–1.

(51) It would be pointless to cite parallels at every point; but see Rudwick, *op. cit.* (21).

(52) Lyell, *op. cit.* (3), Vol. 1, p. 385; Wilson, *op. cit.* (3), p. 357.

(53) The resolution (Council minute book 'B', meeting of 18 May 1832) was 'That the attendance of ladies be not permitted at any course of lectures to be hereafter read in the College, and that the several Professors be made acquainted with this regulation;' but Copleston (who was in the chair) afterwards wrote in the margin: 'N.B. This resolution was not passed at the Council on May 18th but was postponed.' I have found no further reference to it in the minutes. Presumably the Council agreed that it should not come into force until the following academic year, i.e. that women should not be debarred from Lyell's lectures in the middle of his course.

(54) Martin J. S. Rudwick, 'Lyell on Etna, and the antiquity of the earth,' *in* Cecil J. Schneer (ed), *Towards a History of Geology* (Cambridge, Mass., 1969), pp. 286–304.

(55) Wilson, *op. cit.* (3), p. 359.

(56) Lyell, *op. cit.* (3), Vol. 1, p. 387.

(57) *ibid.*, Vol. 1, p. 395.

(58) *ibid.*, Vol. 1, pp. 396–7.

(59) Hearnshaw, *op. cit.* (1), p. 109; King's College, Council minute book 'B', minutes of 11 October 1833 and 17 January 1834; John Phillips, *Syllabus of a course of eight lectures on geology* [London, 1834].

(60) Royal Institution Archives: Lecture Index, 1833, p. 41; Subscribers' Book, No. 1. I am indebted to Mr J. R. Friday for sending me copies of these documents.

(61) Charles Lyell, *Syllabus of a course of twelve lectures on Geology* (London, [1833]); Royal Institution Archives, Managers' Minute Books, Vol. 8, p. 66 (meeting of 14 January 1833), and account books. The Council minutes at King's do not specify the amount of the 'College Deduction' in Lyell's case; but others were charged one-fifth of the fees (see minutes of 8 June 1832). The number of subscribers and total income reported by Lyell are consistent with the inference that his deduction was the same.

(62) Published in W. D. Conybeare, 'Report of the Progress, Actual State, and Ulterior Prospects of Geological Science,' *Rep. Br. Ass. Adv. Sci., 1st & 2nd Meetings, 1831–2* (London, 1833), pp. 365–414.

(63) For a critique of Thomas Kuhn's notions of 'paradigm' and 'revolution' as applied to Lyellian geology, see M. J. S. Rudwick, 'Uniformity and progression: reflections on the structure of geological theory in the age of Lyell,' *in* Duane H. D. Roller (ed.), *Perspectives in the History of Science and Technology* (Norman, Okla., 1971), pp. 209–227.

(64) The influence of Lyell seems to have been instrumental in leading Darwin into what he later called a 'gigantic blunder' on this point: see Martin Rudwick, 'Darwin and Glen Roy: a "great failure" in scientific method?' *Stud. Hist. & Phil. Sci.,* 5, 97–185 (1974).

(65) Lyell, *op. cit.* (8), Vol. 3, p. 7.

Charles Darwin in London: The Integration of Public and Private Science

C HARLES DARWIN'S PRIVATE WORK on the problem of the origin of species, in the years following his return from the voyage of the *Beagle*, has become one of the most intensively studied episodes in the history of science.[1] This is hardly surprising, for Darwin's conception of evolutionary change is one of outstanding historical significance by any criterion, from the most "internalist" to the most "externalist." For at least two reasons, however, the interest of the episode extends far beyond the small circle of Darwin scholars, or even beyond that of historians of biology. First, it is an almost unparalleled example of a major scientific theory that can be followed in detail through all the phases of its development, and it thereby throws light on the dynamic processes of cognitive growth on the individual level. Second, it offers an important opportunity for studying the role of direct personal interactions—as opposed to merely bookish "influences" —in the construction of a theory, or more generally the link between the individual and group levels of scientific practice, or between "private" and "public" science.[2] For in one crucial respect Darwin's early work on the species problem is highly atypical of theory construction in science: it was not developed in *mutual* interaction with other scientists. As is well known, his theory remained almost totally private for some twenty years. (Darwin discussed it with a very few colleagues, such as Joseph Hooker, but not until after he had formulated it in most of its essentials; the fact that he was working on a theory of speciation was more widely known, but its content was not.) The theory was eventually forced into the public realm only by external circumstances, namely by the unexpected arrival of Alfred Russel Wallace's paper in 1858, which closely paralleled Darwin's work.

*Science Studies Unit, University of Edinburgh, 34 Buccleuch Place, Edinburgh EH8 9JT, Scotland.

I am grateful to Howard Gruber for discussion of the interpretation I propose and to Gerald Geison and David Kohn for their comments and assistance.

[1]Michael Ruse, *The Darwinian Revolution* (Chicago: Univ. Chicago Press, 1979) includes a useful bibliography; see also John C. Greene, "Reflections on the Progress of Darwin Studies," *Journal of the History of Biology*, 1975, 8:243–273.

[2]On the sequential aspect see David Kohn, "Theories to Work By: Rejected Theories, Reproduction, and Darwin's Path to Natural Selection," *Studies in the History of Biology*, 1980, 4:67–170. On the wider cognitive objective, see Howard E. Gruber, *Darwin on Man: A Psychological Study of Scientific Creativity*, 2nd ed. (Chicago: Univ. Chicago Press, 1981); also Gruber, "The Evolving Systems Approach to Creative Scientific Work: Charles Darwin's Early Thought," in Thomas Nickles (ed.), *Scientific Discovery: Case Studies* (Dordrecht/Boston: Reidel, 1980, pp. 113–130). My approach in this article differs radically from the project of reconstructing a cultural circle for Darwin that is wholly bookish and virtually nonsocial, as in Edward Manier, *The Young Darwin and his Cultural Circle* (Dordrecht/Boston: Reidel, 1978).

Yet Darwin was much concerned with his public reputation, as his reaction to Wallace's paper amply demonstrates; and in other fields he had no hesitation about exchanging ideas with his scientific colleagues and publishing his research promptly.

This paradox can be resolved by relating the private construction of Darwin's species theory more closely to the science that he was doing publicly at the same time. In this paper I outline the social setting of Darwin's public science in his most creative period, emphasizing the opportunities it offered for the development of his powers as a theorist. I argue that the relation between his public and his private science should not be regarded as a simple dichotomy, but rather as a continuum or scale of "relative privacy." This scale of relative privacy can be combined with the time scale of his years in London and the development of his work plotted on the two-dimensional field thereby generated. Figure 2 is the resulting diachronic diagram; Figure 1, a synchronic map of the scientific community with which Darwin was most closely associated in his London years, also illustrates my interpretation. These diagrams are intended not only as visual summaries of my argument, but also as provocative stimuli for further debate. They exemplify my belief that historians of science need to develop a much greater visual awareness and to make much more use of visual communication. This might lead to a sharp reduction in the verbosity and obscurity that characterize some current writing, just as the development of a "visual language" had that desirable effect on many of the natural sciences themselves.[3]

The kind of analysis that I apply to Darwin in this paper should be quite generally applicable to the study of the work of scientists in any branch of science at any period of history, provided that adequate documentary evidence is available. Such analyses, and the "maps" that represent them visually, have only modest explanatory pretensions; but as ways of reframing what is already known descriptively, they have important heuristic value as a source of fresh questions to ask about the pattern and significance of a person's scientific work in its immediate social context.

DARWIN'S LONDON YEARS

It is well known that Darwin's *Origin of Species* of 1859 was a hurriedly improvised abstract of a much longer unfinished work, *Natural Selection*, only a part of which was later developed into book form. Both these articulations of his species theory follow closely the argument he had outlined many years before in the short manuscript "Essay" of 1844 and the even briefer manuscript "Sketch" of 1842. These earliest attempts to formulate the theory were based in turn on a series of notebooks that Darwin kept in the years 1837–1842. This sequence of texts has recently been extended still further back in time: the "Red notebook" of 1836–1837 contains what are probably Darwin's very first recorded conjectures on the mutability of species, and manuscripts from the *Beagle* voyage itself (1831–1836) show that at that time he was not yet an evolutionist in any sense at all.[4] Within the

---

[3]For the case of geology, see Martin J. S. Rudwick, "The Emergence of a Visual Language for Geology, 1760–1840," *History of Science*, 1976, *14*:149–195.

[4]*On the Origin of Species* (London: Murray, 1859); the extant portions of Darwin's "Big Book" have been edited by Robert Stauffer, *Charles Darwin's Natural Selection* (Cambridge: Cambridge Univ. Press, 1975); the part later published was Darwin, *Variation of Animals and Plants under Domestication* (London: Murray, 1868) (represented in Fig. 2 by "Origin," "Natural Selection," and

quarter century spanned by all these texts, scholarly attention has been focused chiefly on the record of Darwin's early and private speculations, from his first evolutionary conjectures to his first coherent formulation of a theory, that is, for the six-year period from 1837 to 1842.[5] Although several important conceptual developments still lay ahead after the completion of the "Sketch" of 1842, this narrower focus on the earlier years has ample justification, since it reflects the intrinsic interest of what was arguably the most innovative period in Darwin's life.

It has not sufficiently been noticed, however, that this creative period was precisely framed by two decisive changes in Darwin's personal and social circumstances. His return to England in the fall of 1836 was marked by the emergence in his work of a new style of theoretical note making, and his move to London a few months later coincides closely with his first recorded evolutionary notes.[6] Six years later his removal in the fall of 1842 from London to Down House, in the Kent countryside, came shortly after he had completed his first coherent formulation of his theory of evolution (i.e., the 1842 "Sketch"), if not yet the more connected prose of the "Essay." I suggest that these are not in fact coincidences.

Darwin's social environment in his London period differed radically from his years on the *Beagle* beforehand and from his years at Down House afterwards. While on the voyage he was, relatively speaking, an isolate. He was not totally isolated, of course: he had the intelligent and scientifically literate Robert Fitzroy as his companion, he had a few scientific books with him, and he wrote and received letters at various ports.[7] But apart from Fitzroy, his contacts with other men of science were at a distance and restricted in range and variety. Similarly, when he moved to Kent he became, relatively speaking, a recluse. Again, he was not totally isolated: he was near enough to London for scientific friends to visit him quite easily and near enough to go there occasionally himself, but also just far enough away to give him a socially acceptable excuse for declining unwanted

---

"Variation"). For the "Sketch" and "Essay" see Charles Darwin and Alfred Russel Wallace, *Evolution by Natural Selection* (Cambridge: Cambridge Univ. Press, 1958), pp. 41–88, 91–254 (represented in Fig. 2 by "1842 Sketch" and "1844 Essay"). (An allegedly earlier and even briefer sketch has now been identified as a part of a draft of the "Essay": See David Kohn, Sydney Smith, and Robert C. Stauffer, "New Light on *The Foundations of the Origin of Species*: Darwin's 1839–1844 Notebooks and Essays, a Reconstruction of the Archival Record," *J. Hist. Biol.*, 1982, forthcoming; cf. P. J. Vorzimmer, "An Early Darwin Manuscript: the 'Outline and Draft of 1839,' " *J. Hist. Biol.*, 1975, 8:191–217.) The Species notebooks (B–E) were published by Gavin de Beer *et al.*, "Darwin's Notebooks on Transmutation of Species," *Bulletin of the British Museum (Natural History) Historical Series*, 1960/61, 2:27–200; 1967, 3:129–176; the Red notebook by Sandra Herbert, "The Red Notebook of Charles Darwin," *Bull. Brit. Mus. (Nat. Hist.) Hist. Ser.*, 1980, 7:1–164; the later torn-apart notebooks have only recently been identified: see Kohn, Smith, and Stauffer, "New Light" ("Species notebooks," "Red notebook," and "torn-apart notebooks" in Fig. 2). (A comprehensive edition of these and other notebooks is in preparation at Cambridge as *Darwin's Theoretical Notebooks*, ed. Paul Barrett, Sandra Herbert, David Kohn, and Sydney Smith.) For the *Beagle* notes, see Sandra Herbert, "The Place of Man in the Development of Darwin's Theory of Transmutation, Part I: To July 1837," *J. Hist. Biol.*, 1974, 7:217–258; and Kohn, "Theories to Work By."

[5]There are few if any major changes of content between the "Sketch" and the "Essay," so that 1842 may be regarded as marking the successful achievement of this phase of Darwin's work. For 1842–1859 (publication of the *Origin*), see esp. Dov Ospovat, "Darwin after Malthus," *J. Hist. Biol.*, 1979, 12:211–230; and Ospovat, *The Development of Darwin's Theory: Natural History, Natural Theology, and Natural Selection, 1838–1859* (Cambridge: Cambridge Univ. Press, 1981).

[6]He left the *Beagle* 2 Oct. 1836, moved to Cambridge 13 Dec. 1836 and settled in London 13 Mar. 1837. Herbert ("Red Notebook") discusses the changing character of the Red notebook and concludes that Darwin's earliest evolutionary notes probably date from around Mar. 1837.

[7]See esp. his letters to John Henslow in Cambridge, in Nora Barlow, ed., *Darwin and Henslow: The Growth of an Idea*; *Letters 1831–1860* (London: Murray, 1967).

commitments and invitations.[8] And as on the voyage, he could of course communicate by letter with whomever he chose. Nonetheless, both environments limited sharply the interactions with other men of science to which he was exposed on a daily or weekly level; and they were environments in which he could *choose* with whom he wished to interact. By contrast, when, after three months in Cambridge, he decided to settle in London in the spring of 1837, he placed himself at the center of scientific activity in the English-speaking world. He was quickly accepted into scientific circles, and for six years—but especially before his marriage and the onset of his illness— he played an active and quite prominent role in the scientific life of the metropolis. In contrast to his life before and after, this was an environment in which face-to-face scientific interactions, chosen or unchosen, planned or unplanned, were frequent, varied, and often intensive. This environment was of crucial importance, not only for Darwin's public science but indirectly for his private science too.

Darwin's brief, as unofficial gentleman-naturalist with the *Beagle*, had been to observe and collect anything of interest throughout the realm of the natural history sciences. When he returned, he farmed out his zoological and botanical materials with efficiency and tact, but above all with professionalism. If that term be used in the sense of a strong commitment to high, collectively held standards of work, then Darwin's aspirations towards professional science were well formed even before the voyage and unmistakable by its end.[9] But his ambitions were not diffuse: he wanted to become a geologist. His first projects on his return were to write up his journal in publishable form, as a scientific travel narrative in the Humboldtian manner, and to get the best relevant specialists to work on his collections of specimens. These projects brought him at once into contact with a wide range of men of science from various disciplines. But he reserved the geological part of the *Beagle* results for himself; and even before he arrived back in England he took steps to join the society of gentlemen-specialists that was the chief forum for geological debate.[10]

In his London years Darwin followed the example and advice of his older mentor Charles Lyell and gave his main loyalty and commitment to the Geological Society of London.[11] In the 1830s the Geological Society was recognized interna-

---

[8]The origin and nature of Darwin's illness (a major factor in his decision to move out of London) is not relevant here: but his move to Kent did *in fact* have this consequence for his social life. For an exhaustive discussion of his illness, see Ralph Colp, *To Be an Invalid: The Illness of Charles Darwin* (Chicago: Univ. Chicago Press, 1977).

[9]Herbert, "Place of Man, Part I." For this conception of professionalism in science, based on perceived excellence rather than on standardized training and certification or on financial remuneration, see Susan F. Cannon, *Science in Culture: The Early Victorian Period* (New York: Science History, 1978). For Darwin's unusual gentlemanly position on the *Beagle*, see Keith Stewart Thomson, "Why Was Charles Darwin on Board the HMS *Beagle*?" *Discovery* (Peabody Museum), 1979, *14*:2–10; also Herbert, "Place of Man, Part I," on pp. 250–252.

[10]Herbert ("Place of Man, Part I," on pp. 226–227) rightly stresses that Darwin's 1839 *Journal of Researches* ("Journal" in Fig. 2), although a book for the "general reader," was also a serious scientific work in the Humboldtian tradition (on Humboldtian science see Cannon, *Science in Culture*). Darwin wrote to Henslow from St. Helena about joining the Geological Society of London: see Barlow, *Darwin and Henslow*, on p. 115. "Gentlemen-specialists" is my own term for the active members of the metropolitan specialist societies at this period; the term is adapted from Jack Morrell and Arnold Thackray, *Gentlemen of Science: Early Years of the British Association for the Advancement of Science* (Oxford: Oxford Univ. Press, 1981). See also Roy Porter, "Gentlemen and Geology," *Historical Journal*, 1978, *21*:809–836.

[11]See esp. J. B. Morrell, "London Institutions and Lyell's Career," *British Journal of the History of Science*, 1976, *9*:132–146.

tionally as the premier body for the earth sciences, and by common consent it was also the liveliest of all the scientific societies in the metropolis. When he returned from his voyage, Darwin was already known to its leaders as a young geologist of great promise, owing to the geological letters he had sent home from South America. He was also a man of independent means and of impeccably gentlemanly status. With such credentials he was admitted to the society without delay, elected to its ruling council at the first opportunity, and appointed one of its two secretaries a year later. He served a three-year stint in that quite onerous position, and thereafter remained on the council until after his move away from London (see Fig. 2).[12] In other words, Darwin was at the center of the Geological Society, and therefore of world geology, throughout his years in London. Like Lyell, he made by contrast very slight commitments to the other metropolitan specialist societies and to the Royal Society.

THE SOCIAL WORLD OF GEOLOGY

To talk of a "scientific community" of geologists in the 1830s is misleading, because it suggests anachronistically a strong-boundaried professional group marked by standardized training and certification, with only the uninitiated public outside. It is more accurate to depict the social topography of the science at that period as a series of graduated zones of "ascribed competence" that shade insensibly into one another (see Fig. 1). (Similar topographies could be drawn for other sciences at the same period; they would overlap in a complex manner with that for geology, because many individuals operated in more than one science.) The competences referred to here are primarily those ascribed by the leaders of the science to themselves and others; but there was generally a fair degree of consensus, among those in the outer zones of lesser competence, that that was where they belonged. The mapping of given individuals onto this kind of topography can be recovered from the historical record by noting, particularly from their correspondence, the ways in which they treated each other, referred to each other, and used each others' work. Along the continuum of ascribed competence, it is convenient to distinguish three main zones; it should be noted however that the boundaries between them are more like contour lines than stone walls.

At the center or peak of ascribed competence was the elite of the science. These were men with a strong, indeed primary, commitment to geology rather than any other branch of science; they were active in its affairs and productive in publication; they interacted intensively with each other, whether in cooperation or in rivalry and controversy; and above all they regarded themselves as competent arbiters of the most fundamental matters of theory and method within the science. They were those whom Darwin, soon after joining the Geological Society, recognized as the "great guns" or "great scientific men" whose approbation he hoped to earn.[13] Examples of elite geologists in the society in the 1830s include Henry De la Beche, Charles Lyell, Roderick Murchison, and Adam Sedgwick; and, elsewhere, Léonce Élie de Beaumont in Paris and Leopold von Buch in Berlin.

[12]Darwin was elected a Fellow of the Geological Society 30 Nov. 1836 (see "FGS" in Fig. 2); elected to the Council at the annual (Feb.) "anniversary" meeting for 1837; appointed a secretary at the 1838 meeting, serving until the 1841 meeting; and remained on the council, though in a less and less active role, until the 1851 meeting.

[13]Francis Darwin, ed., *The Life and Letters of Charles Darwin* (London: Murray, 1887), Vol. I, pp. 274, 277, 280.

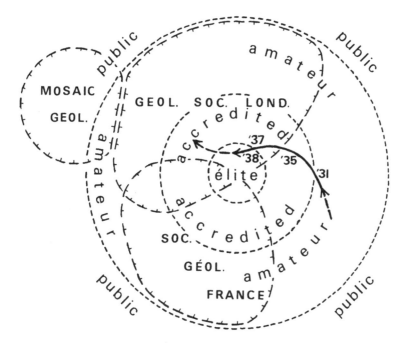

Figure 1. *The social and cognitive topography of geology in the 1830s (drawn as a Venn diagram), showing the three weak-boundaried concentric zones of ascribed competence (elite, accredited, amateur) surrounded by the general public; the two (strong-boundaried) major geological societies with overlapping membership, and the marginal Mosaic geologists; and Charles Darwin's trajectory (thick line) from amateur to elite status during the 1830s and his later reversion to the accredited zone. The relative areas of the various envelopes give a rough qualitative impression of the relative numbers of individuals populating those areas.*

Grouping such men into one zone does not imply any cognitive consensus among them; on the contrary, the zone of the elite was precisely the arena for all major conflicts of theory and method. But even in cases of profound disagreement, members of the elite showed by their actions that they acknowledged the need to treat the opinions of their opponents with the utmost seriousness, even if their rhetorical strategy was to deride those opinions and to detract from their opponents' reputations.[14]

Outside the zone of the elite was a zone of what I term "accredited" geologists. These were men who were less active in geology and less exclusively committed to it, but whose work was highly respected and accepted by the elite within certain limits. Accredited geologists included men of science whose primary commitments and competence lay in other fields of science (where they might belong to an analogous elite), but whose range of expertise impinged in an auxiliary manner upon geology, for example by providing geological debate with authoritative

[14]See, e.g., M. J. S. Rudwick, "The Devonian: A System Born from Conflict," in M. R. House *et al.*, eds., *The Devonian System* (London: Palaeontological Association, 1979), pp. 9–21. I am preparing a full-length account of the Devonian controversy, with a more substantial illustration of the use of sociocognitive topography.

opinions on other sciences. Examples here include men such as William Whewell and John Herschel for the physical sciences and Louis Agassiz and Adolphe Brongniart for the biological sciences (including paleontology). This zone also included men whose primary scientific interest was in geology, but whose first-hand experience was limited to a particular group of strata or geographical region. Examples here include John Phillips for Carboniferous strata and their fossils, Robert Austen for Devonshire, and André-Hubert Dumont for the Ardennes. The opinions of such men would be highly respected by the elite, even in matters of theory, provided those matters pertained directly to the accredited geologists' particular areas of expertise. But the elite geologists did not regard them as competent to judge more fundamental or global questions of theory or method, and accredited geologists were generally content to accept that assessment.

Still further down the continuum of competence lay a zone of those who can properly be termed amateur geologists. These were men—and here at last a few women too—whose ascribed expertise was even more limited, usually in geographical extent, than that of many accredited geologists. In the 1830s they included country gentlemen and ladies, physicians, lawyers, and clergymen with an intimate knowledge of their home areas; and government officials, service officers, and professional and business men whose duties took them to unfrequented parts of the world. Such people might have impressively accurate knowledge of, for example, the strata and fossils of their local area, but the elite and accredited geologists would only rely on that knowledge at a quite factual level. Most of the amateurs accepted this limited role; if they presumed to offer opinions on theoretical matters those opinions would generally be ignored by those in the higher zones of competence.[15]

Finally, outside the zone of the amateurs lay the general public, whose reports on even quite factual matters such as the provenance of anomalous specimens or the effects of earthquakes would be regarded with caution and skepticism, until checked and corroborated, by those in *all* the higher zones of competence.

The well-defined boundaries of membership of the Geological Society (and of similar bodies elsewhere, such as the Société Géologique de la France) cut across this informal topography of social interaction and ascribed cognitive competence. When Darwin joined the Geological Society it had over 800 members, the great majority of whom were clearly amateurs.[16] (Almost by definition, anyone elected to the society would be regarded by the more competent members as a reliable informant on a factual level, and hence as more than an uninformed member of the public.) The society also included a minority of accredited geologists, including

[15]For a brief but suggestive study of one amateur whose work had an important effect on an elite geologist, see J. C. Thackray, "T. T. Lewis and Murchison's Silurian System," *Transactions of the Woolhope Naturalists' Field Club*, 1979, *42*:186–193. For a more general discussion, see David Elliston Allen, *The Naturalist in Britain: A Social History* (London: Allen Lane, 1976). From a slightly later period see, e.g., Darwin's reaction to Robert Chambers's *Ancient Sea Margins, as Memorials of Changes in the Relative Levels of Sea and Land* (Edinburgh: W. R. Chambers, 1848), summarized in Martin Rudwick, "Darwin and Glen Roy: A 'Great Failure' in Scientific Method?" *Studies in the History and Philosophy of Science*, 1974, *5*:97–185, on pp. 145–146. My use of the term "men" is historically correct: women were not eligible for membership of the Geological Society and were not represented on the topography of the science above the amateur zone.

[16]At the end of 1836 the total of Fellows was 810, in the following categories: 97 Compounding (i.e., life members), 251 Contributing, 361 Non-Resident (i.e., living outside the London area), 43 Honorary (a declining group of early provincial members), 55 Foreign, and 3 "of royal blood": see "Annual Report for 1836," *Proceedings of the Geological Society of London*, 1837, *2*(49):462. The Société Géologique de la France was similar in character, but had only 302 members in 1836.

many whose primary affiliations were to other societies and other sciences. The elite was a very small group containing, according to the stringency of definition, perhaps five to ten persons. That small group would however recognize others elsewhere in the world as belonging equally to the elite of the science.

It should be noted in parenthesis that the so-called Mosaic or scriptural geologists of the 1830s, although not banded together in any formal organization, were in effect an almost equally strong-boundaried group. They constitute the most striking case of a radical dissensus about the cognitive topography of geology, because they and the elite geologists were in direct rivalry as dispensers of "authentic" interpretations of the Earth. In effect, there was a mutually hostile boundary between them and the Geological Society (so depicted in Fig. 1); as perceived by the members of the society, they were mainly beyond the limits of even amateur competence, though a few did contribute information that could be adopted on a factual level while ignoring its heterodox theoretical pretensions.[17]

In reality, of course, this relatively static topography was populated by individuals who would tend to shift their location in the course of their careers. Especially at the start of such a career, the trajectory of any individual might move quite rapidly across the map, as Darwin's does in Figure 1. Shortly before he went off on the *Beagle* in 1831, the elite geologist Sedgwick took him on a research trip to Wales, inducted him into the field practice of geology, and thereby effected his transition from a merely amateur status into the role of an accredited if still inexperienced geologist.[18] Darwin's location in that zone during the voyage is shown by the reception of the scientific letters that he sent to John Henslow from South America in 1835. The excerpts that were read to the Cambridge Philosophical Society and to the Geological Society were valued not only for their factual information but also for Darwin's theoretical inferences, since they were limited to South America.[19] After his return to England he joined the Geological Society but at first remained within the zone of accredited geologists. The first two papers he read to the society in 1837 contained explicitly theoretical inferences, but they were still based on and limited to his areas of firsthand experience. His third paper, containing his theory of oceanic crustal oscillation as indicated by coral reefs, was more ambitious, since by implication it could be extended beyond the Pacific and Indian oceans.[20] By the following year (1838) he was clearly being

---

[17]See Milton Millhauser, "The Scriptural Geologists: An Episode in the History of Opinion," *Osiris*, 1954, *11*:65–86; Charles Coulton Gillispie, *Genesis and Geology: A Study in the Relations of Scientific Thought, Natural Theology, and Social Opinion in Great Britain, 1790–1850* (Cambridge, Mass.: Harvard Univ. Press, 1951). On the Mosaic geologists as a strong-boundaried group, see Martin Rudwick, "Cognitive Styles in Geology," in Mary Douglas, ed., *Essays in the Sociology of Perception* (London: Routledge & Kegan Paul, 1982), pp. 219–241. Mosaic geologists are shown on Fig. 1 only in relation to the Geological Society, because they were a much more vocal group in Britain than elsewhere in Europe.

[18]For Darwin's amateur status on the eve of this trip, see the revealing anecdote recorded in *The Autobiography of Charles Darwin, 1809–1882*, ed. Nora Barlow (New York: Harcourt, Brace, 1959), on pp. 69–70. The trip itself is described in detail, though without adequate analysis of Darwin's transition, in Paul H. Barrett, "The Sedgwick-Darwin Geologic Tour of North Wales," *Proceedings of the American Philosophical Society*, 1974, *118*:146–164.

[19]Charles Darwin, *Extracts from Letters Addressed to Professor Henslow* (Cambridge: Cambridge Philosophical Society, 1835); and "Geological Notes Made During a Survey of the East and West Coasts of South America . . . ," *Proc. Geol. Soc. London*, 1835, 2(42):210–212, read (by Sedgwick) 18 Nov. 1835; both rpt. in Paul H. Barrett, ed., *The Collected Papers of Charles Darwin* (Chicago: Univ. Chicago Press, 1977), Vol. 1, pp. 3–19 ("Henslow" in Fig. 2).

[20]Charles Darwin, "Observations of [i.e., "and"] Proofs of Recent Elevation on the Coast of Chili . . . ," *Proc. Geol. Soc. London*, 1837, 2(48):446–449, read 4 Jan. 1837; Darwin, "A Sketch of the

accepted into the company of the elite geologists, for his last paper in this early series embodied much more explicitly a major causal theory of crustal mobility with obvious global pretensions, and it was recommended (by Sedgwick) for full publication in the society's prestigious *Transactions*.[21] Darwin remained an elite geologist for the rest of his time in London and for several years afterwards, while his geology from the *Beagle* was being published in full. Only slowly did his increasing preoccupation with biological science and his weakening commitment to geology cause him to sink back—in the estimation of other geologists—into the merely "accredited" zone; this change is movingly illustrated by the long saga of the waning plausibility of his interpretation of Glen Roy.[22]

DARWIN AS A GEOLOGICAL THEORIST

It will be clear from this outline interpretation of the sociocognitive topography of geology in the 1830s, and of Darwin's trajectory across it, that I do not fully agree with Sandra Herbert's contention that there was no available social role for a theorist in the natural history sciences at that period in England, and that this constrained the form and presentation of Darwin's scientific work. For this is hardly compatible with the assertion, which in my view she is right to make, that Darwin was able to express his geological ideas fully in public at this time, without dissimulation or loss of content, and that they were published quite rapidly.[23] We must, however, first ask what it would or could have meant, in this particular science and period, for a person to be a "theorist" *tout court*. Analogies with theoretical and experimental physics at a later period, or with pure and applied mathematics, are obviously inappropriate. Such a clear division of labor or sharp disciplinary differentiation would have been as unthinkable in geology then as it is now. The concept of a theorist in the geology of the 1830s is still valuable, however, if it is used to denote a person who publicly attached relatively high-level theoretical inferences to his observational descriptions. In this sense all those

Deposits Containing Extinct Mammalia in the Neighbourhood of the Plata," *Proc. Geol. Soc. London*, 1837, 2(51):542–544, read 3 May 1837; Darwin, "On Certain Areas of Elevation and Subsidence in the Pacific and Indian Oceans, as Deduced from the Study of Coral Formations," *Proc. Geol. Soc. London*, 1837, 2(51):552–554, read 31 May 1837; all rpt. in Barrett, *Collected Papers*, Vol. I, pp. 41–49 ("Chili," "Plata," and "Oceans" in Fig. 2). (Note that the title of the last paper states clearly that it concerns primarily the question of crustal mobility and not, as is often assumed, coral reefs for their own sake; see D. R. Stoddart, "Darwin, Lyell, and the Geological Significance of Coral Reefs," *Brit. J. Hist. Sci.*, 1976, 9:199–218.)

[21]Charles Darwin, "On the Connexion of Certain Volcanic Phaenomena, and on the Formation of Mountain Chains and Volcanos, as the Effects of Continental Elevations," *Proc. Geol. Soc. London*, 1838, 2(56):654–660, read 7 Mar. 1838; "On the Connexion of Certain Volcanic Phaenomena in South America; and on the Formation of Mountain Chains and Volcanos, as the Effect of the Same Power by Which Continents are Elevated," *Transactions of the Geological Society of London*, 1840, 2nd Series, 5(3):601–631 ("Elevation [*Proc.*]" and "Elevation [*Trans.*]" in Fig. 2). Note the more ambitious title of the original paper, without the qualifying phrase "in South America" (only the later, full version is reprinted in Barrett, *Collected Papers*, Vol. I, pp. 58–86).

[22]Charles Darwin, *The Structure and Distribution of Coral Reefs* (London: Smith Elder, 1842), *Geological Observations on Volcanic Islands* (London: Smith Elder, 1844), and *Geological Observations on South America* (London: Smith Elder, 1846) ("Coral Reefs," "Volcanic Islands," and "S. America" in Fig. 2). Darwin, "Observations on the Parallel Roads of Glen Roy, and of Other Parts of Lochaber in Scotland, with an Attempt to Prove That They are of Marine Origin," *Philosophical Transactions of the Royal Society of London*, 1839, pp. 39–81, read 7 Feb. 1839; rpt. in Barrett, *Collected Papers*, Vol. I, pp. 87–137 ("Glen Roy" in Fig. 2). For the eclipse of his interpretation, see Rudwick, "Glen Roy."

[23]Sandra Herbert, "The Place of Man in the Development of Darwin's Theory of Transmutation, Part II," *J. Hist. Biol.*, 1977, 10:155–227; on pp. 157–178.

who were tacitly acknowledged to belong to the elite and accredited zones of competence could be regarded as potential theorists.

Herbert rightly comments about the Geological Society that "the practical situation allowed the necessary latitude for the expression of theoretical views, and Darwin, like others, was free to publish within the limits of the rules."[24] But the rules were less restrictive than she infers. It is true that the society's official policy remained severely fact-oriented or "Baconian," as it had been from its foundation thirty years earlier; but social realities had long diverged from official statements and traditions, as they have a way of doing in any institution. It was not for nothing that, just at the time when Darwin joined the society, its outgoing president (Lyell) persuaded one of the most distinguished theorists of science in England (Whewell) to take over the presidency, and also persuaded the council to approve that nomination.[25] Although there was undoubtedly a strong tendency within the elite (Lyell excepted!) to regard theorizing as sharply separable from observation, theoretical inferences and conclusions were in practice readily accepted by the society's own referees when judging papers for publication, provided that the theorizing was demonstrably based on sound observations and that the author was deemed to have shown his competence to operate on the relevant theoretical level. For example, Darwin's geological mentor Sedgwick, who was well known for his emphasis on field observation and for his reluctance to indulge in theorizing, treated respectfully the "theoretical opinions" in Darwin's highly speculative paper on elevation, only asking that they be made "more definite and unequivocal." Likewise he praised the "great ingenuity" of the theorizing in Darwin's paper on Glen Roy, criticizing only his "far too diffuse" style.[26] The extant network of scientific correspondence among the elite and accredited members of the society likewise reveals that the theoretical implications of current research were matters of continual and lively discussion at this period.[27] Most significant of all, such personal interactions were facilitated by a major change in procedure that had marked, and perhaps partly caused, the remarkable intellectual blossoming of the society around the end of the 1820s. This was the introduction of the custom —exceptional at that period in any scientific society—of following the formal reading of papers with free and informal discussion under conditions of relative confidentiality, behind the closed doors of the society's meeting room.[28] In this

[24]*Ibid.*, p. 169.

[25]For the early period see M. J. S. Rudwick, "The Foundation of the Geological Society of London: Its Scheme for Cooperative Research and Its Struggle for Independence," *Brit. J. Hist. Sci.*, 1963, *1*:325–355; Rachel Laudan, "Ideas and Organizations in British Geology: A Case Study in Institutional History," *Isis*, 1977, *68*:527–538. Whewell was president from Feb. 1837 (when Darwin joined the council) to Feb. 1839. The nomination of a new president was by custom the prerogative of the retiring president, but he always checked that his nominee would be acceptable to other elite members and, more formally, to the council.

[26]Sedgwick's report on Darwin's paper on elevation is quoted in Herbert, "Place of Man, Part II," p. 168, n. 27. That Sedgwick regarded theorizing as personal opinion does not imply disapproval of it. His report (to the Royal Society) on Darwin's Glen Roy paper is printed in full in Rudwick, "Glen Roy," pp. 181–183. Any historian who reads these two Darwin papers will, I believe, agree with Sedgwick that at this stage of his career Darwin fully deserved criticism for his obscure and convoluted style.

[27]For Darwin, this should be more apparent in the forthcoming *Collected Letters of Charles Darwin* than it is in the selective and unreliable *Life and Letters* or in F. Darwin, ed., *More Letters of Charles Darwin* (London: Murray, 1903). My forthcoming study of the Devonian controversy will also illustrate this point fully for other elite members of the society.

[28]See Horace B. Woodward, *The History of the Geological Society of London* (London: The Geological Society, 1907), p. 76; also Charles Babbage, *Reflections on the Decline of Science in England and on Some of its Causes* (London: Fellowes, 1830), p. 45. No account of discussions at the

way, the society had in effect found an institutional means for allowing the uninhibited expression of theoretical disagreements, while maintaining towards the wider public a politic stance of corporate theoretical neutrality. In such an arena there was indeed an accepted social role for the would-be theorist, provided only that he had proved his theoretical competence in relation to firsthand observational experience; in other words, provided that he had been tacitly admitted to the higher zones of ascribed competence.

It follows, however, that the true theoretical character of a cognitive enterprise in this social milieu cannot be evaluated by looking only at the finished products of publication. In line with the norms that were customary in the Geological Society, Darwin's geological papers were presented and published with their theoretical component in an apparently subordinate position. But his "piecemeal" mode of publication and his apparent failure to present "a single coherent argument" do not mean that he was inhibited from the full expression of his theoretical views or that they lacked a receptive and appreciative audience.[29] It shows only that his papers were regarded—by Darwin and by other geologists—as contributions to an ongoing and *collective* enterprise within geology.[30]

In the geological science of the 1830s, as it was being practiced internationally, several loosely linked cognitive enterprises of this kind can be discerned beneath the surface of individual publications. One, for example, concerned the reduction of the apparent chaos of the old Transition strata to the kind of sequential order that by this time was well established for the younger and less confusing Secondary strata. Darwin might well have been recruited by Sedgwick into this enterprise, had he not been offered a place on the *Beagle*.[31] But in the event his voyage took Darwin to a part of the world that gave him outstanding opportunities to contribute to another current collective enterprise, only remotely related to the first. This enterprise, which I term "crustal mobility," was concerned with establishing the spatial and temporal pattern of vertical movements of the earth's crust. Within it, attention was concentrated at this time on one particularly contentious focal problem, namely that of the nature of such movements within the Recent or historical period of the history of the earth.[32] Controversy on this point was

---

society was allowed to be published and no official record of them was kept. An unofficial account of one famous discussion is printed in Woodward, *Geological Society*, pp. 138–142.

[29]Quoting Herbert, "Place of Man, Part II" p. 169, n. 30. But it was Herbert who first recognized the importance and theoretical unity of Darwin's geological work on the *private* level and the extent of his ambitions for it: see *The Logic of Darwin's Discovery* (Ph.D. diss., Brandeis University, 1968), Ch. I; also Michael T. Ghiselin, *The Triumph of the Darwinian Method* (Berkeley/Los Angeles: Univ. California Press, 1969), Ch. 1.

[30]Gruber (*Darwin on Man*) uses the term "enterprise" in the context of Darwin's work, but I extend it here to cover the totality of activities directed towards a major cognitive goal in science, whether shared or not. The Lakatosian term "research programme" is focused too exclusively on the formal features of theories to be appropriate in the present context. The Kuhnian term "paradigm" and its successors refer to larger-scale features of scientific practice, and in any case would be inadequate to convey the dynamic, developing, and purposive character of an enterprise.

[31]Herbert, "Place of Man, Part I," p. 222. The Transition strata were those lying beneath the Carboniferous strata but above the unfossiliferous Primary rocks. A similar stratigraphical enterprise at this time, more closely related to Darwin's work, was directed towards finding sequential order in the patchy Tertiary deposits lying above the Secondary strata: see, e.g., M. J. S. Rudwick, "Charles Lyell's Dream of a Statistical Palaeontology," *Palaeontology*, 1978, *21*:225–244.

[32]I use the term "focal problem" to denote any particularly intractable, anomalous, or controversial problem *within* a broader cognitive enterprise. For another and roughly contemporary focal problem, within the parallel enterprise of Transition stratigraphy, see Rudwick, "Devonian." Two of Darwin's later papers from his London period (rpt. in Barrett, *Collected Papers*, Vol. I, pp. 145–171) were

already well under way before Darwin returned to England. His reports to Henslow of the evidence for the elevation of Chile within the Recent period, for example, were therefore received immediately as welcome support by those, led by Lyell, who were arguing for the reality and significance of such movements.[33] Darwin's early papers after his return were likewise seen as important observational and theoretical contributions to this problem, since they presented evidence not only for the continued elevation of the whole continent of South America, but also for the corresponding subsidence of the broad oceanic tracts marked by coral reefs. The significance of Darwin's work was publicly emphasized, for example, in Whewell's presidential address to the society at the meeting at which Darwin took up his secretaryship: Whewell summarized it as the main contribution of the year to what he termed "Geological Dynamics."[34] The last and most ambitiously theoretical of this series of papers on continental elevation, read three weeks later, was seen as extending the Lyellian view on crustal mobility to the level of causal explanation, since Darwin suggested a possible mechanism by which major crustal blocks might have oscillated on a fluid substrate, like drift ice on a polar sea.[35] Darwin's paper on Glen Roy a year later—his first full-length research paper and in effect his masterpiece for entry to the Royal Society—crowned the series by seeking to demonstrate the validity of that theory, using as an example a puzzling phenomenon that was already well known to his geological colleagues. The paper embodied concepts of good scientific theorizing (*vera causa*, consilience of inductions) that Darwin had derived from Herschel and—more importantly—from Whewell.[36] Darwin probably read Whewell's philosophically oriented *History of the Inductive Sciences* (1837) at about the time that he was writing and revising his Glen Roy paper; but he had no need to rely exclusively on this bookish mode of learning, since he was working closely with Whewell on society business at just this same period.[37]

The implication of this interpretation is that Darwin's public science in his London years was not just the unfolding presentation of an individual theoretical enterprise. It was consciously a part of a shared or collective enterprise centered on the focal problem of Recent crustal mobility. To emphasize this is not to detract

---

directed towards a third problem, the possible reality of widespread recent glaciation ("Erratics" and "Glaciers" in Fig. 2). Of all Darwin's geological papers from his London years, only that on the formation of mold by earthworms ("Mould" in Fig. 2; rpt. in Barrett, *Collected Papers*, Vol. I, pp. 49–53) was unrelated to any contemporary focal problem.

[33]For Lyell's enduring commitment to this position see Martin J. S. Rudwick, "The Strategy of Lyell's *Principles of Geology*," *Isis*, 1970, *61*:4–33, on p. 16.

[34]Whewell, "Address to the Geological Society, Delivered at the Anniversary, on the 16th of February, 1838," *Proc. Geol. Soc. London*, 1838, 2(55):624–648, on pp. 643–645.

[35]Lyell's account of the discussion that followed the reading of this paper is in K. M. Lyell, ed., *Life, Letters, and Journal of Sir Charles Lyell, Bart,* (London: Murray, 1881), Vol. II, pp. 39–41.

[36]See Rudwick, "Glen Roy," on pp. 167–175. Darwin submitted his Glen Roy paper (*via* Lyell) 17 Jan. 1839 and was elected a Fellow of the Royal Society a week later ("FRS" in Fig. 2), even before it was read. The importance of Herschel and Whewell for Darwin's *biological* work is stressed by Michael Ruse, "Darwin's Debt to Philosophy: An Examination of the Influence of the Philosophical Ideas of John F. W. Herschel and William Whewell on the Development of Charles Darwin's Theory of Evolution," *Stud. Hist. Phil. Sci.*, 1975, *6*:159–181.

[37]Ruse ("Darwin's Debt") dates his reading of Whewell's *History* to the later part of 1838; certainly he had read it by Apr. 1839 (see his letter to Whewell, 16 Apr. 1839, Trinity College, Cambridge, MS. Add. C. 88[4]). From Feb. 1838 to Feb. 1839 Whewell as president of the Geological Society was assisted by Darwin, whom he had persuaded to become one of the two secretaries. Darwin's contacts with Herschel at the society were less close, but Herschel had been on the council until 1832 and was again quite active in the society after his return from South Africa in 1838.

from Darwin's originality as a geologist, but simply to point out that his theoretical powers were directed towards cognitive goals that he shared with others —whether or not they agreed with his views on the problem. In other words, the cognitive enterprise aimed at understanding crustal mobility developed through a complex process of personal interactions—including frequently those of vehement controversy—among the group of geologists who concerned themselves with it in any way. Darwin returned to England a fairly marginal member of that group, but during his London years he was at its core.[38]

Darwin's social milieu in the Geological Society during his London years gave him a valuable, perhaps even indispensable *training* as a theorist in science. Not only did it give him the opportunity of face-to-face discussions with writers like Whewell and Herschel, who were explicitly concerned with the structure of theorizing in science. Even more importantly, it taught him to articulate his theoretical ideas in an environment that included both the friendly criticism and encouragement of those who shared his views and the more bracing criticism of those who did not.[39] It taught him how to present his theories in a way that would most effectively meet the objections of his opponents and persuade the uncommitted. I suggest that this training may have been vital for the more solitary theorizing on other themes that he was pursuing more privately during those same London years.

A SCALE OF RELATIVE PRIVACY

It is customary to distinguish between the "public" production of a scientist's finished publications and the "private" manuscript evidence of his or her unpublished works. This simple dichotomy is sometimes modified by the recognition of a "semiprivate" area represented for example by scientific correspondence. Even this modification, however, is too crude, and any such scheme is too individualistic. In its place we need the concept of a dimensional scale or continuum between public and private, which is also a scale between the individual and group levels of scientific practice. Such a scale can be illustrated first from the antecedents of Darwin's public science in his London years (see Fig. 2).

While he was on the *Beagle*, Darwin kept a series of rough field notebooks, which he amplified at the time into a series of fair-copy notes and diaries. Although these documents record his wide interests throughout the natural history sciences, their content is heavily weighted towards geology.[40] They were an immediate quarry for the compilation of his published *Journal* and for his early papers to the Geological Society. On the conventional definition they were clearly "private"; but that term can only be relative, since it is difficult to believe that he would have withheld them from his scientific colleagues if asked for a record of his observations on some specific feature. Conversely, the information embodied in his notes included not only his own firsthand observations but also reports that he actively elicited from others (for example, "amateurs" such as local witnesses

[38]The sociocognitive role of those few individuals who are at the center of any focal problem in science is discussed by Harry Collins, "The Place of the 'Core-Set' in Modern Science: Social Contingency with Methodological Propriety in Science," *Hist. Sci.*, 1981, *19*:6–19; nothing in his analysis limits it necessarily to *modern* science.

[39]See also Murchison's remarks to this effect in his presidential "Address to the Geological Society, Delivered on the Evening of the 15th of February, 1833," *Proc. Geol. Soc. London*, 1833, *1*(30):438–464, on p. 464.

[40]Howard E. Gruber and Valmai Gruber, "The Eye of Reason: Darwin's Development During the *Beagle* Voyage," *Isis*, 1962, *53*: 186–200.

to the effects of earthquakes) and material culled from what he chose to read (for example, the narratives of earlier travels in the same regions). In this way even the "zone of field note making," as I term it, represents a level of scientific practice that transcends the individual and embodies the social element.

On his homeward journey, Darwin transformed the last of his rough field notebooks (the Red notebook) into an agenda for the research he planned to do. After his return it was further transformed into a record of highly theoretical thinking, again primarily on geological topics.[41] This change represented a significant plunge away from the public realm of knowledge to be shared with others, into a "zone of theoretical note making." While settling down to life in England, Darwin in effect devised for himself a new genre of note making, at a "deeper" level within the space of his private science. This gave him a medium in which to reflect on issues that could not be shared with others without further intellectual work. One of two notebooks to have this character from the start (notebook A) was specifically devoted to geology, and it was used by Darwin as a theoretical resource in the production of much of this subsequent geological work.[42] But although this notebook was fully "private," in the sense that it was intended for no eyes but his own, it was also oriented towards the public realm, in the sense that it formed a private substrate for the formulation of ideas that he intended to make public as soon as practicable.

His geological projects therefore moved back towards the public realm through the "shallower" level of field note making; here he used retrospectively his notes from the voyage and, in the case of Glen Roy, a new field notebook.[43] From there, his projects shifted in due course into the stage of outlining the form of persuasive arguments. For Darwin, as for any creative scientist, this "zone of drafting" represented an important shift in modality, since the sequence of argument that would have the greatest rhetorical force would never, except fortuitously, be the same as the sequence that was taken in the cognitive growth of the author. Any draft, however sketchy, would involve the reordering of materials to give an argument persuasive power, and would thereby be oriented towards an imagined public audience.[44] This would bring the project into a state in which it could be shown to colleagues for their comments. In this "zone of discussing" the work would be a stage nearer the public realm, but the audience would be small and —more importantly—selected by the author. In the case of Darwin's geological work it is safe to assume that his draft papers went through such a stage of friendly criticism; but his opportunities to talk face to face with friends like Lyell were such as to leave little documentary evidence of this stage. The importance of the stage is indicated, however, by the character of the correspondence between the two men when they were *not* together in London.[45] Beyond such discussions there lay, for

<hr>

[41]For its changing character, see Herbert, "Red Notebook." The predominantly *geological* character of the contents is perhaps obscured by the editor's understandable focus on the short passages dealing with the possibility of transmutation.

[42]Notebook A ("Geology notebook" in Fig. 2) is being prepared for publication by Sandra Herbert.

[43]Darwin's Glen Roy notebook ("Glen Roy notebook" in Fig. 2) is his only *field* notebook from his London years. It has not been published, but its contents are analyzed in Rudwick, "Glen Roy," pp. 116–118.

[44]For his first full-length paper (on Glen Roy), Darwin described this work as "one of the most difficult and instructive tasks I have ever engaged on," Gavin de Beer, ed., "Darwin's Journal," *Bull. Brit. Mus. (Nat. Hist.) Hist. Ser.*, 1959, 2:1–21, on p. 8; also Rudwick, "Glen Roy," p. 129.

[45]See Leonard Wilson, *Charles Lyell: The Years to 1841: The Revolution in Geology* (New Haven/London: Yale Univ. Press, 1972), Ch. 13.

Figure 2. The temporal development of Charles Darwin's three cognitive enterprises (crustal mobility, species origins, Man and Mind) in his London years, plotted against the scale of relative privacy. Notebooks are shown as thick lines (dashes are uncertainties of dating), papers as small black diamonds, books as larger black diamonds; circled points 1 to 5 refer to episodes of theorizing about species. Only the last part of Darwin's Beagle period is shown; for his years at Down House the time-scale is compressed and only a few publications are shown. Tailed arrows show filiation of theoretical notebooks; all other arrows indicate general direction of projects through the scale of relative privacy towards publication. Note that these arrows are schematic: they connect dated documents by the simplest lines and should not be taken to indicate dated transitions through specific zones en route, or that all projects passed through all zones. Abbreviations for drafts, articles, and books are identified in the footnotes; those not closely related to any of Darwin's major enterprises are marked by open diamonds and named within brackets.

Darwin's early geological projects, yet another stage short of the fully public realm, namely the "zone of expert debating" in the restricted semipublic arena of the Geological Society. To present a paper there was to expose it to criticism in a quite different way from any prior discussion with chosen friends: the effective audience of other elite and accredited geologists—the amateurs present hardly counted in this context—might include more hostile critics with quite different theoretical convictions. Darwin's views were recorded in summary form in the society's *Proceedings*, circulated to the membership within a few weeks of the meeting; but his arguments may have been modified substantially in response to the critical interactions on the debating floor of the society's meeting room, not to mention the suggestions of those appointed subsequently to referee the work for publication.[46] Only after that would the project move finally towards the fully public "zone of publishing."

It should be emphasized that this scale of relative privacy is a scale of diverse activities, not a scale from thought to action. The "deepest" levels of the most private zone are no nearer the "true thoughts" of the person concerned (in this case, Darwin) than any other level or zone. Processes of thinking, unrecorded and therefore inaccessible to direct historical inspection, lie behind *all* the activities along the scale: behind the revision of a text for publication in the light of critical expert debate, no less than behind the jotting down of an ephemeral conjecture in a notebook. No one part of the scale has a privileged status for the understanding of the individual's work.[47] Furthermore it should be clear that the precise form of the activities that stretch from private to public science is contingent on particular historical circumstances. The scale just outlined is that which characterized Darwin's science in his London years, but its form altered even in his later years (for example, his books were published without passing through any zone of expert debating), and for other scientists at other periods its form might differ still more. Nonetheless *some* form of the scale of relative privacy must characterize the work of any scientist at any time.

[46]Summaries for the *Proceedings* were generally made after the meeting by one of the secretaries, though they were shown to the author for approval; the printed version often already embodies modifications made in the light of the discussion. Full publication in the *Transactions* was invariably slow—see, e.g., Fig. 2, "Elevation (*Proc.*)" and "Elevation (*Trans.*)"—and allowed ample time for still further revision.

[47]For a vigorous defense of the use of laboratory notebook evidence, comparable to field notebook evidence, see F. L. Holmes, "The Fine Structure of Scientific Creativity," *Hist. Sci.*, 1981, *19*:60–70. The historical inaccessibility of *any* evidence of a scientist's thoughts is claimed in somewhat extreme form by Steven Shapin and Barry Barnes, "Darwin and Social Darwinism: Purity and History," in Barry Barnes and Steven Shapin, eds., *Natural Order: Historical Studies of Scientific Culture* (Beverly Hills/London: Sage, 1979), pp. 125–142. For criticism of this view, see Martin J. S. Rudwick, "Social Order and the Natural World," *Hist. Sci.*, 1980, *18*:269–285, on pp. 278–280.

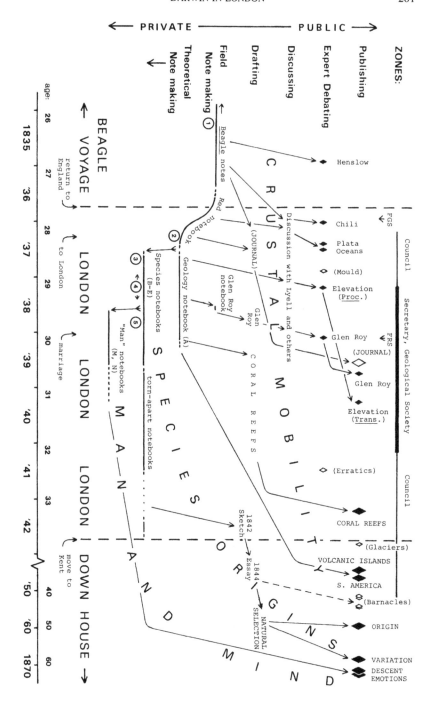

DARWIN'S PRIVATE THEORIZING

Returning to the specific case of Darwin, this scale can now be extended to include those other and better known aspects of his private science that never reached the public realm until long after his London years. Darwin's "private" theorizing was not of course totally isolated from any discussion with others, but in the earlier years he did not disclose to any of his colleagues the full content of the enterprise to which their knowledge or opinions were to contribute: the cognitive flow, as it were, was essentially one way, from them to Darwin.

At the same time that Darwin was contributing to the collective enterprise of crustal mobility, he was becoming increasingly involved in a related but distinct cognitive enterprise of his own, namely that of finding an adequate explanation for the origin of species. This first appears in the documentary record as notes scattered within the mainly geological materials of his *Beagle* notes and the Red notebook.[48] In some geological notes of February 1835, Darwin commented on the "births" and "deaths" of species; his views were "creationist"—though in no naive sense—and derivative from those of Lyell (Fig. 2, Point 1). Around the time he settled in London, he recorded his first transformist conjectures in the context of his geological notes in the Red notebook; he speculated here on a saltatory mode of origin of species, and on a mode of extinction in terms of racial senescence (Fig. 2, Point 2). When in the summer of 1837 he began a new theoretical notebook (A) to continue his geological theorizing, he also began in parallel a similar one (B) devoted specifically to the problem of species. He quickly formulated a "first theory" for the nonsaltatory origin of species, in which geographical isolation and sexual reproduction were central elements (Fig. 2, Point 3). In the subsequent period of about a year, he consolidated this theory and sought to extend its explanatory range, while rejecting various elements that might have "destabilized" its growing coherence and scope (Fig. 2, Phase 4). Then came his famous insightful moment of rereading Malthus, which precipitated a major reconstruction of his theory, with natural selection now as its centerpiece (Fig. 2, Point 5). This new theory was likewise extended and improved subsequently, up to and beyond the end of his notebook period.

This documentary record of Darwin's early work on the species problem suggests that his private cognitive enterprise emerged from a preexisting matrix composed primarily of his contribution to the collective enterprise of crustal mobility in geology. In comparison with this, his previous concerns in the rest of natural history were relatively subordinate. The immediate occasion for his first transformist conjectures in the spring of 1837 (Fig. 2, Point 2) may well have been the natural history work that he was doing for his *Journal* and for the zoology of the *Beagle* voyage.[49] But the problems of biogeography raised by those projects were intimately related—for Darwin, if for no one else—not only to the possibility of geographical isolation as a mechanism of speciation, but also to the underlying geological events that might make that mechanism possible. They were related, in

---

[48]In the following summary I rely especially on Kohn, "Theories to Work By." The black circled points 1 to 5 on Fig. 2 represent his Episodes 1 to 5; Episode 4, unlike the others, was a protracted period of about a year. See also Gruber, *Darwin on Man*; Herbert, "Place of Man, Part II"; and George J. Grinnell, "The Rise and Fall of Darwin's First Theory of Transmutation," *J. Hist. Biol.*, 1974, 7:259–273. Gruber uses Darwin as an example in "On the Relation Between 'Aha Experiences' and the Construction of Ideas," *Hist. Sci.*, 1981, 19:41–59.

[49]Herbert, "Place of Man, Part I," p. 249; Grinnell, "First Theory."

other words, to the concept of continuous crustal oscillation that formed the focus of his theoretical work at this time. His later decision to start a notebook devoted specifically to the species problem may therefore be taken to reflect his growing awareness that by then it deserved to be treated as a fully distinct enterprise, notwithstanding its substantial and enduring links with his geological work, particularly in the area of biogeography.

A few months later, when he began the third (D) of his species notebooks, he took the opportunity to begin in parallel another new one (M) devoted to issues related more specifically to human beings. Such concerns had been apparent in the notebooks from the beginning of his work on the origins of species, but the start of a new series surely reflects his growing awareness that yet another cognitive enterprise was emerging in his work, namely that concerned with the problems of "Man and Mind," a somewhat inadequate label for a wide-ranging network of related issues, including metaphysical and ethical themes.[50] This too deserved to be treated as distinct, notwithstanding its important and enduring links with the enterprise on the origins of species, particularly in such areas as the study of instinctive behavior. In other words, this third enterprise differentiated out of a matrix of the second, just as the second had emerged from the first.

This reconstruction of the filiation (or phylogeny!) of Darwin's theoretical notebooks into three parallel series, with distinct though interlinked contents, is borrowed from the meticulous research of Darwin scholars into their relationships and dating. In incorporating their results, as summarized in Herbert's diagrams,[51] into my own visual scheme (Fig. 2), I have, however, tried to suggest how the successive differentiation of Darwin's three enterprises is at the same time a record of varying degrees of privacy. All the notebooks display equally Darwin's style of theoretical note making, and in that sense they are all equally "private." But even within this most private zone, different degrees of privacy must be distinguished. Darwin's geological notes could be used without delay as a theoretical resource in the construction of drafts for his geological papers and later books (see, e.g., arrows from Red and Geology notebooks in Fig. 2). The start of a new series of notebooks on the species problem, on the other hand, represented a further move away from the public realm, for Darwin recognized that this new enterprise, unlike his geology, could not in the immediate future be articulated even in draft form, let alone as a publication. His approach to the problem was so innovative that, on the evidence of the notebooks themselves, his theorizing had to go through a lengthy process of construction, maturation, and reconstruction before it even began to seem adequate to him. Furthermore, in contrast to his geological work, there was no ready-made potential audience for his particular style of theorizing about the problem of species. For while the definition and delineation of species and varieties were ever-present problems in the routine taxonomic practice of the zoologists and botanists among his contemporaries, Darwin's project to find a purely naturalistic explanation of the *origin* of species was one that his most respected col-

---

[50]The relevant notebooks are M and N ("Man notebooks" in Fig. 2), edited by Paul H. Barrett and Howard E. Gruber, *Metaphysics, Materialism and the Evolution of Mind: Early Writings of Charles Darwin* (Chicago: Univ. Chicago Press, 1980). See also Herbert, "Place of Man, Part II," pp. 196–226; Manier, *Young Darwin*; and Sylvan S. Schweber, "The Origin of the *Origin* Revisited," *J. Hist. Biol.*, 1977, *10*:229–316.
[51]Herbert, "Place of Man, Part II," pp. 178–188, esp. Fig. 3; and Herbert, "Red Notebook," esp. Fig. 2.

204

leagues regarded as intractable, insoluble, or even beyond the scope of science altogether.[52]

Only after perhaps four or five years of work with his notebooks (from about 1837 to 1841 or 1842) did he feel ready to reorder the elements of his theory into a first outline of a persuasive argument. This move towards drafting may be reflected in the way he switched his working method to a system of collecting loose notes in a series of portfolios, and perhaps also in his method of excising pages from his notebooks and, presumably, rearranging them too according to topic.[53] In any case, by the summer of 1842 he was at last ready to write a first sketch of his theory, only three months before he finally left London. Thereafter the project remained for many years in the zone of drafting, in the successively enlarged and improved versions of the 1844 "Essay" and the uncompleted full-length *Natural Selection* (see Fig. 2). Not only did Darwin retain the same basic structure of argument throughout this long process of development; but even in its earliest form (in the "Sketch") that argument mirrors the form that he had used in his interpretation of Glen Roy.[54] The form of theorizing that he had first put into practice in the context of the collective enterprise of understanding crustal mobility was thus carried over and applied again to the more private and more difficult enterprise of understanding the origin of species.

The differentiation of a new series of notebooks on Man and Mind out of the matrix of the species origins enterprise represented a further plunge away from the public realm (see "Man and Mind" in Fig. 2). The problems he addressed here were not only regarded by many of his colleagues as being of dubious scientific legitimacy, but also raised sensitive personal and social issues about the nature and destiny of human beings.[55] But the early differentiation of this enterprise out of a preexisting matrix implies that Darwin's separation of the specifically human issues from his nascent theory of evolution was not just a tactical move made at a late stage in order to divert or postpone personally unwelcome criticism. Rather it reflects Darwin's judgment that the two enterprises could be pursued most effectively, and be brought up into the public realm most easily, by being worked on in parallel from the beginning.

[52]The lack of a receptive audience in the 1830s is argued convincingly by Herbert, "Place of Man, Part II." She is too restrictive however when she suggests (p. 190) that even in the 1850s Darwin's potential audience consisted only of Wallace; for the taxonomists just mentioned did constitute a kind of *latent* audience which Darwin was anxious by that time to reach with his own theory. Darwin's own taxonomic monographs of 1851–1854, on living and fossil cirripedes ("Barnacles" in Fig. 2), can be regarded as *tacit* contributions to the public presentation of his theory, which only became explicit with the *Origin* of 1859 (hence the dashed arrow on Fig. 2).

[53]The portfolio method is known to have been in use by 1844 at the latest, but the excisions from the notebooks are as yet undated: see Kohn, Smith, and Stauffer. "New Light."

[54]In my opinion Gruber (in *Darwin on Man*) exaggerates the social and personal consequences that a man in Darwin's position would have experienced had he published his naturalistic species theory in the 1840s. But the effect on his professional reputation, had he published it without the massive supporting evidence planned for *Natural Selection*, would surely have been damaging. Significantly, he wanted the "Essay" published only in the event of his early death; its lack of substantive documentation would then have been readily excused by his colleagues. For the continuity of structure from the "Sketch" (or "Essay") to the *Origin*, see Ruse, "Darwin's Debt;" also M. J. S. Hodge, "The Structure and Strategy of Darwin's 'Long Argument,' " *Brit. J. Hist. Sci.*, 1977, *10*:237–246. For the anticipation of that structure in the Glen Roy paper, see Rudwick, "Glen Roy," pp. 170–175.

[55]See Herbert, "Place of Man, Part II," and Gruber, *Darwin on Man*. Here Gruber's emphasis on Darwin's reluctance to face the imagined personal consequences of his theorizing is more convincing. Parts of the Man and Mind work were eventually published as *The Descent of Man* (1871) and *The Expression of the Emotions in Man and Animals* (1872) ("Descent" and "Emotions" in Fig. 2).

CONCLUSION

I have argued that the filiation of Darwin's theoretical notebooks represents a graded structure of relative privacy, lying as it were beneath the series of zones that leads to the fully public realm. When the sequence of notebooks is plotted in relation to events and documents representing those more public activities, a fairly simple and intelligible pattern emerges, as depicted in Figure 2.

Leaving aside Darwin's early project for publishing his journal of the *Beagle* voyage, which was descriptive in the Humboldtian sense rather than truly theoretical in any explanatory sense, Darwin's work in his London years (and indeed afterwards) can be seen to have developed into three interlinked cognitive enterprises, differentiated from each other in turn and then running in parallel. The first, focused on the geological problem of Recent crustal mobility, found a ready-made and receptive audience within the milieu of the Geological Society of London. Indeed, Darwin was here contributing to a process of collective theory construction; and as he rose rapidly into the highest zone of ascribed competence, he joined others in the accepted role of geological theorist. He published his work as rapidly as contingent constraints allowed, and its theoretical content was accepted as important and legitimate even by those who were left unconvinced by his conclusions. Out of the matrix of this collective enterprise, including particularly the issues it raised for the relation between living beings and an ever-changing physical geography, Darwin found himself developing a second and more private cognitive enterprise, concerned with the problem of the origins of organic species. This had to take second place on his long-term agenda, not only because of the temporal priority of his geology but also because the more innovative character of his species work demanded more time and more effort before it could be brought towards the public realm. Darwin's third cognitive enterprise, concerned with the relation of human beings to the natural world, differentiated out of the second at an early stage, but was even more private because of the problematical and sensitive issues it raised. It is therefore not surprising that it took third place on Darwin's agenda; and even after the enterprise on the origins of species had been forced up into the public realm (as the *Origin of Species*) he was still somewhat reluctant to present this most private of his enterprises in published form.

In conclusion, these maps of Darwin's work during his London years suggest how his success as a theorist within the social world of geology can be related to his other, more private cognitive enterprises. In the social milieu of the Geological Society he gained valuable training and experience in the practice of theorizing in science, by participating in a *collective* but controversial theoretical enterprise. In doing so, he not only had the opportunity to try out for himself the models of good scientific theorizing that he learnt from others, but also earned esteem and recognition as a *competent* theorist. I suggest therefore that his public success as a theorist in geology not only provided him with models of theorizing that he could apply to other, less recognized fields, but also gave him the sense of legitimation that he may have needed in order to sustain himself while pursuing that more lonely course. If this interpretation is correct, Darwin's work in his London years constituted a *network* of enterprises that were linked by even more ties than those of overlapping problems and common methods.[56]

---

[56]The notion of a *network* of enterprises within a creative individual's life work is being developed

I put forward that conjecture, however, simply as an illustration of the sort of historical question that comes into prominence in the light of the kind of mapping that I have outlined in this article. I have shown how Darwin's ascribed competence in his most creative period can be mapped as a personal trajectory across the social and cognitive topography of the science to which he gave his primary allegiance (Fig. 1), and how the development of his work in that period can be mapped on the scale of relative privacy (Fig. 2). These visual representations, together with their complementary verbal exposition, do not in themselves explain anything in any strongly causal sense. But they do explain the pattern of Darwin's work in the sense that they make intelligible the sequence of his activities; and at the very least they have heuristic value as a way of highlighting the problems that need further historical explanation. More generally, I hope that this reframing of Darwin's public and private science in his London years will suggest a model for a similar understanding of the innovative work of other scientists, in a way that does justice both to the individual particularities of that work and to its immediate social matrix. This could help to bridge the current gap in historical understanding between the individual and group levels of scientific practice.

---

by Nancy Ferrara in collaboration with Howard Gruber. I borrow the notion of a long-term agenda from Gruber's current work in this field.

X

DARWIN AND GLEN ROY: A "GREAT FAILURE" IN SCIENTIFIC METHOD?

"It is by an attention to circumstances which at first appear trivial,
that abstruse truths are often discovered."

John MacCulloch, *On the Parallel
Roads of Glen Roy.* (1817)

CONTENTS

I	*Introduction*	99
II	*The Lake Hypothesis*	
	1 The problem stated: lakes and barriers	102
	2 The barriers located	109
III	*The Marine Hypothesis*	
	1 Darwin in Lochaber	114
	2 The marine hypothesis proposed	118
IV	*The Glacial Lake Hypothesis*	
	1 The possibility of glacial lakes	130
	2 The lake hypothesis improved and defended	140
	3 The marine hypothesis restated	145
	4 The glacial lake hypothesis improved and accepted	147
V	*Origins of Darwin's Hypothesis*	
	1 Darwin and crustal elevation	153
	2 Darwin and the Glacial Theory	157
VI	*Darwin's 'Tenacity'*	
	1 Personal and social factors	159
	2 Theoretical factors	161
	3 The influence of Lyell	165
	4 The influence of Herschel and Whewell	167
	5 Glen Roy and the Species Theory	170
VII	*Conclusion*	175
	Appendices	179

PLATE I. "View of the Parallel Roads of Glen Roy", from Darwin's paper of 1839. The Roads, seen as fine "lines" on the hillsides, are those termed in this article R1 (the highest), R2, and R3 (the lowest). The view appears to have been taken from a point on R3 on the side of the Hill of Bohuntine (see Figure 1), looking up Glen Roy.

I

Introduction

R E C E N T debates on the structure and development of scientific knowledge have suffered from the paucity of available detailed case-studies of the processes by which scientific knowledge comes into being. Some might even urge that a moratorium on all such general debates should be called, until a larger number of detailed studies is available. Without subscribing to such an extreme (and unenforceable) view, I believe that detailed analyses of what were, in their time, difficult but genuine scientific *problems* may help to clarify much broader philosophical and historical issues. More specifically, now that philosophers of science are increasingly aware of the historical relativity of the view that physics is the paradigm science, it is important to provide the general debate with a wider *variety* of detailed studies. The positivist 'peck-order' of the sciences, which continues to form the background assumption to much writing on the philosophy of science, allows no place for large areas of science as it has been (and is) actually practised. Geological science is one such area. In this article I shall analyse in detail one scientific problem in nineteenth-century geology.[1] It may appear at first sight strictly technical in interest and narrowly circumscribed in content. Even if this were true, it might still be useful as a casestudy in scientific debate. In fact, as I hope to show, it had important implications which gave it significance and interest up to a high theoretical level. It is also a problem with considerable intrinsic historical interest, in that the most distinguished individual involved in the debate put forward an explanation which failed to stand the test of time.

In 1838, during the most creative phase of his life, Charles Darwin made a major excursion to the Scottish Highlands in order to study the celebrated Parallel Roads of Glen Roy. His conclusions formed his first full-length scientific paper. But when writing his *Autobiography* some forty years later, Darwin condemned this paper as "a great failure", and added "I am

[1] I am grateful to the University of Connecticut for inviting me as a visiting Professor for part of a semester in 1972, during which I was able to write a major part of this article. Thanks to Professor A. J. Frueh, I was able to try out the argument on the graduate geology seminar there, and on my return to Cambridge I had a similar opportunity in the History and Philosophy of Science seminar. I am grateful for many constructively critical comments received on these occasions. I am particularly indebted to Gerd Buchdahl, not only for helpful comments on the text of this article, but even more for the stimulus of his Tarner Lectures on "Science and Rational Structures", delivered in Cambridge in 1973. My debt to his methodological schema will be apparent to all who attended the lectures.

ashamed of it." Of no other part of his life's work was he ever so out-spokenly self-critical. This in itself would suggest that Darwin's work on Glen Roy merits more attention than it has received from historians of science. But its interest is highlighted by Darwin's own diagnosis that his "failure" had been *philosophical*, for he maintained that "my error has been a good lesson to me never to trust in science to the principle of exclusion."[2]

This judgment has been accepted at face value by most commentators ever since,[3] but to do so raises several intriguing problems. Darwin himself explained his "great failure" in the following terms:

"Having been deeply impressed with what I had seen of the character of the land in South America, I attributed the parallel Lines [*i.e.*, the 'Roads' of Glen Roy] to the action of the sea; but I had to give up this view when Agassiz propounded his glacier-lake theory. Because no other explanation was possible under our then state of knowledge I argued in favour of sea-action."[4]

In fact, however, Darwin had had to attack another hypothesis, which was already in existence, and which had substantial explanatory power. It is therefore not true that "no other explanation was possible" in the 1830s. Furthermore Darwin was aware of the glacial theory when he wrote his paper; and when in 1840 Agassiz applied it specifically to Glen Roy, Darwin did not in fact "give up" his own view. On the contrary, he held on to it with great tenacity for more than twenty years, despite the gradual erosion of its plausibility. When finally he abandoned it, he did so, in his

---

[2] Nora Barlow (ed.), *The Autobiography of Charles Darwin 1809–1882* (London, 1958), 84.

[3] Soon after Darwin's death, Archibald Geikie called it "a curious example . . . of the danger of reasoning by a method of exclusion in Natural Science": quoted in Francis Darwin (ed.), *The Life and Letters of Charles Darwin* (London, 1887) 1, 291. [This will be referred to hereafter as *LLCD*]. The same acceptance of Darwin's own diagnosis is found in more recent writing by both pro-Darwinians and anti-Darwinians. For the former, see for example Gavin de Beer, *Charles Darwin: Evolution by Natural Selection* (London, 1963), 76–7; for the latter, see Gertrude Himmel-farb, *Darwin and the Darwinian Revolution* (London, 1959), 87–9. One recent full-length analysis of "the Darwinian Method" contains, surprisingly, no mention of Glen Roy at all, although the geological component in Darwin's work is rightly given greater prominence than in more traditional accounts: Michael T. Ghiselin, *The Triumph of the Darwinian Method* (Berkeley and Los Angeles, 1969). The "principle of exclusion" is put into better historical perspective by David L. Hull, who, however, does not analyse the sequence of events between the Darwin of the 1830s writing the Glen Roy paper and the Darwin of the 1870s condemning it: 'Charles Darwin and Nineteenth-Century Philosophies of Science', R. N. Giere and R. S. Westfall (eds.), *Foundations of Scientific Method: The Nineteenth Century* (Bloomington, 1972), 115–32, see p. 128; *Darwin and his critics: The reception of Darwin's Theory of Evolution by the Scientific Community* (Cambridge, Mass., 1973), 13–14, 25–6. The only study specifically devoted to Darwin's work on Glen Roy adopts the traditional judgment: Paul H. Barrett, 'Darwin's "Gigantic Blunder" ', *Journal of Geological Education*, 21 (1973), 19–28. This article is valuable, however, for making generally available a previously unpublished document recording Darwin's views at a crucial moment in the controversy.

[4] *Op. cit.* note 2.

own words, "with more sighs and groans than on almost any other occasion in my life".[5]

It is therefore at least worth considering whether Darwin's philosophical criticism of his own work may not contain an element of myth-making. Like the well-known comment in his *Autobiography*, that on the problem of species he had "worked on true Baconian principles, and without any theory collected facts on a wholesale scale,"[6] his comment on his work on Glen Roy may be a recollection that bears little resemblance to what actually happened.

This article is divided into two parts: the first (Sections II–IV) narrative, the second (Sections V–VII) interpretative. The division is merely one of convenience. The form of the narrative is governed by the philosophical interpretation I later expound, and I have not hesitated to add explicit interpretative comments where appropriate. But the narrative form is imperative in the first instance, because my interpretative conclusions will emerge only from a consideration of the Glen Roy controversy as a *dynamic* scientific problem. This apology for the narrative form of the major part of the article is the more necessary, in that it may appear to be excessively detailed in content. In part, this is an unavoidable feature of any debate in an 'unrestricted' science like geology.[7] More generally, however, I hope that this case-study will bear out my belief that a serious philosophical engagement with the problems of scientific knowledge will need to take the *details* of scientific argument as seriously as the scientists themselves have done. The formalizations that are conventional in much current philosophy of science are intellectual tools far too crude to do justice to the subtle diversity of reasoning and judgment that is found in real scientific argument.

One final introductory point concerns my use of thematic maps and

---

[5] Francis Darwin (ed.), *More letters of Charles Darwin, A record of his work in a series of hitherto unpublished letters* (London, 1903) [hereafter abbreviated to *MLCD*] **2**, 193.

[6] *Op. cit.* note 2, 119.

[7] C. F. A. Pantin, *The Relations between the Sciences* (Cambridge, 1968), chs. 1, 5. My interest in the Glen Roy problem was first aroused years ago by some brief comments by the late Professor Carl Pantin, F.R.S., in the course of a talk to the British Society for the Philosophy of Science. After his death, the late Dr. A. M. Pantin kindly allowed me to search his papers for any publishable material on the subject. Although I found a short incomplete typescript on Glen Roy, it became clear that a fuller historical study would be both necessary and worthwhile. I am however greatly indebted to Pantin's work for drawing attention to the way in which Agassiz, as much as Darwin, 'saw' what his theory required, even though Agassiz's interpretation was later pronounced 'correct' and Darwin's 'incorrect'. Pantin's notes on his historical fieldwork on the problem also amplified in various ways my own first-hand study of the Parallel Roads in 1968, and confirmed my belief that such a study of original phenomena is an indispensable part of any historical analysis of a complex scientific problem of this kind.

other diagrams for the analysis of successive interpretations of the Parallel Roads. The linguistic and logical traditions of conventional philosophy of science have led to a neglect of the various non-verbal and non-mathematical languages that are important components of the discourse of many scientists. In particular, the special properties of the visual languages of illustrations, maps and diagrams have generally been neglected by philosophers of science, or at best treated as second-class adjuncts of mathematical, logical and verbal languages. In this article I have merely followed the usage of the relevant science (both in the nineteenth century and today) in using the visual language of maps and diagrams as an integral part of scientific discourse. My visual interpretation of the Glen Roy debate is complementary to my verbal analysis, and neither can be understood without the other.

II

The Lake Hypothesis

1. *The problem stated: lakes and barriers.* Many years before Darwin visited Scotland, the origin of the Parallel Roads of Glen Roy had been discussed in great detail in two published papers. Darwin himself studied these papers closely, and in proposing an alternative explanation he was obliged to treat their arguments very seriously. I shall do the same.

The Parallel Roads had first become known to the world of learning in the eighteenth century through a brief description by the traveller Thomas Pennant.[8] They are situated in Lochaber, an area of the Scottish Highlands that was still remote even in the early nineteenth century (Figure 1). But they became a common feature in the itinerary of those who travelled through the Highlands for a variety of purposes—politico-economic, natural-historical, or in search of the romantic and picturesque. For example, George Greenough included them in his extensive travels in 1805, two years before he became the first President of the Geological Society of London, and he recorded his impressions and conclusions about them, along with observations on the economic life of the Highlands.[9] Pennant had noted the local inhabitants' belief that the Roads were ancient hunting rides (a notion that would have seemed more plausible before the de-afforestation of the area), but Greenough rejected this ex-

---

[8] Thos. Pennant, *A Tour in Scotland, MDCCLXXII. Part II* (London, 1776), 394–6.

[9] M. J. S. Rudwick, 'Hutton and Werner compared: George Greenough's geological Tour of Scotland in 1805', *British Journal for the History of Science*, x (1962), 117–35; see pp. 130–2.

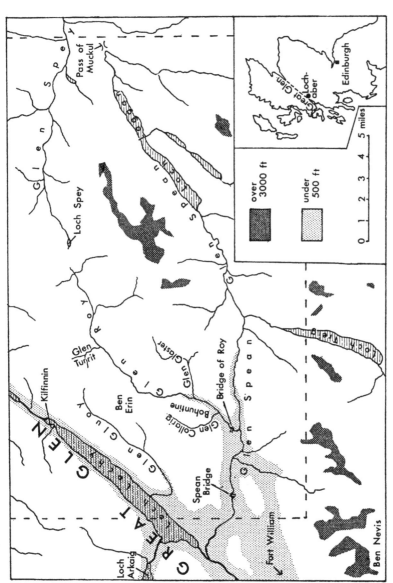

Fig. 1. General map of Lochaber to show the topographical setting of the Parallel Roads. Place-names are given in their nineteenth-century anglicized forms (see Appendix D). The Roads are in Glens Spean, Roy and Gluoy, and around Loch Treig. This map also serves to illustrate MacCulloch's lake hypothesis. If the Roads were former lake shore-lines, then *whatever* the precise limits of the lake or lakes, barriers to contain the water would be required *at least* at two points in the west of the area (mouths of Glen Spean and Gluoy, *or* both ends of the Great Glen) *and* somewhere in the east of the area (at one or more points in Glen Spey). Other maps illustrating this article (Figures 3, 4, 7, 8) show a slightly smaller area (within dashed lines).

planation with scorn and asserted that they were certainly natural in origin. This was one point on which there was to be no further controversy. The term 'Roads' can therefore be used here without risk of misunderstanding.

Yet the Roads seemed astonishingly regular for a natural feature of the landscape. From a distance they appear as fine lines running with perfect horizontality along the steep sides of certain valleys; and in one valley, Glen Roy, there are three such lines running parallel to each other (Plate 1). Close up, the Roads are found to be narrow terraces, sloping gently outwards. Greenough suggested that they were the beaches of a former lake which had stood at three successive levels, the impounding barrier or barriers being broken down in stages. This was the core of what I shall term the *Lake Hypothesis*. Although Greenough's *Journal* was unpublished it appears to have circulated in fair-copy form among his friends, so that his suggestion may have become widely known, at least within the Geological Society circle.

John MacCulloch, a member of the Society, first visited Lochaber in 1814, and again in the two subsequent years. In the introduction to his paper,[10] which was read to the Society in 1816, he stressed the "geological consequences of the first importance" that would follow from a satisfactory explanation of the Roads as *natural* features. Thus Darwin, as we shall see later, was following an accepted tradition when he used the unique phenomenon of the Roads in the service of a theory of much wider, indeed global, implications.

But in attempting a causal explanation, MacCulloch was running counter to the strongly 'Baconian' policy of the Society. He therefore defended himself at the outset by stating that "it is the duty of the philosopher to investigate causes," although at the same time he promised to "keep clear of all speculations purely hypothetical." It was preferable to leave causes in their "natural obscurity," if necessary, rather than invoking "imaginary causes". This was no mere window-dressing, for MacCulloch was in fact prepared, as we shall see, to admit that his own interpretation was incomplete and to urge its acceptance merely as a provisional and

[10] J. MacCulloch, 'On the Parallel Roads of Glen Roy', *Transactions of the Geological Society of London* [1st series], 4, part 2 (1817), 314–92, Pls. 15–22. John MacCulloch (1773–1835) studied medicine at Edinburgh; practised for some years at Blackheath; and worked as a consultant chemist for the Board of Ordnance and, later, as a consultant geologist to the Trigonometrical Survey of Scotland. He was elected to the Geological Society in 1808, soon after its foundation, and was President in 1816–17, at the time this paper was read. See also V. A. Eyles 'John MacCulloch, F.R.S., and his geological map: an account of the first geological survey of Scotland', *Annals of Science*, 2 (1937), 114–29.

temporary theory. In this he forms a striking contrast to Darwin, for whom a satisfying explanation of every detail seems to have been an inner necessity as much in his Glen Roy work as in his work on the origin of species.

The structure of MacCulloch's paper also seems to reflect the 'empiricist' climate of the Geological Society. The first part was devoted to a description of the Roads, as theory-free as he could make it; the second, to an assessment of three or four hypotheses in the light of these observations.

His description of the Roads—for which he used the deliberately noncommittal term "lines"—was indeed in some respects uncontroversial. He noted that the Roads were about sixty feet wide, sloping gently outwards; just above and below each terrace the general slope of the hillside was slightly steepened. They maintained precise horizontality throughout their length, curving round every spur and into every re-entrant and lateral valley. Their continuity was broken only by minor gullies that had evidently been formed at a subsequent period, and also in places where a flattening of the general slope of the hillside made the course of a terrace indistinguishable.

The localized distribution of the Roads was more of a problem. The highest, which on levelling and barometric readings MacCulloch estimated at 1278 feet above sea level, was confined to Glen Gluoy (I shall term this Road G).[11] The next, at 1266 feet, was confined to the adjacent Glen Roy (Road R1). Another, at 1184 feet, was also confined to Glen Roy, paralleling the previous one around the sides of the valley (Road R2). The lowest and most extensive, at 972 feet, likewise paralleled the previous two around the sides of Glen Roy, but also emerged into Glen Spean (to which Glen Roy is tributary) and extended right to the head of that Glen and also into another tributary valley, Glen Treig (Road R3/S).

In addition to these Roads, MacCulloch maintained that "Many marks or fragments of *lines* are seen between the principal ones", but were relatively short and obscure. He reported two minor Roads in Glen Gluoy. He also noted a broad terrace merging into Road R3 at one point in Glen Roy (at the mouth of Glen Turrit). One further detail in MacCulloch's description was to become important for the interpretation of the Roads: he reported that, at the very head of Glen Roy, Road R1 was sixty-three

---

[11] Almost every author used a different scheme of terms to refer to the various Roads. To simplify exposition, I shall use a standardized scheme throughout this article: its equivalents in some of the nineteenth-century papers are given in Appendix C.

feet above Loch Spey, the source of the River Spey, and also above the level of the col separating Glen Spey from Glen Roy.

MacCulloch first dismissed the traditional local explanation that the Roads were ancient hunting rides, asserting that their "real cause" lay in the action of water. He then considered three alternative hypotheses, all of which were actualistic, *i.e.*, based on natural processes observable in action.[12] He conceded that these three hypotheses all had their difficulties, but he said he could think of no others.

The first was an application of Sir James Hall's theory of 'Huttonian diluvialism', in which the sudden elevation of one part of the earth's crust could have sent a transient wave or deluge sweeping over adjacent land areas, on the actualistic analogy of tsunamis ('tidal waves') caused by present-day earthquakes.[13] But MacCulloch saw that this was unsatisfactory for several reasons. The causative crustal movements would have had to be far from Lochaber, or else they would have disturbed the perfect horizontality of the Roads. More seriously, it was impossible to conceive how three such movements could have produced equal effects (*i.e.*, Roads of identical breadth, etc.) in such unequal circumstances (*i.e.*, having such different extents): for by the Newtonian rule the "almost absolute equality of their dimensions . . . proves that the causes which produced these *lines*, have been similar and equal". Furthermore, extra movements and waves would then have to be invoked to account for the fragmentary minor Roads, and anyway a violent transient wave flowing up or down a valley of irregular width would not leave perfectly horizontal traces on the valley sides. For these and other reasons, the hypothesis was rejected.

The second hypothesis depended on the visual resemblance between the Roads and the ordinary river terraces seen on the floors of many valleys. But the resemblance was slight, and the hypothesis was only supported by the broad terrace coincident at one point with the Road R3. This hypothesis also could be rejected.

The third, and in MacCulloch's view, "only other" hypothesis was that

[12] Lyell's writing constantly implied that at the time he wrote his *Principles of Geology* (1830–3), the science was a mass of speculation and wild theorizing unrelated to observable "actual causes". It is therefore worth stressing that MacCulloch's actualistic method was normal procedure at this period; "extraordinary causes" were invoked only as a last resort. For actualism and the meanings of "uniformity" in early nineteenth-century geology, see R. Hooykaas, *Natural Law and Divine Miracle: A Historical-critical Study of the Principle of Uniformity in Geology, Biology and Theology* (Leiden, 1959); also M. J. S. Rudwick, 'Uniformity and Progression: Reflections on the Structure of Geological Theory in the Age of Lyell', *Perspectives in the History of Science and Technology*, Duane H. D. Roller (ed.) (Norman, Oklahoma, 1971), 209–27.
[13] See Leroy E. Page, 'John Playfair and Huttonian Catastrophism', *Actes de XIe Congrès international d'Histoire des Sciences* (1969), 4, 221–5.

the Roads had been the shore-line beaches of a lake. This at once explained their absolute horizontality. Moreover, their form was closely similar to the shore-lines of modern lakes (Loch Rannoch was given as an example); and the steepening above and below each terrace could be attributed to slight erosion on the landward side of the beach and deposition of material on the side towards the lake. Furthermore, the broad terrace at the level of Road R3 at one point in Glen Roy could then be interpreted as a *delta*, situated in an appropriate position where a stream would have entered the putative lake.

But this hypothesis involved a formidable difficulty, which MacCulloch fully admitted. It required a barrier or barriers to contain the lake or lakes at the levels recorded by the roads. Nevertheless, "it is proper", he said, "to commence by assuming the least difficulties, assigning no more causes than are strictly necessary for the production of the desired effect." This principle of economy led him to try postulating only a single lake, with three or four successive shore-lines. (The minor Roads would then be explicable as the traces of briefer pauses in the intermittent lowering of the lake-level.) MacCulloch was clearly concerned to 'economise' on the number of necessary partial ruptures of the barrier, for he suggested (not unreasonably) that the apparent twelve-foot difference in height between Road G (in Glen Gluoy) and Road R1 (in Glen Roy) might be due to instrumental error. If they were regarded as identical in height only two partial ruptures of the barrier would be necessary, before the final draining of the lake.

At least *three* barriers had to be postulated in order to contain this lake over the extent indicated by the Roads. Two would be needed to block the lower ends of Glen Spean and Glen Gluoy, or alternatively two to block both ends of the Great Glen to which they are tributary. At least one further barrier would be needed beyond the head of Glen Spean, *i.e.*, somewhere in Glen Spey, to account for the Road S in Glen Spean and for the reported extension of Road R1 into the head of Glen Spey (see Figure 1).

Whatever the precise positions of the barriers there were further difficulties about their origin, composition and ultimate removal, for there were no obvious traces of any such barriers. MacCulloch tried to minimize the difficulties by postulating that only one of the barriers had, by repeated partial ruptures, been responsible for the successive levels of the lake and for its final draining. The disappearance of the other barriers, "if by any causes of a nature similar to those which we see in daily action", was then attributed to river erosion (he seems to have been thinking in terms of barriers of rock débris). Yet there was an obvious difficulty about postulating the complete removal of the barriers during the period subsequent to the lake, for during this same period the Roads, although exposed to erosion, had suffered remarkably little and were extremely well preserved. (He thought the Roads were of "very high antiquity" and therefore that erosion of the land surface must be very slow). In the end he said he thought it was "fruitless to speculate" on the nature and cause of removal of the barriers. This then was one important point at which he was content to leave his interpretation incomplete.

A further problem which he left unexplained was the curious distribution of the Roads. If the same three or more barriers had been in existence throughout, it might be expected that all three principal Roads (accepting the identity of G and R1) would appear on *all* the hillsides within the area enclosed. At the very least, R1 and R2 should have extended above R3/S throughout Glen Spean (assuming Glen Gluoy had been cut off from the rest after the first fall in level). This objection was left unsolved, though MacCulloch was aware of it.

MacCulloch's whole hypothesis was thus incomplete and imperfect; but despite the aspects left in obscurity he felt it deserved acceptance "for the present" as the most probable explanation available.

A fourth hypothesis was mentioned at the very end of the paper, and may have been suggested to him after the original version had been read. This was that the Roads were the successive shore-lines, not of a lake but of the sea. Although MacCulloch discussed it only briefly, this hypothesis is of great importance as the one that Darwin was to develop later. MacCulloch thought it was "undoubtedly possible", but that it, too, involved great difficulties. In fact he mentioned only one difficulty, namely the absence of marine fossils among the pebbles and other débris on the Roads. He did not mention the greatest explanatory virtue of this hypothesis, that it completely eliminated all the difficulties about barriers that he had wrestled with so inconclusively. Perhaps this was because he saw that it would simultaneously aggravate another difficulty, namely the curiously local distribution of the Roads. In any case it is significant that he assumed that on this hypothesis it would have been the sea that had been lowered in level intermittently, not the land that had been elevated.[14]

[14] This was in line with the highly 'stabilist' attitude that was general in geological theorizing at this time. Large-scale physical changes in the earth's surface features were readily suggested, but they were generally explained in terms of eustatic movements, *i.e.*, changes in sea-level relative to unmoved land-masses. This is in sharp contrast to the attitude that was becoming common

2. *The barriers located.* Thomas Dick Lauder visited Lochaber and wrote his paper[15] at almost the same time as MacCulloch. Although it was published later there is little reason to doubt his assertion that his work was independent.[16] His 'factual' description was in most respects identical. It was followed by a general statement of "infinitely the most probable" hypothesis, the only one that Lauder considered to be worth discussing, namely that the Roads were shore-lines of a former lake. Like MacCulloch, he noted the similarity between the Roads and the shore-lines of modern lakes (he mentioned that he studied the shores of Loch Aven in the Cairngorms *before* he visited Lochaber). Lake Subiaco near Rome was cited as an analogical example of a former lake with similar shore-lines, which had been drained since classical times by the breaching of its natural barrier of rock.

The Roads were then described in detail, the comments being clearly directed towards the support of the lake hypothesis. Lauder was sceptical about the reality of the 'minor' Roads, since the hillsides are covered with roughly horizontal sheep-tracks and other features that often look shelf-like from a distance: he issued a "caution to future observers, not to decide too hastily as to such faint appearances . . . For, aided by fancy, which is always alive in an investigation of this kind, the eye is very apt to lead the judgement into error." The caution was to be much needed in later studies of the Roads: in this respect, even the most basic 'facts' were to prove elusive. Lauder doubted the authenticity of the minor Road or Roads in Glen Gluoy, and also the similar "pseudo-shelves" in Glen Roy, which he said were not horizontal. This reduced the explanatory task to the four principal Roads.

No "marine exuviae", *e.g.*, sea-shells, had been found anywhere in

by the time of Darwin's work, which was prepared to envisage large-scale tectonic movements of the land-masses themselves.

[15] Thomas Lauder Dick, 'On the Parallel Roads of Lochaber', *Transactions of the Royal Society of Edinburgh*, 9 (1821), 1–64, Pls. 1–7. Sir Thomas Dick Lauder (1784–1848) was the author of successful Scottish historical novels in the manner of Walter Scott, and of works on Scottish topography. He also published several scientific papers. He was a popular and influential Whig, an active promoter of schemes for educational reform, and in his later years Secretary of the Board of Scottish Manufactures.

[16] He first visited Glen Roy in 1816, and returned in June 1817, accompanied by a surveyor, to make a full study of the Roads. He could have heard of MacCulloch's paper, which had been read in London in January 1817, before his fieldwork. More probably, he heard of it, or even saw it, later in 1817, and therefore attempted to establish the independence of his own work by outlining it in a "few hasty remarks" to the Royal Society of Edinburgh that winter. The full version of his paper was read in March 1818 but not published until 1821. A postscript Note (pp. 62–4) asserts that it was written in 1817 and that he did not read MacCulloch's paper until January 1818.

Lochaber, not even in the cuttings made for the Caledonian Canal through the Great Glen (which, being only a little above present sea level, would be most likely to contain marine fossils if there had been a period of higher relative sea level). Like MacCulloch, Lauder evidently felt that this was sufficient to rule out the marine hypothesis.

The most important new element introduced by Lauder was the recognition of three cases of coincidence between the levels of the Roads and the heights of certain cols separating adjacent valleys (Figure 2). He maintained, on the basis of careful levelling by a civil engineer who had accompanied him, (i) that Road G was exactly on a level with the col (Col G) at the head of Glen Gluoy (into Glen Roy), and that the twelve-foot difference between this Road and Road R1 was genuine and *not*, as MacCulloch had suggested, due to instrumental error; (ii) that Road R1 was exactly on a level with the col (Col R1) at the head of Glen Roy (into Glen Spey), and *not*, as MacCulloch had reported, well above that col; and (iii) that Road R3/S was on a level with the low col (Pass of Muckul; Col S) at the head of Glen Spean (into Glen Spey). He did not, however, state these observations in the form of a *rule* of coincidence; had he done so, he would have had to note that Road R2 was an apparent exception to the rule, for he did not find any col on that level (Figure 3.) This anomaly became, as we shall see, one of Darwin's major explicit objections in the lake hypothesis.

Lauder then elaborated the lake hypothesis to take account of his particular observations. Like MacCulloch he had to face the difficulties of locating and accounting for the barriers that must have dammed the lake or lakes, but his task was substantially easier in two respects. The three cols coincident in level with three of the Roads were interpreted as the *sills* or overflows by which the lakes had been drained at certain times: this not only explained the coincidence, but also eliminated the need to postulate the barrier or barriers to the east, which had been necessary in MacCulloch's interpretation. Nevertheless, Lauder still found it necessary to invoke no less than three barriers. An "immense bulwark or barrier" was required across lower Glen Gluoy (Barrier G), to dam a deep "Loch Gluoy": this had formed a shore-line at Road G and had overflowed by Col G into Glen Roy. Another vast barrier (Barrier S) was needed across lower Glen Spean, to dam a long "Loch Spean" extending into Glen Roy and Glen Treig: this had formed a shore-line at Road R3/S and had overflowed by the Pass of Muckul (Col S) more than twenty miles away. These two barriers were relatively straightforward.

Darwin and Glen Roy

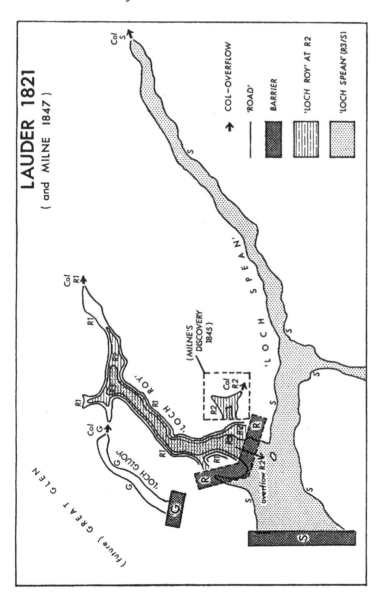

Fig. 2. Map to illustrate Lauder's lake hypothesis. Note the three presumptive barriers impounding lakes, placed at the points where the Roads end "abruptly", barrier R being more temporary than the others. Note also the anomaly that "Loch Roy" at level R2 had no known col-overflow, and that Lauder *failed to see* the lateral valley (Glen Glaster) at the head of which Milne subsequently discovered the missing col. Barriers are depicted in purely diagramatic form. (This and subsequent maps are based partly on the maps and other diagrams published at the time, and partly on the texts of the papers.)

The third was more problematical, and Lauder ran into difficulties in explaining its history. It was the barrier (Barrier R) required across Lower Glen Roy, to dam a "Loch Roy". At first this lake had formed a shore-line at Road R1, overflowing by the col (Col R1) at the head of Glen Roy into Glen Spey. The barrier had to be situated near the mouth of Glen Roy in order to explain why Road R1 ended "abruptly" at that point; but later in the sequence of events it had to be eliminated completely, so that a unified "Loch Spean" could form a single shore-line at Road R3/S ex-

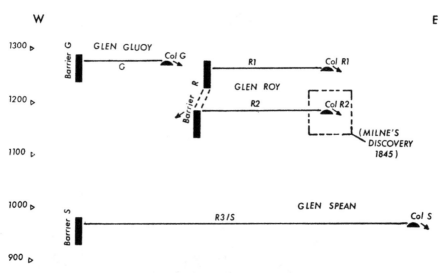

Fig. 3. Diagram to illustrate Lauder's lake hypothesis, showing the heights of the Roads above present sea-level (by early nineteenth-century measurements) against their relative lateral extent (horizontal scale arbitrary). Note the col-overflows coincident in level with three of them, and the anomalous level R2 draining over the barrier (R) itself (later replaced by the overflow by Col R2 discovered by Milne).

tending past the site of the erstwhile barrier. Furthermore, the anomalous Road R2, for which no col-overflow could be found, but which ended abruptly at about the same point, had to be explained in terms of a partial rupture of this barrier, lowering the lake to a level at which it had overflowed across the barrier itself (the same mechanism as that invoked by MacCulloch). Lauder was clearly uneasy at this point. On his hypothesis the other three Roads were the shore-lines of lakes that had had reversed drainage and overflows over rocky sills, so that long stationary lake-levels were not implausible. At the level of Road R2, on the other hand, the overflow would have been over the barrier itself, which *ex hypothesi* had to be composed of loose débris (in order to explain its complete and relatively rapid removal). This would be very unlikely to maintain a uniform level for long, since the outflowing stream would quickly erode the loose débris to a lower level. Lauder was therefore forced to argue that the level might not have been at Road R2 for long; but then that left unexplained the striking uniformity of Roads R1, R2 and R3, which (as MacCulloch had pointed out) naturally suggested uniform causes for all three.

Finally, Lauder had to explain the ultimate disappearance of both "Loch Gluoy" (which he thought had persisted throughout the changes in the other valleys) and the enlarged "Loch Spean". In other words, the two barriers damming these lakes had to be eliminated. He assumed without supporting argument that the two barriers (G and S) had been ruptured simultaneously, and he identified this event with the "extraordinary and powerful convulsion" that he thought had opened the Great Glen right across Scotland. The "terrible convulsions" that must have accompanied this event would have weakened the barriers, and the weight of water behind them would then have broken them "almost instantly".

If he had simply suggested the weakening of the barriers by one or more earthquakes, this relatively paroxysmal interpretation would in fact have been respectably actualistic. But even so it would have left his reconstruction with the unsatisfactory feature that the rupture of the "Loch Roy" barrier (R) had been somehow accomplished at an earlier time without affecting the others. It is not clear why he correlated the presumptive earthquakes with so gratuitously 'catastrophic' an event as the opening of a previously non-existent Great Glen, since in all other respects he was clearly concerned to interpret the phenomena actualistically. However, he did not leave the barriers totally hypothetical: he thought that the barrier (S) to "Loch Spean", at least, had left substantial remains in the hills of loose débris (in modern terms, of glacial morainic material) at the foot of Glen Spean. (He also postulated that a string of three small lakes had remained in lower Glen Spean after the rupture of Barrier S, and had later become silted up, leaving patches of flat ground on the valley floor.)

III

The Marine Hypothesis

1. *Darwin in Lochaber.* Before he left on the *Beagle* voyage, Darwin had a season of geological training from Adam Sedgwick, one of the best field-

geologists in Britain.[17] He also took with him the first volume of Lyell's *Principles of Geology*, and was sent the two subsequent volumes in South America. The scientific value of his letters to Henslow from South America was regarded highly enough for extracts, mostly on geology, to be printed and circulated in his absence.[18] But it is unlikely that he was aware of the Glen Roy problem in any detail before his return from the voyage. Probably his first knowledge of it came from a passing allusion in Lyell's book—an allusion of great significance for Darwin's later work.

In the third volume of the *Principles*, which Darwin received at Valparaiso in 1834, Lyell commented on the 'Parallel Roads of Coquimbo' in Chile and explicitly compared them with those of Glen Roy.[19] Basil Hall, who had first described these Chilean 'Roads', had interpreted them as lake beaches,[20] by analogy with MacCulloch's and Lauder's hypothesis for Glen Roy. Lyell felt unsatisfied by this, however, and on further enquiry discovered that Hall had privately had doubts about his interpretation and that he had been inclined to regard the Chilean Roads as marine in origin.

In May 1835 Darwin was able to study the Roads of Coquimbo for himself.[21] There were five or six major terraces, extending many miles up a valley eastwards from the coast. In addition, he noted some minor "intermediate" terraces. He was in no doubt about their marine origin, not only because they extended to the present coast-line but also because in places marine shells were abundant on them. The highest terrace at Coquimbo was 360 feet above sea-level; when traced inland up the valley they all rose gradually in level.[22] This accorded with Darwin's hypothesis,

---

[17] *The Life and Letters of the Reverend Adam Sedgwick*, J. W. Clark and T. McK. Hughes (eds.) (Cambridge, 1890), **1**, 379–81; Paul H. Barrett, 'The Sedgwick-Darwin Geologic Tour of North Wales', *Proceedings of the American Philosophical Society*, **118** (1974) 146–64.

[18] C. Darwin, *Extracts from Letters addressed to Professor Henslow* (Cambridge, 1835); 'Geological notes made during a survey of the East and West Coasts of South America . . .', *Proceedings of the Geological Society of London*, **2**, No. 42 (1835), 210–12.

[19] Charles Lyell, *Principles of Geology, being an Attempt to explain the former Changes of the Earth's Surface, by reference to Causes now in Operation*, **3** (London, 1833), 131–2.

[20] Basil Hall, *Extracts from a Journal, written on the coasts of Chili, Peru and Mexico, in the years 1820, 1821, 1822* (London, 1824), **2**, 6ff.

[21] His copy of the third volume of the *Principles* is now in the University Library, Cambridge. The scarcity of marginalia should not be taken to indicate lack of use, as the example given here shows.

[22] University Library, Cambridge, Darwin MS, 37 (i), "Notes on the geology of the places visited during the voyage, Part IV", p. 182. He attributed the terraces to "at least five distinct efforts" of the elevatory force, which had raised the land by an average of seventy-two feet at a time. Later, in working over these Notes, he put a question-mark beside "five distinct efforts", and wrote in the margin, "Reconsider this. Mem: Patagonia", where he attributed elevation solely to a gentle movement without earthquakes.

which he was formulating at this time, that the elevation of continents occurred along linear "axes of elevation", one of which was presumed to lie some way inland, parallel to the coast.

The Coquimbo Roads were only one of many phenomena that Darwin attributed to the gradual (*i.e.*, stepwise) elevation of the continent from beneath the sea. He had earlier found a series of broader terraces up to 1300 feet above present sea-level, which he interpreted as evidence of the geologically recent emergence of the whole of the Chilean coast-line. His investigation of the results of the great 1822 earthquake, and his first-hand experience of that of 1835, convinced him that the elevation of the Andes was still continuing.[23]

In this context, Lyell's passing reference must surely have pre-disposed Darwin to follow him in regarding the Glen Roy Roads as a direct analogy to those in Chile, and hence to adopt a marine hypothesis to account for them. It would have been clear to Darwin—as to any geologist who read the two published accounts with care—that the lake hypothesis was not entirely satisfactory in either version. MacCulloch and Lauder were agreed that the Roads were shore-line beaches. In the complete absence of marine fossils, they both went further in inferring that the beaches must have been *lake* beaches. They were therefore forced to invoke barriers to contain the lake waters; yet the nature, position, origin and removal of these barriers were highly problematical. The Roads clearly constituted an unsolved problem. It is hardly surprising that Darwin should have suspected that the mysterious barriers had never existed at all. His vivid experiences in South America would have been sufficient to ensure that when he saw Mac-Culloch's and Lauder's excellent drawings, Darwin would immediately have seen the Roads *as* marine beaches.

This much is a reasonable inference from what we know of Darwin's work at this time. The only documentary evidence of his views on Glen Roy *before* his visit is a MS note (published here as Appendix A) entitled "Chief Points to be Attended to." Although undated, its contents leave no doubt that it was an 'agenda', written after reading MacCulloch and Lauder, but before his own visit to Lochaber. It is of great interest as providing direct confirmation of what I have already inferred as probable on general grounds, namely that he expected to find evidence of the *marine* origin of the Roads.

His agenda is brief and in note form, but the points listed show that before setting foot in the area he had a remarkable grasp of the crucial

[23] *Op. cit.* note 10.

features that might help either positively to support his marine hypothesis or negatively to undermine the lake hypothesis. He would search for fossil barnacles and adherent tube-worms on the rocks that—on his hypothesis—would have been situated in intertidal positions.[24] He would tackle "The great problem, why lines [are] absent in other parts": that is, why the Roads were so curiously localized if the elevation of the land had been general to the whole area. This would require a study of "The relative preservation of the shelves", to see whether he could explain their preservation in some areas and non-preservation in other parts on the same level. More particularly he would need to look at the allegedly "abrupt termination of shelves" (Roads R1 and R2) at the outlet of Glen Roy: if their non-extension beyond this point was due simply to differential preservation, a more gradual 'dying away' might be expected. If the Roads had been formed during pauses in an intermittent elevation of the land, the minor Roads could be attributed to shorter pauses: it was therefore in the interests of the marine hypothesis to establish the reality of as many minor Roads as possible (notwithstanding Lauder's note of caution about accepting them). If the Roads were marine beaches, the cols coinciding with them could not be overflow sills, as Lauder maintained, but would need some alternative explanation: accordingly, Darwin noted that he would have to check whether the coincidence of heights was genuine. He would also need to see whether there was any "lip of escape" where Lauder had failed to find one, on the level of Road R2: since the rule of "col-coincidence" appeared to fail altogether in one of the four instances, it might perhaps be an illusion in the others. With these and other items on his agenda, Darwin travelled to Scotland in June 1838 fully prepared, if possible, to establish the marine hypothesis.

His field notebook[25] shows that he was indeed testing the phenomena against the two rival hypotheses even before he reached the Lochaber region.[26] He spent eight days in Lochaber; he studied Glen Roy and

---

[24] He had already used barnacles successfully as indicators of elevation in: Charles Darwin. 'Observations of proofs of recent elevation on the coast of Chili, made during the survey of His Majesty's ship *Beagle*, commanded by Capt. Fitzroy, R.N.', *Proceedings of the Geological Society of London*, **2**, No. 48 (1837), 446–9.

[25] Glen Roy Notebook, 1838 (University Library, Cambridge, Darwin MS, 130). This notebook is very small, and the notes are rough, faint, and often difficult to decipher even by the usual standards of Darwin manuscripts. Though mainly geological, the notebook also contains a few isolated observations relating to animal breeding. I am indebted to the Syndics of the University Library for permission to quote from this and other Darwin manuscripts.

[26] For example, he wrote, of irregular hillocks of débris (in modern terms, of glacial origin) near Tyndrum, south of Rannoch Moor: "Rivers could not have deposited it. Barrier of lake very lofty and no trace of it: to the sea more probable."

lower Glen Spean (as far as the mouth of Glen Trieg) fairly thoroughly; but he saw Glen Gluoy only from a distance (from Ben Erin) and he did not visit the Pass of Muckul (Col S), the channel at the head of Glen Spean which was to be important in the later debates. He said many years later that "for my first two days I was a convert to the Lacustrine theory"[27] and that MacCulloch's and Lauder's conclusions were "impregnable", but his notebook shows few signs of any such wavering.[28] Even his descriptive language, written on the spot and for no eyes but his own, is slanted towards a marine interpretation of the features he was observing. For example, he used the terms "flat-bottomed strait" and "tidal channel" for the col between Glen Collarig and Glen Roy, and "tidal plain" and "isthmus" for other cols or areas of level ground.

Nevertheless, even if he scarcely wavered in his conviction that his marine hypothesis was correct, he may have found the contrary indications unexpectedly strong. In particular, contrary to his experience at Coquimbo, Darwin could find no trace whatever of marine fossils on the Scottish Roads, although (unlike earlier investigators) he was certainly searching for them with particular care. In fact, he admitted the problem to Lyell soon after his visit to Lochaber:

"I have fully convinced myself (after some doubting at first) that the shelves are sea-beaches, although I could not find a trace of a shell; and I think I can explain away most, if not all, the difficulties."

It was candid of him to admit that it had been an effort to retain the hypothesis he had gone there hoping to prove, and that the difficulties were such as to need explaining *away*. Perhaps surprisingly, for one who had quite recently returned from seeing some of the most spectacular geology in the world, Darwin told Lyell that Lochaber was

"far the most remarkable area I ever examined. . . . I can assure you Glen Roy has astonished me."[29]

[27] Darwin to Robert Chambers, undated letter (probably 1848). University Library, Cambridge, Darwin MSS, 50, 'Glen Roy notes and scraps'.

[28] For example, when he first entered the area, he wrote, of the "buttresses" below Road S in Lower Glen Spean: "In all cases [*inserted*: "I urge"] deposition marine—because if not chain of lake[s] and if so there would be barriers." Noting the flat patches on the floor of Glen Spean (Lauder's residual lakes), he wrote, "NB lake gradually draining off would form plains such as those near Bridge [of] Roy (and other cases) but then if gradually drained where is barrier." Of the lateral ending of the Roads near the mouth of Glen Roy, he wrote, "Shores die away only where slopes less, best developed on *steep earthy* slope, two circumstances rarely united.—die away also without cause, must be tides etc." One of the few possible moments of wavering I have noticed is: of certain areas of flat ground where Glen Turrit joins Glen Roy, "From this point plain appears like one uniform slope slightly bending up each main valley, and that river alone had modified it—perhaps however sea also." (See also second quotation in note 30.)

[29] Darwin to Lyell, 9 August [1838]. American Philosophical Society, Darwin MSS; *LLCD*, **1**, 293.

Darwin's notebook also provides direct confirmation that while in the area he was actively drawing analogies between what he saw in Scotland and what he had seen in South America.[30] At the end of the notebook, and apparently written after he had finished his fieldwork, he reminded himself to "Speculate on Beagle Channel";[31] and it is clear that in writing up his work for publication he did indeed relate the landscape of Lochaber, with the sea restored in his mind's eye, to the labyrinthine tidal channels that the *Beagle* had negotiated in South America (see Figure 4). The Great Glen itself, filled with an arm of the sea, would have been exactly like the Beagle Channel. Of a spot near the mouth of Glen Trieg (on the level of Road R3/S), he wrote:

"It required little imagination to go back to former ages, and to behold the water eddying and splashing against the steep rocks."

And as he summarized the matter in his paper, in language of almost biblical cadence:

"Whoever walks over these mountains, and believes that each part has been successively occupied by the subsiding waters of the sea, will understand many trifling appearances, which otherwise, I believe, are unintelligible."

2. *The marine hypothesis proposed.* Like MacCulloch and Lauder, Darwin began his paper[32] with a descriptive section (I). Much of this was uncontroversial, but theory inevitably seeped into the way he described the phenomena. Lauder's col-overflows were re-named "flat land-straits" and said to be "precisely what might be expected from *straits*" of the sea. Darwin noted the three cols concerned, and commented that "the coincidence of land must be ultimately connected with the origin of the shelves"—as on Lauder's interpretation it was; but he immediately undermined the force of this connection by pointing out that Road R2 was an anomaly and had no known col on its level, so that "such relation is not absolutely necessary" (Figure 5). He reported an "obscure line of terrace" above and beyond the col (R1) at the head of Glen Roy, and thought this might have misled MacCulloch into thinking that Road R1 passed over the col at this height.

---

[30] For example, he wrote, of the non-extension of coastal raised beaches (*not* Roads) into Glencoe: "Mem Coast of Chile—? is not Mica Slate too hard and uneven to be impressed." Noting the "buttresses" of débris at Roy Bridge, he wrote (with a possibly wavering moment about his hypothesis) "lake required to deposit this Remember however the first Chilian Valley [place name illegible], must have deposited much—on other hand remember Modelling power of sea N of Valparaiso."

[31] The first words of a brief memorandum written in ink.

[32] Charles Darwin, 'Observations on the Parallel Roads of Glen Roy, and of other parts of Lochaber, with an attempt to prove that they are of marine origin', *Philosophical Transactions of the Royal Society*, vol. for 1839, 39–81.

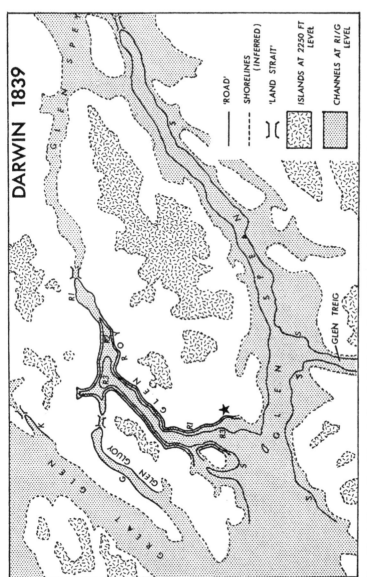

Fig. 4. Map to illustrate Darwin's marine hypothesis. The lateral extent of the principal Roads was accepted without change from Lauder, but they were given a radically different interpretation (of Darwin's minor Roads, only that near Kilfinnin is shown here). The stippling shows (as an example) the marine channels that Darwin envisaged at Level R1/G, illustrating his argument that the Roads were shore-lines formed (or preserved) only in specific circumstances (in protected cul-de-sac channels). Lauder's col-overflows were re-interpreted as tidal shallows ("landstraits") separating certain channels. Darwin too failed to *see* Glen Glaster (note the Roads scarcely indented at the point marked with a star, and compare with Figure 2 or 8). The 'islands at 2250-feet level' are to illustrate the geography that Darwin envisaged at the (earlier) period at which the highest erratic boulders had (he believed) been rafted on icebergs into their present position at this altitude.

He himself interpreted it as a minor Road (R_1^+) *above* the col and Road R_1, which therefore supported his belief that the sea had stood at even higher levels than the highest principal Road. This was a good example of the features, "trifling" in themselves, that he said would be intelligible on the marine hypothesis. Since that hypothesis was strengthened by every new Road discovered anywhere in the area, Darwin was evidently pleased to have discovered a completely new Road (K) in a short valley above

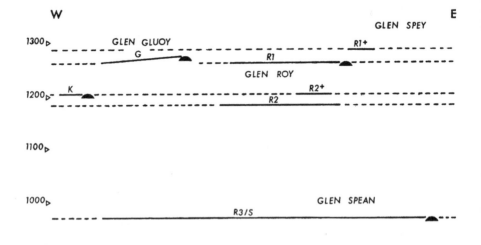

Fig. 5. Diagram to illustrate Darwin's marine hypothesis, for comparison with Figure 3. The Roads (continuous lines) are shown as localized traces of successive shore-lines (dashed lines) that existed at the time throughout the area. Note Darwin's explanation of the small vertical difference between Roads G and R_1 as due to tidal rise up the less sheltered Glen Gluoy. Note also how his minor Roads increased the number of successive shore-line traces, and effectively reduced the frequency of association between Roads and "land-straits" (shown here by black 'hump' symbols).

Kilfinnin in the Great Glen. Its very "want of continuity and shortness"—which might have given Lauder grounds for doubting its reality—was characteristically turned by Darwin into an advantage: the same causes as in Glen Roy, operating in a different situation, had produced little effect, so that in still other circumstances "all evidence would have been obliterated."

This ingenious way of making an explanatory virtue out of observational necessity was further developed in the next section (II) of the paper, which was ostensibly a critique of MacCulloch's and Lauder's hypotheses. The disappearance of the Roads at their outer ends—particularly that of

Road R1 and R2 near the outlet of Glen Roy—had been characterized by Lauder as "abrupt", and this had led him to reconstruct barrier R in that position. To Darwin, on the contrary, their "disappearance is so extremely gradual" that another explanation was needed.[33] Darwin argued that the circumstances unfavourable to their formation or preservation had been the widening of the valley at this point, corresponding to greater exposure of the shores of the former marine "sound" (Figure 4). (He also noted the curious fact that Road R1 faded away slightly further up the valley than Road R2 below it, though this was not explained satisfactorily until many years later.) Lauder's barrier R at this point was therefore criticized both on account of the *gradual* fading of the Roads that (on Lauder's reconstruction) had been enclosed by it, and on account of its complete removal before the formation of the lowest Road, R3/S, across its site; "more convincing proof of the non-existence of the imaginary Loch Roy, could scarcely have been invented." Darwin argued that, with similar criticisms applicable to Lauder's other lakes, one was driven towards MacCulloch's hypothesis of a *single* lake, only to find still greater difficulties. Three or four barriers were needed, of unexplained composition, formation and removal; and the col-coincidences were left a complete puzzle. By this manoeuvre of elimination, Darwin maintained that "the conclusion is inevitable", that *no* theory involving barriers, and therefore no theory of lakes, was possible. Here was the operation of the "principle of exclusion" that he was later to regret.

Having cleared the ground of competing hypotheses, Darwin then began the development of the marine hypothesis in earnest. His main "proofs" of the existence of the sea in the valleys of Lochaber (Section III) were however nothing to do with the Roads. He called attention to the vast accumulation of "alluvial" gravels and other débris (in modern terms, glacial outwash deposits and morainic till), often forming flat-topped "buttresses" below the level of Road S, in Glen Spean. He attributed them to deposition by rapid currents, but not by any "débâcle." Through them the river Roy has cut a gorge, which most observers would have attributed to a period *after* the deposition of the "alluvial" material. But this reversal of the action of the river (*i.e.*, first depositing, later eroding) seemed unsatisfactory to Darwin. Characteristically he sought to explain the two apparently opposite phenomena as effects of a single continuous cause. He

---

[33] To a modern eye, both would seem correct, depending on the scale chosen: on the scale of the whole Road-system their disappearance *is* abrupt, but following them on the ground they certainly fade away gradually.

argued that if the valley had been filled with an arm of the sea, into which the river brought its load of débris, and if the relative level of the sea was gradually falling (*i.e.*, in a succession of short steps), the observed effects would be explained. The river would have deposited its débris in deltaic form on entering the arm of the sea; but as the sea-level fell, it would have eroded a gorge through these deposits and simultaneously deposited more material further downstream and at the new, lower, level of the sea. In time, a continuous tract of deposits, with "buttresses" on many different levels, would have been formed all the way down the valley, cut through by an equally continuous gorge. This hypothesis depended on an analogy between the terraces extending at certain points outwards from the level of Road R3/S, which were agreed to be deltas, and these (much less regular) "buttresses" at lower levels. But Darwin, believing the analogy was valid, saw no reason to reject the deltaic explanation as the "*vera causa*" for the "buttresses". With this interpretation, he thought he had clear evidence of a gradual fall in the level of the water in Glen Spean, from Road S down towards present sea-level (Figure 6). He concluded with characteristic confidence "that it is satisfactorily proved" that Glen Spean had contained a gradually retreating body of water. Equally characteristically, he immediately extrapolated this conclusion, asserting that it could "be urged with only a little less force" for *all* the valleys in the area.

This interpretation was also supported by a brief reference to the intermediate (minor) Roads, which Darwin firmly believed were genuine, and which also suggested a slowly and almost continuously falling relative sea-level. Without the marine hypothesis, one would be forced into the "startling assumption" of a mysterious barrier; and to invoke a barrier for each valley was nothing short of "monstrous", because no trace of them could be seen. Having again dismissed the lake hypothesis in these emotive terms, Darwin concluded that there was "but one alternative, which we are compelled to admit", namely that the gradual retreating body of water had been the *sea*.

The next section of the argument (IV) cited collateral evidence of the plausibility of the marine hypothesis. There was good fossil evidence, recently discovered,[34] to prove that elsewhere in Scotland the sea had stood in geological recent times at about 350 feet above the present sea-level: on Darwin's hypothesis this would represent a relatively late stage in the much longer process of emergence that he was postulating on the basis of evidence

[34] James Smith, 'On the Last Changes in the relative Levels of the Land and Sea in the British Islands', *Edinburgh new Philosophical Journal*, 25 (1838), 378–94.

in Lochaber. He pointed out that even at a 350-foot sea-level the Great Glen would have been an "open strait" like the Beagle Channel in Patagonia (see Figure 1); and he asked, "may we not . . . deliberately affirm it proved" that the sea could have produced *all* the effects at *every* level in Lochaber. Furthermore, Lyell's recent work in Sweden[35] indicated

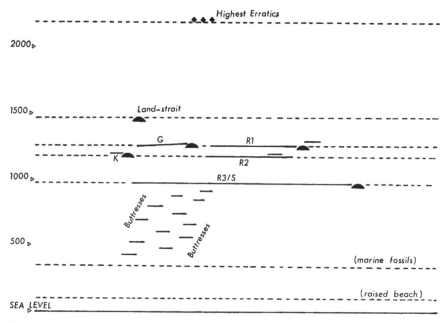

Fig. 6. Diagram for comparison with Figure 5, showing Darwin's assimilation of the Roads into a 'consilience of inductions' from other phenomena, all indicating formerly higher relative sea-levels. Successive levels (dashed lines) are shown for the principal Roads and for a selection of other features: the highest erratic boulders and highest isolated "land-strait" mentioned by Darwin; and the highest (geologically recent) marine shells and the highest unambiguous marine 'raised beach' known elsewhere in Scotland (*not* in Lochaber). Note how Darwin's "buttresses" (shown here at arbitrary levels) bridged the gap between the lowest Road (R3/S) and the highest uncontroversial evidence of higher sea-levels. Vertical scale in feet above present sea-level (on measurements accepted by Darwin).

an "*equably progressive* elevation" of the land there; and his own work in Chile, with his interpretation of the close causal connection between eleva- tion and vulcanism,[36] implied that the action of elevation was generally

[35] Charles Lyell, 'On the proofs of a gradual rising of the land in certain parts of Sweden', *Philosophical Transactions of the Royal Society*, vol. for 1835, 1–38; *Principles of Geology*, 4th edn. (London, 1835), **2**, 331–49.
[36] Charles Darwin, 'On the connexion of certain volcanic phenomena, and on the formation of mountain chains and volcanos, as the effects of continental elevations', *Proceedings of the Geological Society of London*, **2**, No. 56 (1838), 654–60.

intermittent and stepwise. Thus the effects predictable on this view of elevation were exactly those observed in Lochaber.

Having set out the case for the marine hypothesis in these very confident terms, Darwin now had to face the two most obvious objections to it (Section V). Both of them, significantly, obliged him to explain *away* the non-existence of certain classes of evidence.

The first objection was that of the "non-extension of the shelves". If the sea had stood successively at the various levels recorded by the Roads, it might have been anticipated that each of the Roads would be found in every valley in Lochaber—and indeed throughout Scotland—along the relevant contours. But they were not. This objection could have been raised in more limited form (*i.e.*, for the valleys within Lochaber) against MacCulloch, whereas Lauder had avoided it completely by postulating that the various lakes had been co-extensive with the corresponding Roads. On Darwin's hypothesis the localized distribution of the Roads had to be explained in terms of their preservation (or formation) in certain valleys, and non-preservation (or non-formation) elsewhere; and he admitted that this was "a very extraordinary circumstance". Once again his South American experience provided him with what seemed a relevant analogy. Characteristically he inverted the norm of expectation by asserting

"that it would be more proper to consider the preservation of these ancient beaches as the anomaly, and their obliteration from meteoric agency as the ordinary course of nature."

Thus the burden of explanation was shifted from the non-existence of the Roads to their existence. His hypothesis here was that the Roads would only have formed in sheltered embayments, not in more open "straits" where the tides and currents would have been stronger. The only evidence brought in support of this was the questionable assertion that the Roads were more the product of deposition than of erosion, so that in less protected situations no beach of loose débris could have accumulated. (MacCulloch's careful, measured cross-sections of the Roads had in fact suggested that they had been formed as much by erosion on the landward side as by deposition on the other.) But with this hypothesis, Darwin was able to explain quite ingeniously the observed distribution of the Roads. Roads R1 and R2, for example, faded away at the mouth of Glen Roy because at this point the narrow sheltered arm of the sea in that valley would have opened out into the wider and more exposed channel occupying Glen Spean (Figure 4). When, however, the sea-level had fallen to the level of the Pass of Muckul (Col S) at the head of Glen Spean, this channel

X

would have become an almost closed and therefore more sheltered arm of
the sea; with the sea-level pausing at this point, beaches could thus accumulate not only in Glen Roy (Road R3) but also in Glen Spean itself
(Road S). (He failed, however, to comment on the fact that, on this type
of explanation, Road R1 and R2 should have formed not only in Glen Roy
but also in the equally narrow and sheltered Glen Treig.)

The "trifling" features of the minor Roads—of whose very existence
Lauder had been justly sceptical—were also brought in to support the
hypothesis. In particular, Darwin thought he had detected a faint terrace
(R2$^+$) intermediate in height between Roads R1 and R2, extending for
three-quarters of a mile along a hillside (Tombhran) near the head of Glen
Roy (Figure 5). Without any supporting argument, he asserted that it
was "an incontestable fact" that even so limited an intermediate Road
must represent almost as long a pause in the process of elevation as the
well-marked principal Roads. The highly-localized occurrence of this
minor Road was then characteristically turned to explanatory advantage,
for Darwin asserted that it proved that the extent of preservation of a
shore-line was no guide at all to its original extent. Indeed, it showed that
a very "slight difference of circumstances" could determine whether a
beach was fully preserved or completely obliterated. All arguments against
the marine hypothesis on the grounds that the beaches are only preserved
in certain places and not others were therefore "valueless".

The second major difficulty about the marine hypothesis, raised even by
MacCulloch, was the non-existence of any marine fossils anywhere in
Lochaber. Darwin conceded that "this may at first be thought a strong
objection", but then set himself to explain it away. He himself had not been
successful in finding any marine shells, barnacles or tubeworms anywhere,
although following his agenda he had doubtless looked very hard for them.
He circumvented the problem by a manoeuvre exactly analogous to the
preceding one: the preservation of fossils was the anomaly, their disappearance the norm. The work of Lyell, Murchison, and others elsewhere was
cited in support of this, and he thought it "easy to imagine" circumstances
that would determine the preservation of fossils in certain areas and their
disappearance in others. The acidic peaty groundwater in a region such
as Lochaber was suggested (not unreasonably) as a possible reason for the
absence of fossils on the Roads.[37] Such reasoning, Darwin maintained,

[37] Lyell had told him of the patchy preservation of shells on raised shore-lines in Scandinavia,
explicitly to help "the Glen Roy case". Lyell to Darwin, 6 September 1838, *Life, Letters and
Journals of Sir Charles Lyell, Bart.*, Mrs. Lyell (ed.) (London, 1881), **2**, 46.

was "quite sufficient" to show that the absence of fossils was no insuperable objection to the marine hypothesis. "The proposed theory," he now said, "explains every essential point in the phenomenon of the parallel roads."

This remained to be shown, however, for a number of features not so far discussed (Section VI). Of these the most important were the three cols coincident in level with three of the Roads. From being highly significant evidence in Lauder's interpretation, as overflows from his lakes, they became in Darwin's scheme minor anomalies that required to be explained away. If the sea had gradually fallen, with pauses at many successive levels, why should some of those levels have had cols at exactly the same height connecting different arms of the sea? Darwin explained this by arguing that, with the sea-level stationary at any particular level, any gap that happened to be at approximately the same height would be eroded down or built up to *exactly* the same height, by tidal erosion or deposition, and kept open as flat tidal shallows, forming a strait between the arms of the sea on either side. This interpretation was reinforced (as already mentioned) by re-naming the cols by the theory-loaded term "land-straits". He failed to explain or even to mention the fact that in each case a Road had been formed (or preserved) only on *one* side of the "land-strait" and not on the other side. He was able, however, to point out that his interpretation was indirectly supported by the fact that the Roads were not invariably associated with a land-strait: Road R2 had no known col at the same height.

But while this anomaly seemed to be in Darwin's favour, there was another that required explaining away. The very small (twelve-foot) difference in height between Road G and Road R1 could no longer be regarded (as MacCulloch had done) as an illusion due to instrumental error. But on Darwin's hypothesis it seemed peculiar that the sea-level should have paused at the higher level, forming a beach (Road G) *only* in Glen Gluoy, and then sunk by a mere twelve feet and formed a beach (Road R1) *only* in Glen Roy. The local "circumstances" of physical geography, on which he relied to explain the patchy preservation of the Roads in general, could hardly be held to have changed much with such a minor fall in sea-level. Darwin conceded that two pauses at such a close interval were "improbable in the highest degree". But he immediately—and characteristically—undermined the force of this admission by adding that the localized preservation of other Roads showed that such a situation would not in any case falsify his hypothesis. But he evidently felt the problem strongly enough to try to suggest an alternative explanation. This was that the two Roads represented the same period of stationary sea-level, but that the beach had formed at a slightly higher level in Glen Gluoy that in Glen Roy, because Glen Gluoy had had more direct access to the open sea than Glen Roy (since the former opens directly into the Great Glen) and would therefore have had greater tidal rise. Once again, the channels around the Straits of Magellan were cited as analogies.

Darwin recapitulated his main argument at this point, and concluded that "the theory of the marine origin of the 'parallel roads of Lochaber' appears to me demonstrated." This was the statement whose confident assertion was later to embarrass him. I hope I have shown that it was no isolated or careless remark, but rather an epitome of an argument that is marked throughout by an almost aggressive confidence in his own hypothesis and a scornful dismissal of the alternatives.

This ended the main part of the paper, but four accessory topics were discussed after the main conclusion. The "erratic boulders" strewn over the hills of Lochaber, as elsewhere in Britain, clearly pre-dated the Roads (Section VII). Darwin thought the older "débâcle" explanation of erratics was full of difficulties, and he preferred Lyell's suggestion that the boulders had been transported on floating icebergs at a period when the climate of northern Europe had been somewhat colder. Since he believed the hills of Lochaber had been submerged below sea-level to at least the level of the highest Roads, Darwin was not disturbed at finding erratics even higher, up to 2200 feet above the present sea-level, at which height only the summits would have stood out as islands (see Figure 4). He was not unaware of the new and third alternative to account for erratics, namely the glacial theory of Venetz, Charpentier and Agassiz. He referred in a footnote to their early papers on the subject.[38] But he did not discuss their theory at all, and merely implied that the phenomena of striated bedrock surfaces, which they explained in terms of glacial movement, would be covered by Lyell's iceberg hypothesis. It was this failure to consider the full implications of the glacial theory that he was later to regard as having led him into an unwarranted trust in the "principle of exclusion".

Although he noted that geologists believed "the so-called 'erratic block period' is recent" in geological terms, his own hypothesis of a very slow elevation of the area implied that the Roads were nevertheless extremely ancient (Section VIII). The smallest features of these ancient beaches had remained almost perfectly preserved, "during a period which cannot be reckoned in thousands of years."[39] Their age could only be measured "by

[38] *Op. cit.* note 32, 72, footnote.
[39] It should perhaps be explained that here, as elsewhere in his work, Darwin's conjectural

those great revolutions of nature which are the effects of slow and scarcely sensible changes." But by assigning the Roads to stages on this *extremely* lengthy process of crustal elevation, he was thus forced into concluding that the erosive power of subaerial agents (rain, soil-creep, rivulets) was almost neglibible.[40] This implied that the significant forces in the sculpturing of the land were those of *marine* erosion—a conclusion to which Lyell had already turned, in a remarkable *volte-face* from his earlier emphasis on the power of subaerial erosion.[41] Such a view was relevant to the theme of the paper (though not stated explicitly), because it made Darwin see the present landscape of Lochaber even more vividly as an area that had been sculptured into its present form during its submergence beneath in the sea.

Another objection to Darwin's hypothesis, not mentioned previously in the paper, was the extremely precise horizontality of the Roads (Section IX). Was it possible, critics might ask, that this crustal block should have been elevated with such astonishing uniformity that no tilting could be detected within the limits of observational error? Darwin countered this by citing again Lyell's work in Sweden and his own in South America. Both of these had in fact involved gently tilted elevation, but he presumably thought that in Scotland the tilting might have been too slight to be detectable within the relatively small area of Lochaber. He therefore argued that the horizontality of the Roads, "marvellous though the fact be", could not be counted as evidence against his hypothesis. Indeed, by his characteristic move of making a virtue out of necessity, he argued that "on the contrary, a most important geological fact is established", namely that crustal elevation *can* be extremely "equable"!

As for the mechanism of such elevation, he admitted that this was to "enter on speculative grounds" (Section X). Rejecting the Playfairian hypothesis of expansion due to deep-seated heating, he outlined his own— highly Lyellian—hypothesis of the simultaneous elevation and subsidence

time-scale for earth-history was far *in excess* of even the modern radiometric time-scale. It was not based on any even roughly quantitative measurement, but only on the requirements of quite separate theoretical speculations—in this case, what he assumed must be an extremely slow rate of elevation. (On the modern interpretation the age of the Roads *can* be estimated in terms of the few thousand years of post-glacial history.)

[40] He wrote to Lyell: "At some future time I shall be extremely curious to talk over with you, the inferences about the small amount of Alluvial action [*i.e.*, subaerial erosion] in Lochaber, which has taken place, since the sea retired.—No one point of interested me more, & though it cost me no small effort to swallow the inference that I have given, I can see no sort of loop-hole to escape from the result." (Darwin to Lyell, [January 1839], American Philosophical Society, Darwin MS.)

[41] Charles Lyell and R. I. Murchison, 'On the excavation of valleys, as illustrated by the Volcanic rocks of Central France' *Edinburgh new Philosophical Journal*, 7 (1829) 15–48; compare Lyell *Principles of Geology*, 3 (1833), chs. 21, 22.

of the vast blocks into which he believed the earth's crust to be divided. Referring to the theoretical paper on the subject which he had read to the Geological Society only a few months before visiting Lochaber,[42] he explained how he had inferred that the subcrustal material must possess "tolerably perfect" fluidity. On this hypothesis, a crustal block would be somewhat like a tabular iceberg, and would indeed be expected to rise with perfect "equability". The absolute horizontality of ancient shore-lines on such a block would not therefore be surprising, and this feature of the Roads was, he argued, explained as satisfactorily as any other.

Perhaps the toughness of Darwin's explanatory problems with the Parallel Roads of Lochaber explains why he noted, after completing his paper on 6 September 1838, that it had been "one of the most difficult and instructive tasks I have ever engaged on".[43] He told Lyell that the "Glen Roy paper has lost me six weeks", which might be taken to mean that he regarded it as peripheral to the main work—on South American geology and especially the species problem—that was engaging him at the time. But in fact he immediately went on to retract the idea that the time had been "lost", because he pointed out that the concluding inference of the paper was of great theoretical importance. "I cannot doubt", he wrote, "that the molten matter beneath the earth's crust possesses a high degree of fluidity, almost like the sea beneath the polar ice."[44] I hope to show that the importance of this conclusion extended even beyond his geology and into his work on the theory of evolution.

Darwin submitted his paper to the Royal Society on 17 January 1839, and a week later was elected a Fellow. The paper was read on 7 February 1839. The Society then submitted the paper to Adam Sedgwick as a referee. (His report is published here as Appendix B.) Sedgwick made some constructive criticisms about the presentation of Darwin's paper, but declared himself completely convinced by Darwin's main argument, and regarded even the most theoretical parts as "ingenious" and worth publishing. He concluded that the paper contained "much original and important matter", and recommended that it should be published in the *Philosophical Transactions*. Darwin's "Observations on the Parallel Roads of Glen Roy . . ." was his first major scientific paper to appear in print (later in 1839), and the only one of his works that was ever published by the Royal Society.

[42] *Op. cit.* note 36.
[43] *LLCD*, 1, 290.
[44] Darwin to Lyell, 13 September [1838]. American Philosophical Society, Darwin MS; *LLCD*, 1, 297 (the word "polar" is here mis-transcribed as "block").

These details are not unimportant. At a period when the Society was trying self-consciously to reform itself by raising its scientific standards, Darwin's paper was scrutinized by one of the best geologists in England, and found convincing. Although Sedgwick himself had been responsible for initiating Darwin into the science of geology before he embarked on the *Beagle*, he might have been inclined, as a moderate 'catastrophist', to be sceptical about an argument as Lyellian as Darwin's. Yet in fact he reviewed it very fairly and declared himself persuaded by Darwin's marine hypothesis and—by implication—by the conclusion that the Highlands had been elevated gradually from beneath the sea. Thus an experienced and potentially sceptical referee evidently saw no glaring logical error in Darwin's paper.

Sedgwick was fully justified, however, in criticizing the style of Darwin's paper. Darwin told Lyell that every sentence was essential to the argument and that he hoped the Royal Society's editor would not insist on his shortening the paper.[45] In fact it is verbose and repetitive. This is quite understandable, however, since it was his first attempt to write a connected argument of some length and complexity for publication. The paper is also difficult to follow without a detailed knowledge of the geography of Lochaber and of the previously published literature. (It is for this reason that I have paraphrased its argument in some detail.) It is characteristic of Darwin's curiously underdeveloped visual sense that even the inclusion of a map and a general view of the Roads seems to have been an afterthought, and the map itself was virtually copied from Lauder's. Altogether, Darwin was right to admit in later years that the paper had been "obscure".

IV

The Glacial Lake Hypothesis

1. *The possibility of glacial lakes.* As already noted, Darwin was aware of the glacial theory when he wrote his paper. He knew that Charpentier had given wider currency to Venetz's earlier suggestion that the Alpine glaciers had extended much further in the geologically recent past, on the evidence of polished or scratched surfaces of bedrock, and the widely dispersed Alpine erratics. They had attributed this, however, not to a globally cooler climate but to a higher elevation of the Alps. Agassiz had seen that this

[45] Darwin to Lyell, January 1839. Copy in University Library, Cambridge, Darwin MS, 146.

explanation could not cover the similar erratics and scratched surfaces in northern Europe, far from any high mountains. In a frankly speculative essay he attributed all these effects to a geologically recent period of intense cold—the *Eiszeit*—which simultaneously explained the apparent mass extinctions of Tertiary faunas at the same period. Most of the northern hemisphere, in Agassiz's view, had been covered by a sheet of stationary ice; the strictly 'glacial' effects dated not from the 'Ice Age' itself, but from the subsequent period of warming climate, when moving glaciers had been formed from the *remnants* of the ice-sheets.[46]

Agassiz visited Britain in 1840, primarily in connection with his work on fossil fish. (His preoccupation with a highly speculative glacial theory was a digression which many of his scientific friends were embarrassed by and regretted.) Going to Scotland for the meeting of the British Association at Glasgow, he took the opportunity to tour the Highlands with William Buckland to search for corroborative evidence for the glacial theory. They found such evidence in abundance, and returned convinced that the Highlands had indeed been glaciated in the geologically recent past.[47]

Like Darwin, Agassiz consciously and explicitly *saw* the landscape of Lochaber through 'eyes' provided by his previous experience and by the hypothesis he was applying to the area. For Darwin, as we have seen, the image was the tidal channels of Patagonia and Tierra del Fuego: for Agassiz, it was the glaciated valleys around Chamonix. By a coincidence, their most vivid expressions of the way they 'saw' Lochaber related to almost the same place. Agassiz wrote after his visit:

"I shall never forget the impression I experienced at the sight of the terraced mounds of blocks at the mouth of the valley of Loch Treig, where it joins Glen Spean; it seemed to me as if I were looking at the numerous moraines of the neighbourhood of Tines, in the valley of Chamonix."[48]

Darwin at the same spot had seen, in his mind's eye, the former tidal channels, with "the water eddying and splashing against the steep rocks".

Agassiz's first reports on his tour of Scotland merely mentioned the bare fact that he considered the parallel Roads to have been due to glaciers damming the valleys and forming glacial lakes.[49] Nevertheless, this at

---

[46] M. J. S. Rudwick, 'The Glacial Theory', *History of Science*, 8, (1970), 136–57.

[47] G. L. Davies 'The Tour of the British Isles made by Louis Agassiz in 1840', *Annals of Science*, 24 (1968), 131–46; *The Earth in Decay: A History of British Geomorphology, 1578–1878* (London, 1969), 273–87.

[48] 'The Glacial Theory and its Recent Progress', *Edinburgh new Philosophical Journal*, 33 (1842), 271–83, at p. 222.

[49] Agassiz, 'Discovery of the former existence of glaciers in Scotland', *Scotsman*, 7 October 1840;

once complicated the issue by making it in effect a three-sided controversy: Darwin's falling sea-level *versus* Agassiz's glacial lakes *versus* Lauder's ordinary lakes. To anyone who had followed the Glen Roy papers to this point, it must have been obvious that Agassiz's hypothesis potentially lent new plausibility to MacCulloch's and Lauder's interpretation, since it indicated how their hypothetical "barriers" might have appeared and disappeared while leaving so little trace.

Darwin, however, told Lyell:

"I think I have thought over the whole case without prejudice, and remain firmly convinced they are marine beaches."

He cited several reasons for retaining his earlier conclusion. In particular, his ingenious marine interpretation of the "alluvial buttresses" in Glen Spean still seemed attractive, and his claim to a newly-discovered Road (K) above Kilfinnin added to the number of glacial barriers that would be required on Agassiz's hypothesis. But he thought the greatest objection to the glacial interpretation of the Roads related to the way in which the supposed glacial lakes had been drained. The lakes must have stood for a long time at the levels of the principal Roads in order for the beaches to accumulate. That was plausible if—adopting Lauder's interpretation—they had invariably drained over one of the flat cols that Darwin had renamed "land-straits". But Darwin argued that many such cols in the hills around Glen Roy did not coincide with any Road, and—most seriously—one Road (R2) was unconnected with any col. This was the most important anomaly: during the presumptively long time required to form this Road, the glacial lake in Glen Roy "*must* (for there is no other exit whatever) have drained *over* the glacier." It seemed inconceivable that the overflow could have remained constant in level under these circumstances.[50] (The same objection had been raised by Darwin against Lauder's inference that at the same stage the lake had drained over a barrier of *débris*.)

In a later letter, after reading what he termed Agassiz's "capital book" on present-day Alpine glaciers,[51] with its magnificent illustrations, Darwin admitted to Lyell:

'On Glaciers, and the evidence of their having existed in Scotland, Ireland and England', *Proceedings of the Geological Society of London*, 3, No. 72 (1840), 327–32.

[50] Darwin to Lyell [9 March 1841]. American Philosophical Society, Darwin MS; *MLCD*, 2, 173–4. Compare the following note by Darwin, probably written at this time: "Position of Boulders in Glen Roy shows not glaciers.—Middle shelf [*i.e.*, Road R2] must have been drained over glacier—preposterous ice so long at same level, as a stream flowing over rock [*i.e.*, as in cases of Roads R1 and R3/S]—and how great a time!! I anticipated glaciers—and now it seems they extended after [word illegible]." (University Library, Cambridge, Darwin MS 50, 'Glen Roy notes and scraps'.) The meaning of the final sentence is obscure.

[51] Louis Agassiz, *Études sur les Glaciers* (Neuchatel, 1840) 2 vols.

"I made one great oversight, as you would perceive. I forgot the glacier theory."[52]

This of course was the omission at the root of his ultimate repudiation of his original paper. He now modified his position by admitting that the outwash from the putative glacier from Ben Nevis might account for the "buttresses" in lower Glen Spean. But he felt that the anomalous Road R2, without an overflow col, still constituted an "insuperable" objection to the glacial lake interpretation. On the other side of the argument, he admitted that the limited extent of the Roads was still the "most obvious" objection to his marine hypothesis, but he thought that, since his own discovery of a minor intermediate Road (probably Road R2[+]) known nowhere else, this was really no longer a valid objection.[53]

Darwin had no reason in general to be reluctant in accepting Agassiz's reconstructions of glaciers in the highland areas of Britain. Agassiz's arguments for the former extension of glaciers were based impeccably on actualistic comparison with existing glaciers. It is therefore not surprising that both Lyell and his disciple Darwin were early converts to this aspect of the glacial theory (which was readily separable from the much more speculative idea of an *Eiszeit*). Both of them rapidly learnt to look at the landscape of highland areas through new, glacialist eyes. Darwin's enthusiasm for the explanatory power of the glacial theory is evident in his letters:

"What a grand new feature all this ice-work is in Geology."[54]

In the summer of 1841 he studied the hills of North Wales and found the evidence of glacial action overwhelming.[55] He later reported to Lyell that the more conservative Roderick Murchison refused to be converted, and added:

"I confess I am astonished, so glaringly obvious after two or three days did the evidence appear to me."[56]

Furthermore, his own conversion, like Lyell's, did not involve abandoning

[52] This comment is something of a puzzle. It refers to "my note about Glen Roy" in which, Darwin told Lyell, "I merely scribbled what came uppermost." Clearly he was not referring to his long, and far from merely 'scribbled" paper, which in any case he had written two or three years before this letter. The 'note' must either be a document no longer extant, or else it refers to a slightly earlier letter discussing Glen Roy, though in no such letter (among those extant) does he seem to 'forget' the glacial theory.

[53] Darwin to Lyell [1841]. American Philosophical Society, Darwin MSS; *MLCD*, **2**, 148–50.

[54] *Op. cit.* note 53.

[55] Charles Darwin, 'Notes on the Effects Produced by the Ancient Glaciers of Caernarvonshire and on the Boulders transported by Floating Ice', *Philosophical Magazine and Journal of Science*, **21**, (1842), 180–8.

[56] Darwin to Lyell, 7 October 1842. American Philosophical Society, Darwin MSS; *MLCD*, 151–2.

the iceberg hypothesis as an additional explanation for the distribution of erratic boulders. Glaciers and floating icebergs were postulated in conjunction with a formerly higher relative sea-level. All of this could in principle have been transferred by Darwin without difficulty from North Wales to Lochaber; yet in fact it was not.

In 1842, Agassiz reviewed the "recent Progress" of the glacial theory in an essay which included a section on the problem of the Roads.[57] It should be emphasized, however, that even this fuller discussion was very brief and sketchy by comparison with the three earlier major papers on the subject (*i.e.*, MacCulloch's, Lauder's and Darwin's), and was also based on much more cursory fieldwork.

He reported three crucial observations (Figure 7). One involved a re-interpretation of phenomena already noted; one was entirely new, and was to be accepted by most subsequent observers; the third was equally novel, but was ultimately to be dismissed as a figment of Agassiz's imagination. (1) The loose débris forming a confused mass of hills in the lower part of Glen Spean, which Lauder had suggested were the remains of barrier S, were re-interpreted by Agassiz as moraines and outwash gravels from a large glacier that had descended into the valley from the Ben Nevis massif to the south. (2) Further up Glen Spean, opposite the mouth of the tributary Glen Treig, Agassiz reported finding striated rock surfaces which clearly indicated another glacier that had flowed northwards out of Glen Treig and across the floor of Glen Spean. (3) Agassiz also asserted that the two higher Roads of Glen Roy (Roads R1, R2) not only extended out of that Glen on to the north side of lower Glen Spean, but also existed on the *south* side of Glen Spean, opposite the mouth of Glen Roy (between the 'stars' on Figure 7).

With these observations, Agassiz outlined an interpretation of the Roads which was clearly based on his perception of a direct analogy with his own experience of the Alps.[58] If the Glacier de Taconay and the Glacier de Bois, descending from Mont Blanc, had extended further in the waning phases of the Ice Age than they do today, they might have blocked the Arve valley in two places and impounded a glacial lake between them (over the site of Chamonix). Agassiz cited no evidence that such a lake had actually been formed. It was merely an analogy to what he believed *had* happened in Lochaber: two glaciers there had similarly descended into

---

[57] *Op. cit.* note 48.
[58] The analogy was presented visually in two sketch-maps on the same Plate accompanying the paper.

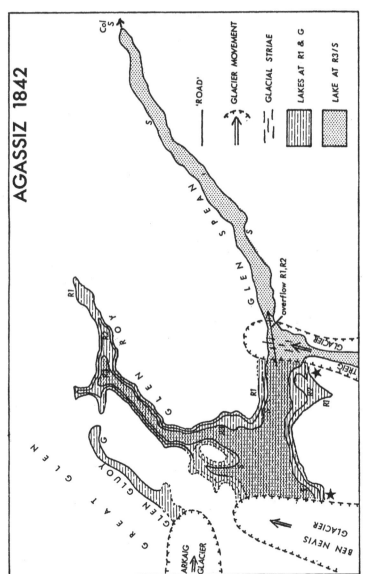

Fig. 7. Map to illustrate Agassiz's glacial hypothesis. A glacial lake was impounded at first (levels R1, R2) between two glaciers; and later (level R3/S) by the Ben Nevis glacier alone, after the retreat of the Treig glacier. A third glacier impounded another lake in Glen Gluoy. Note that the main lake had to drain *over a glacier* while at levels R1 and R2. Note also that Agassiz claimed to have found Roads at these levels on the south side of Glen Spean (between the two stars) where no other observer had seen them (compare with Figures 2 and 4), but where they were 'required' by his hypothesis. The glacial striae (scratched rock surfaces) near the mouth of Glen Treig were Agassiz's main observational evidence for re-constructing a glacier across Glen Spean at that point.

Glen Spean, blocking the valley at two points and impounding a glacial lake between them. This lake had stood at two successive levels, forming the upper beaches in Glen Roy (Roads R1, R2). Later, the retreat of the glacier in Glen Treig (the smaller of the two) had caused the lake to fall to a third level in both Glen Roy and Glen Spean, forming a new and more extensive beach (Road R3/S). Still later, the retreat of the other glacier had drained the lake altogether. The Road (G) in Glen Gluoy was attributed to a separate lake impounded by a third glacier which had crossed the Great Glen at that point (from Glen Arkaig).

Agassiz noted that the Pass of Muckul (Col S) was at the same height as the lowest of the Roads (Road R3/S), and he adopted Lauder's interpretation of it as an overflow channel. But he did not mention the similar col (R1) on the level of the highest of the Glen Roy Roads, and he attributed the drainage at both the highest levels (*i.e.*, at Roads R1 and R2) to overflow over the Glen Treig glacier itself. This laid his interpretation open to the same objection as Lauder's explanation of Road R2: how could a lake draining over the surface of a glacier (or of a detrital barrier) remain at the same level for long enough to form a well-marked beach? Darwin had already picked on this as one of the weakest points in Agassiz's explanation.

Perhaps the most interesting aspect of Agassiz's work on the Roads was his insistence that Roads R1 and R2 extended on to the south side of Glen Spean. Agassiz's hypothesis, like Lauder's, incorporated an explanation of why the Roads were *not* of very wide extent. But the detailed form of his hypothesis required that at this particular point the Roads should have extended further than had previously been reported: if the lake had been impounded initially between two glacial barriers, it must have extended out of Glen Roy and into lower Glen Spean. In correspondence with this expectation, Agassiz was convinced that he had in fact seen the Roads in the positions required.

Darwin at once noted this discrepancy, for he himself had seen nothing but sheep-tracks above Road S on the south side of Glen Spean: "so much, again, for difference of observation."[59] He might have welcomed any such extension of the shelves as equally favouring his own hypothesis, but he seems to have been sceptical of the validity of Agassiz's observation.

In spite of his conversion to a belief in former glaciers in North Wales, Darwin seems to have retained his confidence in his own marine interpretation of the Roads in Lochaber, or at least to have suspended judgment

[59] *Op. cit.* note 56.

on the question. The following year (1843) he said he was planning a trip to Scotland "to hunt for more parallel roads". He added:

"My marine theory for these Roads was for a time knocked on the head by Agassiz ice-work, but it is now reviving again."[60]

That he should even have contemplated such a journey, at a time when he was constructing the first drafts for the presentation of his Species theory, is surely a further indication that the Glen Roy problem remained important to him.

It should be emphasized that during the early 1840s Darwin had strong social support for holding on to his marine hypothesis. Lyell was wise to summarize the position publicly as an unsolved problem, in which there were difficulties about both Darwin's marine hypothesis and Agassiz's "conjecture" of glacial lakes,[61] but in private he supported Darwin more emphatically. Murchison, in his Presidential Addresses to the Geological Society, lent Darwin unequivocal support.[62] In other words, among geologists whose opinions he respected, Darwin's minutely argued interpretation of the Roads as evidence of crustal elevation had by no means been falsified by Agassiz's crude and hasty sketch of an alternative.

Darwin never made a second visit to Lochaber. But neither he nor others forgot the problem. In 1846 Leonard Horner suggested that the Geological Committee of the British Association should apply to the Ordnance Survey for an accurate survey of the Roads, principally to check whether their horizontality was as precise as had been alleged. Darwin clearly recognized that this could yield important evidence for the marine hypothesis. He pointed out to Horner that what was needed was an extremely accurate survey of the level of the Roads on *two* lines at right angles to each other: a single line might happen to run parallel to an axis of elevation, and would therefore be inconclusive. Darwin stated quite explicitly the underlying significance of the whole matter; this apparently local puzzle bore on "one of the most important problems in Geology—namely, the exact manner in which the crust of the earth rises in mass." By this time Darwin had fully recovered his earlier confidence in his own hypothesis for the Roads, and said he was convinced "there never was a more futile theory" than Agassiz's.[63]

---

[60] Darwin to C. D. Fox, 5 September 1843. *LLCD*, **1**, 332–5.
[61] Charles Lyell, *Elements of Geology*, 2nd edn. (London, 1841), 173–9.
[62] R. I. Murchison, 'Anniversary Address of the President', *Proceedings of the Geological Society of London*, **3**, No. 86 (1842), 637–87, see p. 679; *ibid.*, **4**, No. 93 (1843), 65–151, see p. 98.
[63] Darwin to Horner [1846]. *MLCD*, **2**, 174–7.

Table 1. Comparison between the observational 'expectations' associated with (though not strictly *deducible* from) the major alternative explanations of the Roads. The 'expectations' of the lake hypothesis (including the modifications associated with *glacial* lakes) were cumulatively fulfilled over the decades between MacCulloch's paper and Jamieson's, whereas the 'expectations' of the marine hypothesis (including the modifications associated with Chamber's falling sea-level' as opposed to Darwin's 'rising land-mass') almost all required 'secondary' hypotheses to *explain away* their observational failure.

PHENOMENON	EXPLANATION	
	LAKES, on assumption of non-glacial origin (modifications for *glacial lakes* in italics)	SEA, on assumption of rising land-mass (modification for *falling sea-level* in italics)
1. Lateral distribution of Roads	Strictly localized (by extent of presumptive lakes), but continuous within these areas (interruptions within these areas require secondary explanation of special circumstances of non-preservation, *e.g.*, subsequent erosion, or hillside not steep enough)	Throughout region, on appropriate levels, at least as widely scattered traces; if systematically localized, secondary explanation required (*e.g.*, special circumstances for preservation)
2. Lateral limits of Roads	Abrupt; at sites of barriers impounding lakes (*abruptness 'softened' by seasonal freezing near glacial barriers*)	Gradual; in areas where topography induced changes in relative exposure of shore-lines
3. Vertical distribution of Roads	On few but definite levels (principal Roads); minor Roads at intermediate levels absent or rare, requiring secondary explanation (*e.g.*, gradual erosion of barriers) if present	On many levels, some more definite than others (*i.e.*, no absolute distinction between principal and minor Roads)
4. Horizontality of Roads	Absolute (assuming no subsequent crustal movement)	Possibly slight overall tilting, in some directions (but perhaps imperceptible within a restricted region); *no overall tilting if emergence due to falling sea-level*; possibly local inclinations due to tidal gradients

5. General Preservation of Roads	Good throughout (since all geologically recent)	Graded according to height (and therefore age); if good throughout, secondary explanation required (*e.g.*, excessively slow rate of subaerial erosion)
6. Barriers in relation to Roads	At lateral limits of principal Roads; possible traces (after rupture and/or erosion) as low hills of loose débris; *morainic hills and scratched bedrock surfaces etc., if barriers glacial*	No barriers; no special features at lateral limits of Roads
7. Cols in relation to Roads	Cols coinciding in height with every principal Road; Roads only on one (lake) side of each col, with possible channelling on other (overflow) side	Cols may occur on same level as some, but not necessarily all, Roads; Roads on same level on both sides of such cols; no channelling (except possibly on top of cols)
8. Preservation of fossils on Roads	Unlikely; possibly freshwater mollusc shells; *extremely unlikely if lakes glacial*	Likely; marine mollusc shells and possibly adherent organisms; if absent, secondary explanation required (*e.g.*, subsequent leaching by acidic groundwater)

2. *The lake hypothesis improved and defended.* The very next year, however, Darwin's peace of mind on the matter was shattered again. In a paper read (in part) to the Royal Society of Edinburgh, David Milne reported that he had visited Lochaber in 1845 "with a strong conviction that the lake theory was indefensible" and that Darwin was correct; but had come away with the opposite conclusion.[64] The work of "so justly celebrated a geologist, and an accurate observer" deserved the highest respect, and Darwin's interpretation had received the approval of no fewer than three some-time presidents of the Geological Society (*viz.*, Murchison, Lyell, and Horner). Yet Milne felt he must dissent from this view, and argue that the lake hypothesis was more plausible. He promised, however, that "the gradual operation of ordinary courses" would be substituted for the "convulsions of nature" that MacCulloch and Lauder had been obliged to invoke. It was in fact a modified Lauder hypothesis, not the *glacial*-lake hypothesis of Agassiz, that Milne proposed.

Milne described four main points of observation. (1) Not waiting for the outcome of Horner's request for an official survey, he himself had had the Roads re-surveyed. This, he reported, *"leaves no doubt as to the perfect horizontality of the Roads"*. (2) The survey showed that the difference in height between Roads G and R1 was twenty-three feet or twenty-nine feet (by different methods)—in either case, roughly doubling the previous figure of twelve feet and making it clear that the difference had to be explained, and not explained away (as Darwin had done). (3) He described the Pass of Muckul (Col S) as having all the features that would be expected of a major overflow channel, now abandoned. He thought Darwin would have been convinced by this, if he had ever visited the spot. (4) Most importantly, he reported his discovery of a col (R2) coincident in level with Road R2 (see Figures 2, 3). This was at the head of a valley (Glen Glaster) tributary to Glen Roy, which, as he commented, had "oddly enough" not been visited previously by any geologist.

It was indeed odd. Lauder had had a strong motive to find such a col, which would have completed his rule by showing that all four principal Roads were coincident with cols at the same height. His base map, though admittedly crude, did show the lateral valley concerned, yet he drew the

---

[64] David Milne, 'On the Parallel Roads of Lochaber, with Remarks on the Changes of Relative Levels of Sea and Land in Scotland', *Transactions of the Royal Society of Edinburgh*, 16 (1847), 395–418. For his diluvial interpretation of glacial features elsewhere, see his paper 'On Polished and Striated Rocks, lately discovered in Arthur Seat, and other places near Edinburgh', *Edinburgh new Philosophical Journal*, 42 (1847), 154–74. David Milne-Home (1805–90) was a Berwickshire country gentleman, a founder of the Scottish Meteorological Society, and the author of several geological papers.

course of the upper Roads most unrealistically at this point, showing them scarcely entering the valley at all (see Figure 2). Darwin's 'agenda' shows that he too was aware of the need to check whether the anomalous Road R2 had a col or not. But his notebook shows that he never explored Glen Glaster, and he copied the course of the Roads at this point directly from Lauder's map (see Figure 4). Yet the valley is conspicuously deep when viewed from Glen Roy, and the Roads clearly extend back into it. Furthermore, Glen Glaster is directly opposite the Hill of Bohuntine, where Darwin made a detailed study of the way in which the Roads R1 and R2 faded away. Perhaps we must conclude that, being already convinced of the truth of his own marine hypothesis, Darwin lacked the motive to explore the only conceivable area in which the missing col could be located.[65] However that may be, it was in fact Milne who found the col, and moreover found clear indications on the far side that it had indeed been an overflow channel. He stressed that this discovery eliminated what had been an anomaly, and showed that *every* principal Road had a col on the same level (see Figure 3).

This was important evidence in favour of the lake hypothesis. Each of these cols had a Road on the same level on one side but not the other, whereas on Darwin's hypothesis the water would have stood at the same level on both sides and might have been expected to form a beach on both sides. Furthermore, at least two of the cols (R2 and S) preserved clear evidence of having been overflows, water from the side with a Road having eroded a clear channel while flowing *down* the other side of the col. Darwin's evasion of the difference in level between Roads G and R1 was inadmissible; if it were due to greater tidal action in Glen Gluoy, Road G ought to have sloped upwards towards the head of the valley, whereas in fact it was perfectly horizontal. Indeed if the water had been tidal anywhere, the Roads should all have sloped upwards in this way, as in modern tidal inlets. Another serious objection was the astonishingly perfect preservation of the Roads, which Darwin, assuming their very great age, had taken as evidence for the negligible effects of subaerial erosion. Yet on Darwin's hypothesis, they were far older than the much *less* well preserved raised marine beaches known at much lower elevations elsewhere in Scotland (see Figure 6).

[65] It is fair to add that if Darwin had had a more accurate base map of Lochaber he could hardly have missed the significance of Glen Glaster: this is perhaps the only point at which his lack of such a map was a serious handicap. On the other hand, Milne too lacked any accurate topographical map. It is not therefore unreasonable to attribute Darwin's failure to explore Glen Glaster to his lack of motive to find the missing col.

With these and other telling comments, Milne rightly concluded that Darwin's objections to the lake hypothesis "resolve entirely into the difficulty of explaining the disappearance of the barriers." But this, he argued, was no proper reason for rejecting such a hypothesis, and he regretted that Darwin should have asserted so forcefully that his interpretation was "*demonstrated*".

Milne's own interpretation was in effect a gradualist or Lyellian modification of Lauder's. He postulated "a blockage of some sort"—namely of detrital débris—in the same three positions (at or near the mouths of Glens Roy, Spean and Gluoy), but denied that their removal need have been in any way sudden. No "stupendous agent" was required (contrary to what Lauder had suggested), but simply "the continuous working of ordinary and natural agents" like those still operating there, namely the ordinary processes of river erosion. His interpretation thus differed from Lauder's only in two essentials: (*a*) the stable shore-line at Road R2 was now explicable on the same basis as the other principal Roads, as a stage at which the overflow was by a col at the same height; and (*b*) Milne stressed the adequacy of *gradual* erosion of the barriers. The reasons for the existence of barriers at these particular points still remained obscure.

Milne felt it hardly necessary to refute the other rival hypothesis, namely Agassiz's. He had evidently never seen any modern glaciers, and he thought Agassiz's striated rock surfaces and moraines in Lochaber were purely fanciful. He was on firmer ground in denying that Roads R1 and R2 extended into lower Glen Spean as Agassiz had asserted.

Despite his opposition to Darwin's hypothesis for the Roads, Milne did believe in changes in relative sea-level almost as drastic as Darwin's. In the privately printed version of his memoir[66] he cited evidence from all over Scotland to suggest that the sea had been at many higher levels than the present, though he was inclined to think the changes had been due to falls in sea levels, not to a rising of the land. But he believed the Roads were a different class of phenomenon from the raised beaches he detected in other areas, far better preserved and far younger.

The news of Milne's attack, Darwin said, made him "horribly sick": "I entirely gave up the ghost and was quite chickenhearted." But after Robert Chambers had got him a copy of Milne's paper, he told Lyell he had recovered and was "not even staggered".[67] He felt Milne's general hypothesis made no advance over Lauder's, and he still thought barriers

---

[66] David Milne, *On the Parallel Roads of Lochaber, with Remarks on the Change of Relative Levels of Sea and Land in Scotland, and on the Detrital Deposits in that Country* (Edinburgh, 1847).

[67] *LLCD*, **1**, 361–2.

of débris "more utterly impossible than words can express."[68] Despite Milne's attempt to make them more respectably Lyellian (by stressing their gradual erosion by ordinary means), Darwin still felt that they were "monstrous": "I utterly disbelieve in *the barriers*."[69]

Yet he was deeply struck by Milne's discovery of the missing col R2, for he had previously stressed this anomaly as an insuperable objection to any lake hypothesis. Combined with his continuing rejection of any barriers of *débris*, this naturally made him begin to look more favourably on Agassiz's hypothesis of lakes dammed by ice: barriers of ice were "incomparably more probable" than barriers of débris.[70] He told Milne:

"The oddest result of your paper on me is that I am very much staggered in favour of the ice-lake theory. Until I read your important discovery of the outlet in Glen Glaster I never thought this theory at all tenable. Now it appears that a very good case can be made in its favour."[71]

Darwin was clearly exploring the implications of the glacial-lake hypothesis with a very open mind, trying to see whether Agassiz's sketchy version of it could be improved. For example, he thought the glacier from Glen Treig could have extended *further* than Agassiz suggested, spreading sufficiently into Glen Spean to block up all the exits from Glen Roy.[72] This would eliminate the need to postulate the extension of the lake from Glen Roy into Glen Spean, and would therefore explain why the higher Roads (R1, R2) did *not* (*pace* Agassiz) exist in Lower Glen Spean.

But though Darwin told Milne, "I tremble for the result", he also said, "I am not, however, as yet a believer in the ice-lake theory."[73] For he felt that most of the arguments he had used in his own paper still remained strong. Notwithstanding Milne's forceful interpretation of the cols, he felt they were too large to have been overflow channels, and still thought they could be regarded as tidal channels like those he himself had seen in Tierra del Fuego. He was still puzzled by the gradual fading of the higher Roads (R1, R2) where they emerged from Glen Roy into Glen Spean, and thought this irreconcilable with a barrier of *any* kind in that position. He could not believe that a lake draining over the ice itself (as it would have to do between the main levels) could have stopped at one level for long enough even to form the fragmentary minor Road (R2+) he had detected in Glen Roy. Although the valley bottoms contained much débris of various

[68] Darwin to Chambers, 11 September 1847. *MLCD*, **2**, 178.
[69] Darwin to Lyell, "Wednesday 8th" [probably late 1847] *MLCD*, **2**, 181–7.
[70] *Op. cit.* note 68.
[71] Darwin to Milne, 20 [September 1847] *MLCD*, **2**, 180–1.
[72] *Op. cit.* note 69, 182.
[73] *Op. cit.* note 71.

kinds, some of which could well be morainic in origin as Agassiz had suggested, Darwin did not think there were any clear *terminal* moraines in the positions that might have been expected. For these and other "weak but accumulating reasons", he told Lyell he still preferred his original hypothesis of a "gradual rise of a group of islands", which would have been gradually united and the sea restricted to channels and inlets.[74]

Darwin urged Lyell to go to Lochaber himself to settle the matter: it needed "some one who knows glacier and ice-berg action, and sea action well." With perhaps a premonition of the final outcome, he now said, "I enjoyed my trip to Glen Roy very much, but it was time thrown away."[75]

Lyell suggested that Darwin should publish a reply to Milne, and Darwin actually drafted a long and involved letter to the *Scotsman*, which had carried the first report of Agassiz's Scottish tour.[76] In content, this letter shows Darwin inclined still further towards the glacial lake hypothesis. In addition to the points he had already made in correspondence, he now thought a more developed version of Agassiz's hypothesis could explain such apparently trivial observations as the slightly greater extension of Road R_2 beyond R_1 at the mouth of Glen Roy: if the barrier had been of ice, its gradual retreat might have been marked by just such a slight extension of the lake. More important, Lyell had suggested to him how the existence of glacial lakes could explain the *gradual* fading away of the Roads at these points: with the lake frozen over for much of the year near the barrier of ice, wave action would be reduced and the beaches less well developed than in the main part of the lake. Even Milne's evidence of large overflow channels, implying that large volumes of water had flowed over the cols, now seemed plausible to Darwin, since under the climatic conditions that would exist during a period of glacial extension there would be a high volume of surface run-off after the melting of the snow each spring.

Against the glacial lake hypothesis, he reiterated many of his earlier arguments for his marine interpretation. It was easier to imagine the minor Roads being formed as shore-lines during short pauses in elevation than while the overflow from the glacial lakes was over the ice itself. His own Road (K) above Kilfinnin and MacCulloch's minor Road (R_1^+) above Loch Spey (a "crucial case") were problematical on the glacial lake hypothesis. Most of all, the "buttresses" in the bottom of Glen Spean seemed indistinguishable from those he thought were indubitably marine elsewhere.

[74] *Op. cit.* note 69, 185.

[75] *Op. cit.* note 69, 186.

[76] *Op. cit.* note 71. Darwin's letter to the *Scotsman* is printed in Barrett (see note 3), 24-7.

Darwin appended a list of nine crucial points of observation, not unlike his own earlier agenda. They concerned the rival interpretations of the cols as 'land-straits' or overflow channels; the reality of Agassiz's moraines and of his alleged higher Roads in Glen Spean; the exact heights of the principal Roads in relation to possibly analogous features elsewhere in Scotland; and the degree of horizontality of the Roads. It was on the whole a fair list of observations that might serve as *tests* between the two hypotheses. But on the question of horizontality Darwin retained his characteristic inclination to make his own position unfalsifiable: if the Roads were found to be perfectly level, it might seem to favour the lake hypothesis, yet it would not in fact falsify the marine hypothesis, since other regions were known to have been elevated "equably."

Darwin's conclusion to this letter, that he was still inclined to prefer the marine hypothesis, seems at variance with his own weighing of the evidence: his assessment of the points in favour of each side seems clearly to favour the glacial lake hypothesis. Perhaps this inconsistency explains why he never published the letter.[77] He said that if the marine hypothesis was wrong, Agassiz's would prove to be "the true solution". Yet he was still reluctant to accept the failure of his own interpretation.

3. *The marine hypothesis re-stated.* Chambers had told Milne that he for one still believed in Darwin's marine hypothesis for "the Glen Roy mystery".[78] Darwin admitted to Chambers that "if the Roads were formed by a lake of any kind, I believe it must have been an ice-lake." But the crucial features he recommended Chambers to examine were those that still held out hope of rescuing the marine hypothesis. In particular, he suggested a careful study of the horizontality of the principal Roads, of the cols coincident with them, and of the minor Roads between them.[79]

In 1848 Chambers published his volume on *Ancient Sea Margins*, in which the Roads were added to numerous other phenomena to build up a case for the gradual emergence of the whole of Scotland.[80] Superficially this seems similar to Darwin's interpretation, but Chambers (like Milne) thought the emergence was due more probably to a falling sea-level than

[77] The letter was declined by the editor of the *Scotsman* on the reasonable grounds that it would not be intelligible to his readers, who had not been given any abstract of Milne's views. The letter was passed on to Robert Jameson for possible publication in the *Edinburgh new Philosophical Journal*—a far more appropriate medium—but Darwin wrote to Jameson to ask him to destroy the letter, and not to publish it. (Darwin to Lyell, 12 October 1847, copy in Darwin MS, 146.

[78] Chambers to Milne, 7 September 1847, *MLCD*, 2, 177.

[79] *Op. cit.* note 68.

[80] Robert Chambers, *Ancient Sea Margins, as memorials of changes in the relative levels of sea and land* (Edinburgh, 1848).

to the elevation of the land. (Darwin referred to this as a *"heretical* and *damnable* doctrine".) But like Darwin, Chambers put great emphasis on the abundance of traces of shore-lines at many different levels in addition to those of the principal Roads, and claimed many intermediate or minor Roads and other faint horizontal markings. For Lochaber alone he tabulated no fewer than twenty-five "terraces and other markings" ranging from 1495 feet above present sea-level down to 325 feet. He concluded, in a style reminiscent of the *Vestiges* he had published anonymously four years earlier, "How clear and direct is the discourse of Nature, when the true key to the cipher in which she writes has once been discovered!"

The Glen Roy mystery was now, he thought, near solution. The inference to be drawn from all his evidence of raised shore-lines was "majestically simple": since the cold period at which the sea had stood high on the land and the erratics had been deposited from icebergs, there had been a gradual and quiet withdrawal of the sea, with no paroxysmal movements whatever, until the present sea-level had been reached. Agassiz's hypothesis was summarily dismissed as having "fatal objections".

Darwin told Chambers he was pleased to have his support for the marine hypothesis. He said that Milne's work had temporarily "staggered" him to the glacial-lake interpretation, but that he had now recovered his former "positiveness" about his own conclusions. He admitted that he had always felt a deep "personal interest" in the problem, and added, "I should have been more sorry to have been proved wrong on it, than upon almost any other subject." Once again, this indicates the continuing importance of the problem in Darwin's mind. He commented approvingly on some minor Roads that Chambers had newly reported, but complained that Chambers had failed to acknowledge his (Darwin's) priority of discovery in other cases.[81] To Lyell, Darwin was more outspokenly critical. Chamber's book was scientifically valueless: "this book for poverty of intellect is a literary curiosity." He said he believed that Chambers was the author of *Vestiges*, and clearly saw the two books as falling into the same category. Even Chambers's "new" minor Roads were none other than those Agassiz had claimed in Lower Glen Spean, and for the rest, Chambers had merely borrowed Darwin's evidence and Darwin's arguments without acknowledgement.[82] Chambers did in fact later admit that he should have made clearer his debt to Darwin's work.[83]

[81] Darwin to Chambers, undated [probably 1848] University Library, Cambridge, Darwin MS, 'Glen Roy notes and scraps', 50.
[82] Darwin to Lyell, June 1848, *LLCD*, 1, 362.
[83] Darwin to Lyell [June 1848], *LLCD*, 1, 363.

4. *The glacial lake hypothesis improved and accepted.* Two further contributions
to the Glen Roy problem were published the same year as Chambers's
book, but I have found no evidence to show whether Darwin took note of
them. Sir George Mackenzie shared the publication of a new map of the
Roads with Milne and Chambers. His hypothesis would have seemed anti-
quated to the readers of his paper.[84] He called himself "the last survivor
of the old Huttonian school", though he was referring to the paroxysmal
Huttonianism of Sir James Hall (whom he had known personally) not
that of Hutton himself. He had revisited Lochaber after reading Milne's
paper, but had concluded that the evidence supported a "débâcle theory".
His argument, though somewhat obscure, did at least stress the important
differences between the Roads and other forms of terraces, and it did re-
iterate once more the improbability of Milne's (and Lauder's) putative
barriers of rocky débris.

To James Thomson, on the other hand, as to Darwin in his unpublished
Scotsman letter, a modified version of Agassiz's hypothesis seemed attractive.[85] Milne (and Lauder) had been methodologically correct, he said, to
insist—against Darwin—that the assessment of evidence for the *existence* of
barriers must be distinguished from the problem of explaining their nature
and mode of removal. Milne had provided strong evidence for barriers of
some sort; Agassiz had shown how the separate second question could be resolved. Agassiz's own treatment of the problem had been merely a preliminary sketch, and he had been more concerned to demonstrate the
former existence of glaciers in general than to work out all the details
which that hypothesis would explain.

The new feature that Thomson introduced was the application of J. D.
Forbes's concept of glacial motion. This had been the subject of acri-

---

[84] Sir G. S. Mackenzie, Bart., 'An attempt to classify the Phenomena in the Glens of Lochaber
with those of the Diluvium, or Drift, which covers the face of the country', *Edinburgh new Philosophical Journal*, 44 (1848), 1–12. Sir George Mackenzie, Bart. (1780–1848) published an account
of his scientific *Travels in Iceland* (1811), and was the author of several papers on mineralogy,
geology, and zoology, as well as works on aesthetics, phrenology and education.

[85] James Thomson junior, 'On the Parallel Roads of Lochaber', *Edinburgh new Philosophical
Journal*, 45 (1848), 49–61. James Thomson (1822–92), at this time a young graduate of Glasgow
University, later became Professor of Civil Engineering at Belfast and Glasgow. He was the
brother of William Thomson, later Lord Kelvin. He did distinguished work on the physics of ice,
and his paper on Glen Roy—one of his earliest—can be seen as an application of this physical
research to a geological problem. An intriguing possibility is that Thomson heard of Darwin's
improved glacial-lake interpretation of the Roads—which closely resembles his own—from
Robert Jameson, the editor of the *Journal*. Jameson had received Darwin's *Scotsman* letter (see
note 77) and may not have destroyed it as requested, or not before reporting its contents either
to Thomson or, more probably, to J. D. Forbes, whose work on the movement of ice was followed
by Thomson.

monious argument between Forbes and Agassiz; but in this instance Forbes's belief that the ice of glaciers flowed like a viscous fluid could be used to reinforce and improve Agassiz's reconstruction of the glacial lakes in Lochaber. For Thomson argued that, on Forbes's "principles of the viscidity of glaciers", the postulated glacier from Glen Treig could easily have spread right across Glen Spean and up into the mouth of Glen Roy. If initially it extended up Glen Roy beyond even the lateral Glen Glaster, it would have impounded a glacial lake that could have formed a beach (Road R1) and overflowed by the col (R1) at the head of the valley. A slight retreat of the glacier would then have extended the lake into Glen Glaster, when the water level would have fallen to the level of the col (R2) that Milne had discovered there; and a new beach (Road R2) would have been formed, extending slightly further down the valley than the first. The further retreat of the glacier away from the mouth of Glen Roy would have drained the lake down to the level of a third beach (Road R3/S) extending into Glen Spean, this much enlarged lake being dammed by a larger glacial mass remaining in Lower Glen Spean, and having its overflow at the head of the valley (Col S). Meanwhile a lake in Glen Gluoy would have been forming a beach (Road G) and overflowing into Glen Roy (Col G), being dammed by a mass of ice in the Great Glen. The same ice could likewise have dammed a lake that formed the faint Road (K) that Darwin had reported above Kilfinnin.

On such a theory, Thomson asserted, "no gratuitous or unsupported assumption is made." He concluded that the evidence accumulated in the previous years in favour of the glacial lake hypothesis was such that "we may now regard it as an established fact."

It is very curious that Darwin should either have been unaware of Thomson's well-argued paper, or else should not have seen fit to comment on it, even in correspondence (as far as extant letters show). Darwin himself had made the same points about the possible extension of the Glen Treig glacier, with the consequent improvement of Agassiz's original interpretation, though he had not linked this explicitly with Forbes's hypothesis of the viscous flow of glaciers.

Whatever the reason may have been, fifteen years passed before the final blow fell on Darwin's original hypothesis. It was Lyell who suggested to Thomas Jamieson in 1861 that he should make a thorough study of the area. But Darwin too, having "to some extent doubted his own observations" there, "not having glacier action in view" in 1838, had also encouraged Jamieson and lent him his maps and notes.[86]

[86] Cited by Jamieson, *op. cit.* note 87.

Curiously Jamieson also made no mention of Thomson's paper, though his arguments and conclusions were similar and he summarized most of the other earlier papers on the subject with great thoroughness.[87] MacCulloch's and Lauder's work was praised, despite their inability to account for the barriers they had to postulate. After briefly dismissing Mackenzie's débâcle hypothesis, Jamieson commented on the marine interpretation of Darwin, "who urged it so forcibly, and handled so well the difficulties that arose on every side" from the alternative explanation involving barriers, that the marine hypothesis had "met with general acceptance". Agassiz's glacial interpretation of the area had been too hurried to be adequate, but had had the merit of suggesting that the mysterious barriers might have been composed of ice, not rock débris. But the glacial theory as a whole had had a mixed reception at that time, so that it was not surprising that Agassiz's application of the theory to Lochaber should have been rejected by Milne, Chambers and others. Lyell in 1851 had therefore felt the problem was still unsolved. But Jamieson pointed out that it was of far more than local interest, since it bore significantly on the whole question of the nature of recent geological history.

Jamieson pointed out that the Roads were beaches "too fine and neat" to have been formed under tidal marine conditions, and that their fine preservation, and that of the deltas coincident with them in places, showed they had accumulated in sheltered water. Moreover he explicitly saw the question of the alleged overflow channels as a discriminatory *test* between the lake and marine hypotheses. The existence of a col coincident in height with each of the principal Roads constituted a "four-fold proof" in favour of lakes, which was further supported by the evidence that the cols had indeed formed channels for much larger volumes of water than could be explained in any other way. Jamieson also re-stated what Darwin had always recognized as the strongest argument against the marine hypothesis, namely the localized distribution of the Roads; and he argued that this must indicate an equally localized cause. The perfect horizontality of the Roads was likewise an objection to the theory of elevation. "The absence, therefore", he concluded, "of any good positive evidence in favour of the marine theory, and so many considerations urging themselves against it, seemed to me to render it untenable."

[87] Thomas F. Jamieson, 'On the Parallel Roads of Glen Roy, and their Place in the History of the Glacial Period', *Quarterly Journal of the Geological Society of London*, 19 (1863), 235–59, Pl. 10. Jamieson (1829–1913) had become an expert on agricultural matters while working as a Factor on Scottish estates. In 1862 he became a lecturer in agriculture at the University of Aberdeen. He published several important papers on the glacial geology of Scotland.

X

150

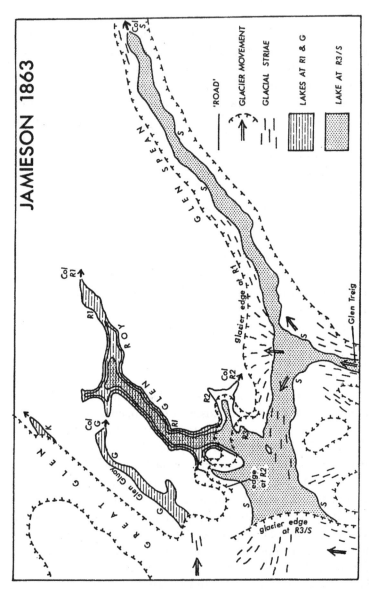

Fig. 8. Map to illustrate Jamieson's developed glacial hypothesis, for comparison with Agassiz's early version (Figure 7). The more widespread glacial striae are those discovered by Jamieson. Note that the ice from the south was envisaged as having at first blocked the whole of Glen Spean, up into the mouth of Glen Roy, impounding a lake at level R1; later, after a slight retreat, Glen Glaster was opened and the lake level fell to R2. Later still the glaciers retreated from most of Glen Spean, so that a much larger lake (level R3/S) was formed (as in Agassiz's interpretation).

For the glacial hypothesis briefly sketched by Agassiz, on the other hand, Jamieson had found abundant confirmatory evidence (Figure 8). There were striated rock surfaces to indicate a glacier in Glen Arkaig, which could have spread across the Great Glen and blocked the mouth of Glen Gluoy opposite. Similar striae in lower Glen Spean confirmed Agassiz's suggestion that a glacier had descended from Ben Nevis to block that valley, but Jamieson thought it likely that it had spread out and filled the whole of this part of the Great Glen in a single mass of ice. This would satisfactorily explain the lowest Road (R3/S) as the shore-line of a vast lake overflowing by the Pass of Muckul (Col S). To explain the upper Roads (R1, R2) in Glen Roy, Jamieson (like Thomson, and indeed Darwin himself in his *Scotsman* letter) argued that the glacier from Glen Treig had spread out into much of Glen Spean, blocking Glen Roy at first enough to force the water to overflow by the high col (R1) at the head of that valley, but later retreating enough to open the slightly lower col (R2) in Glen Glaster. Such an extension of the Glen Treig glacier was no longer merely hypothetical: Jamieson had found moraines and striated surfaces—he said it was some of the clearest glacial evidence in the kingdom—to show that this glacier had indeed spread both up Glen Spean towards the Pass of Muckul, and down the valley into and past the mouth of Glen Roy. "Grant these two ice streams [*i.e.*, from Ben Nevis and Glen Treig]," he concluded, ". . . and the problem of the Parallel Roads can be solved."

The glacial lake hypothesis could now be seen to clear up many other minor features. The absence of deltas on the higher Roads, for example, compared to their common occurrence on the lowest, could be attributed to the smaller amount of fluvial erosion that would have taken place in the more severe climate at that time. Darwin's alleged minor Roads above the highest principal Road were dismissed as probably morainic: they were in any case not horizontal. The only evidence for a higher sea-level in the Lochaber region was a raised beach a mere forty feet above present sea-level; this clearly dated from after the retreat of the valley glaciers, but it contained shells indicating a post-glacial climate still colder than the present. There was no evidence of any higher marine incursion of the area since the period of glaciation. Jamieson did not rule out the possibility of a much greater submergence of the area in "later Pliocene" time; but Darwin had suggested to him that, if so, the later advance of the glaciers would have wiped out all the evidence of it.

The only substantial problem still outstanding was the remarkable height of the ice barriers required on this interpretation. But Jamieson recognized that before such vast dams were rejected as impossible, more must be learnt of the great ice-sheets of (for example) Greenland and Spitzbergen: these, rather than the relatively small present glaciers of the Alps, were the relevant actualistic analogues.[88]

Darwin's first reaction to the news of Jamieson's results, like his first reaction to Milne's years before, was melodramatic. "I am smashed to atoms about Glen Roy", he wrote to Lyell, "my paper was one long gigantic blunder from beginning to end. Eheu! Eheu!"[89] Yet only a few days later, having re-read his own paper, he reiterated what he still felt were "points of very difficult explanation" on the glacial lake hypothesis. These were (1) the existence, even above the highest Roads, of what he believed were similar "land-straits"; (2) the rounded pebbles on the Roads in certain places, which he doubted could be due to erosion on the shore of a glacial lake; and (3) the "buttresses" below the lowest Road.[90]

But even these doubts were rapidly evaporating. About three weeks after hearing of Jamieson's work, Darwin wrote to Hooker, "It is, I believe, true that [the] Glen Roy shelves . . . were formed by glacial lakes":[91] and soon afterwards, to Lyell, "a nice mess I made of Glen Roy!"[92] One more letter from Jamieson removed Darwin's last "difficulties", and Darwin told Lyell:

"Now and for evermore I give up and abominate Glen Roy and all its belongings. . . . I do believe every word in my Glen Roy paper is false."[93]

Darwin's main interests had of course moved away from geology years before, and his comments in these letters are those of one who was no longer an expert in the field concerned. "What a problem you have in hand!", he told Lyell, referring to Jamieson's work, "It beats manufacturing new species all to bits", and he added "It is out of my line nowadays, and above and beyond me."[94] Yet, after making all allowances for his tendency to use extravagant language, Darwin's dismay at the final rejection of the hypothesis he had formed twenty-three years earlier does indicate the strength of his commitment to it. Nearly twenty years later still, in 1880, he recalled to Prestwich:

[88] For his broader views on the glacial period, see his paper 'On the History of the last geological Changes in Scotland', *Quarterly Journal of the Geological Society*, 21 (1865), 161–203.

[89] Darwin to Lyell, 6 September [1861] *MLCD*, 2, 188.

[90] Darwin to Lyell, 15 September [1861] copy in University Library, Cambridge, Darwin MS, 146.

[91] Darwin to Hooker, 28 September [1861] *MLCD*, 2, 190.

[92] Darwin to Lyell, 1 October [1861] *MLCD*, 2, 190.

[93] Darwin to Lyell, 14 October [1861] *MLCD*, 2, 191.

[94] Darwin to Lyell, 22 September [1861] *MLCD*, 2, 188–9.

"As soon as I read Mr Jamieson's article on the parallel roads, I gave up the ghost with more sighs and groans than on almost any other occasion in my life."[95] Jamieson's work was a model of accurate description and cogent reasoning. With the elegant map that accompanied it, this beautifully written paper made a clear, persuasive and virtually watertight case for the glacial lake hypothesis.[96] There is no need to follow here the later history of the interpretation of the Roads. Like Darwin, Lyell accepted Jamieson's argument as settling the problem finally,[97] and the glacial lake interpretation has been retained with only minor modifications to the present day.[98]

V
Origins of Darwin's Hypothesis

1. *Darwin and crustal elevation.* There is strong evidence, as we have seen, that Darwin was pre-disposed to see the Parallel Roads of Glen Roy *as* marine beaches, by his previous experience of the Parallel Roads of Coquimbo and other marks of higher sea-levels elsewhere in Chile. Certainly his agenda of observations to be made, and even more clearly his field notebook, reveal a mental 'set' in favour of *seeing* the features of Lochaber as signs of formerly higher sea-levels. There is nothing surprising or unique about this. MacCulloch and Lauder were probably pre-disposed to interpret the Roads as lake-beaches for the similar reason that they were familiar with the lake-studded Highlands as a whole—it should not be forgotten that some of the valleys of Lochaber *still* contain lakes impounded by barriers of rock or débris (see Figure 1). For Agassiz, likewise, we have his own report of how the sight of glacial traces in Lochaber made him feel he was back at Chamonix. All three rival hypotheses were thus underlain by a strong perceptual influence which made their proponants *see* the phenomena through different 'eyes'.

[95] Darwin to Prestwich, 3 January [1880] *MLCD*, **2**, 193.
[96] For his later improvements in detail, see his 'Supplementary Remarks on Glen Roy', *Quarterly Journal of the Geological Society*, **48** (1892), 5–27.
[97] Sir Charles Lyell, *The Geological Evidences of the Antiquity of Man, with Remarks on Theories of the Origin of Species by Variation* (London, 1863), 252–64. Most other major geologists, *e.g.*, Archibald Geikie, likewise accepted Jamieson's solution, but even at the end of the century there were still some supporters of a marine interpretation: see the Discussion after Jamieson's 'Supplementary Remarks' (Note 96).
[98] See for example J. K. Charlesworth, *The Quaternary Era with reference to its glaciation* (London, 1957), 462–4. The most important modification since Jamieson's time has been the recognition that the glaciers that impounded the lakes were not ordinary valley glaciers, but major ice-streams draining a substantial ice-cap over the present Rannoch Moor to the south. In other words, as Jamieson clearly suspected, the relevant actualistic analogues were not the glaciers of the Alps but those of the Polar regions.

This perceptual factor would have lacked force, however, if it had not been combined with a regulative factor. In Darwin's case, his South American observations were only relevant to Lochaber by virtue of a possible analogical relation between them. Here we can trace the influence of the two works that Darwin conceded were the most important in his intellectual development at the time of the *Beagle* voyage: Herschel's *Discourse* and Lyell's *Principles*. Herschel particularly stressed the value of using a clear instance to illuminate, by analogy, the cause of a more obscure phenomenon.[99] Darwin believed that his South American evidence of crustal elevation was just such a clear instance; while his reading of MacCulloch and Lauder would have shown him that the nature of the Parallel Roads was still relatively obscure. Though less formally expressed than Herschel's, Lyell's whole work was likewise based, of course, on the use of analogy: between present and past, between the relatively clear Tertiary period and the more obscure earlier periods of earth-history, and generally between known and unknown. Such examples would have encouraged Darwin to expect a successful outcome to his use of South American phenomena as analogous to those in Lochaber. Furthermore, his success in applying Lyell's regulative principles of actualism and gradualism in South America would have pre-disposed him to use a similar *type* of explanation in Scotland. In other words, Darwin had already interpreted the elevation of South America as a gradual process still in operation, and not as the result of paroxysmal action in the past. He was therefore likely to favour a similar explanation of the Scottish Roads. Lyell's conception of the principles of uniformity and continuity favoured an explanation of the Roads that linked them, through other signs of higher shore-lines, down to the present sea-level, in a single continuous series reflecting a long-continued uniform process. This is precisely the kind of explanation that Darwin applied to the Roads, as we have seen, for he linked them *through the "buttresses"* down to less controversial evidence of higher sea-levels.

This regulative factor was reinforced and justified by a theoretical factor. Darwin interpreted his South American observations through Lyellian eyes—the perceptual metaphor is Darwin's own[100]—as evidence of large-scale gradual crustal elevation. But this had been only one half of

---

[99] John Frederick William Herschel, *A preliminary discourse on the study of natural philosophy* (London, 1831), 149.

[100] Darwin to Leonard Horner, 29 August [1846], *MLCD*, 2, 117. "When seeing a thing never seen by Lyell, one yet saw it partially through his eyes." This well-known retrospective comment does not refer specifically to Darwin's work in South America, though it is clear from other evidence that he had felt this effect particularly there.

Lyell's steady-state theory of the gradual elevation of continental masses and simultaneous depression of other segments of the earth's crust. Darwin's unpublished notebooks,[101] and the papers he read to the Geological Society after his return to England, show that he was deeply engaged in developing a general causal theory of crustal elevation and subsidence in confirmation and extension of Lyell's work. Lyell himself, subsequent to the first edition of the *Principles*, became convinced that elevation could occur insensibly slowly, without the agency of earthquakes.[102] This highly gradualistic modification was adopted by Darwin. In his first paper to the Geological Society he asserted that earthquakes and abrupt elevations of a coastline were only "irregularities of action in some more widely extended phenomenon."[103] The modification of Darwin's views can be seen in the particular case of the Roads of Coquimbo. When he came to write the relevant part of his *Journal* for publication, he now denied that the Roads marked pauses between "distinct elevations." "On the contrary", he said, they marked "periods of comparative repose during the gradual and perhaps scarcely sensible rise of the land."[104] This was the concept of elevation that he was later to apply to the Parallel Roads of Glen Roy.

In a paper read to the Geological Society in May 1837, he presented his brilliantly original hypothesis on the formation of coral reefs and atolls as a further contribution towards a global tectonic theory.[105] The "main object" of the paper was to use reef phenomena to show that there were wide areas of "continental subsidence", corresponding to those of elevation. The whole crust of the earth was divided into parallel strips, some rising, some sinking. These "linear spaces" were undergoing "movements of an astonishing uniformity": coral reefs proved that subsidence, like elevation, affected vast areas slowly and simultaneously. The pattern of these "symmetrical areas", affected by opposing forces in balance, would give further insight into the (Lyellian) system of the "endless cycle of change" in the earth's crust.

[101] His 'A' Notebook (University Library, Cambridge, Darwin MS, 127) is mainly concerned with the problems of elevation and subsidence. It dates from about the same period as his better-known 'Species' Notebooks 'B'–'E'.
[102] *Op. cit.* note 35.
[103] *Op. cit.* note 24.
[104] Charles Darwin, *Narrative of the Surveying Voyages of His Majesty's ships* Adventure *and* Beagle . . ., **3,** *Journal and Remarks 1832–1836* (London, 1839), 424. See also note 22.
[105] Charles Darwin, 'On certain areas of elevation and subsidence in the Pacific and Indian Oceans, as deduced from the study of Coral formations', *Proceedings of the Geological Society of London,* **2,** No. 51 (1837), 552–4.

In the last of this series of papers to the Geological Society, read in March 1838, Darwin extended the synthetic power of this embryonic theory by interpreting not only earthquakes and volcanic eruptions, but even the elevation of mountain-chains themselves, as mere surface effects of continental elevation.[106] Following Lyell again, the rising of continental masses was attributed to the injection of fluid magma at depth; but elevation was only one side of a global equation of balanced elevation and subsidence.

Darwin's commitment to this theory ensured that he would approach the geology of any new area anticipating evidence either of the rising or of the sinking of that portion of the earth's crust. The Parallel Roads therefore represented a rich prize, if they could be brought within the theory as evidence for the substantial elevation of the crustal block of which Scotland formed a part. The perception of a possible analogy between Chile and Scotland was thus greatly strengthened by its integration into a global tectonic theory.

Perceptual, regulative and theoretical factors thus combined to give Darwin strong grounds for anticipating that a marine hypothesis could be applied successfully to the Parallel Roads, even before he himself had seen them. These positive factors were further reinforced by negative factors that led him to look with disfavour on alternative explanations of the Roads.

Darwin's original notes on MacCulloch and Lauder are apparently no longer extant; but all his references to their work in his own paper, and in his correspondence for many years afterwards, betray a powerful and even emotional antipathy to the lake hypothesis. It would be easy to interpret this in terms of the traditional historiography of nineteenth-century geology as the antipathy of a uniformitarian towards a hypothesis espoused by catastrophists. But as we have seen, MacCulloch's version of the lake hypothesis mentioned no 'catastrophes' at all, and Lauder's invoked such an event only to explain the final rupture of some (not all) of the presumptive barriers.[107] Even this minor use of a paroxysmal event was logically extraneous to his main explanation. This was later recognized implicitly by Milne, who removed all trace of 'catastrophism' from his version of the lake hypothesis. Yet Darwin's opposition to the hypothesis remained implacable.

---

[106] *Op. cit.* note 36.

[107] Barrett tends towards the view mentioned above: "The Glen Roy controversy was in the beginning basically . . . a metaphysical contest between opposing world-views." I cannot agree with this, or with the assertion that "Sir Lauder Dick's basic thesis was catastrophism." (*Op. cit.* note 3, 23.)

This surely suggests that the source of Darwin's antipathy lay elsewhere. It is in fact clear from his comments that he objected not to any supposed 'catastrophist' component of the lake hypothesis, but to the postulate of *barriers*. These remained objectionable even when, in Milne's work, they were treated in a respectably actualistic, non-catastrophist, manner.

I suggest that the reason for Darwin's antipathy to the supposed barriers, and hence by extension to any version of the lake hypothesis, was methodological. Like the invocation of land-bridges in biogeographical explanation, barriers were essentially *ad hoc* devices. Darwin felt that they were invoked purely to save the phenomena prematurely. MacCulloch postulated barriers somewhere on the edges of Lochaber, or even beyond, simply to explain why beaches had formed at high levels. But this was premature and counter-heuristic, in Darwin's view, because it deflected attention from the unstated underlying assumption, that the land (and the sea) had remained stable. Lauder postulated barriers *within* Lochaber, not only to contain the water at high levels, but also to explain the limited extent of the shore-lines. This in Darwin's mind was even worse, because it was not only an *ad hoc* device but also offended against the canons of observation: barriers were inferred in definite positions where no trace of them was reported even by Lauder himself (except for the putative remains of Barrier S in Glen Spean).

The emotive force of Darwin's continuing repugnance for the idea of any barriers in Lochaber reflects, I believe, the depth of his commitment to the explanatory norms represented by Herschel's and Lyell's work. *Ad hoc* explanations were inadmissible because they impeded the search for *verae causae*. I shall take up this point later in this article.

2. *Darwin and the Glacial Theory.* The foregoing discussion indicates why Darwin might have approached the problem of the Roads expecting and hoping that they would yield to a marine interpretation. We have seen that Darwin blamed his "great failure" over Glen Roy on his failure to consider the glacial theory; yet in fact he was not unaware of that theory when he wrote his paper. Nevertheless, it is hardly surprising that the powerful factors influencing him in favour of a marine interpretation should have triumphed completely over a consideration of the possible application of an alternative theory that was, in many respects, scientifically and philosophically questionable.

Like other geologists at the time, Darwin made substantial use of glacial agencies in geological explanation. In his interpretation of erratic boulders, both in South America and in Lochaber, he adopted Lyell's suggestion that a different pattern of land and sea might have caused a colder regional climate in the past. Glaciers might have reached sea-level, even in regions that are now temperate in climate, and might have discharged icebergs that would have been capable of rafting large boulders for long distances before dropping them as they melted. Darwin used this iceberg hypothesis to explain the erratics in Lochaber, and hence to argue that the erratics themselves were further evidence of the former submergence of the whole area.

Darwin was also well aware of another component of the later glacial theory. He knew that Charpentier had recently revived and extended Venetz's suggestion that the Alpine glaciers had been far more extensive in the geologically recent past. They had found scratched bedrock surfaces, distinctive erratics and terminal moraines far beyond the present limits of the glaciers.[108] This hypothesis was based on impeccably actualistic comparisons between the present and past effects of glacial action, and would be expected to appeal to a geologist as Lyellian as Darwin was. At the time he visited Lochaber, however, there were good reasons why Darwin should not have been sensitive to possible traces of former glaciers in that area. All the early versions of the glacial theory stressed the notion of the former extension of glaciers, but in Scotland there were no glaciers to have been extended. Ironically, therefore, only an observer such as Agassiz, with highly 'catastrophist' theoretical commitments, was likely to anticipate, and therefore find, glacial traces so far from any present glaciers. (Even the Scandinavian erratics, although spread hundreds of miles from their source-areas, could be attributed to the vastly greater extension of present, observable glaciers.) Once Agassiz had made the perceptual 'leap' required, however, Darwin, Lyell and others were able to see similar evidence and to adopt the more moderate aspects of Agassiz's interpretation (*i.e.*, valley glaciers), without conceding the validity of the much more radical aspect (*i.e.*, the *Eiszeit*) that enabled Agassiz to make the 'leap' in the first place. Hence although Darwin had seen glaciers at first hand in South America and noted some of their effects, it is not surprising that he failed to see the traces of similar glaciers in the valleys of Lochaber.

The theoretical superstructure of the early glacial theory would also have led Darwin to underestimate its explanatory value. Charpentier attributed the extension of the Alpine glaciers to a former greater elevation of the

[108] See Rudwick, *op. cit.* note 46.

Alps; but he explained the latter in terms that were not only 'catastrophist' but which would have seemed to Darwin as inadmissibly *ad hoc* as landbridges to oceanic islands or barriers in Lochaber. Such objections were heightened in the even more speculative version of the glacial theory that Agassiz proposed. Darwin criticized Agassiz not only for postulating an "assumed sudden elevation" of the Alps, but even more for his suggestion of a drastic global Ice Age followed by "the assumed sudden renewal of life". Once again, it was not only the "sudden" catastrophic element that Darwin rejected, but even more the "assumed" or *ad hoc* nature of the explanation. He asserted that in his own explanation of erratics, by contrast, "only *verae causae* are introduced".[109]

But even if Darwin had had stronger reasons to look for traces of valley glaciers in Lochaber, he would still have been unlikely to think of glacial *lakes* as an explanation of the Roads. Such lakes are unusual features; only someone with extensive first-hand experience of glaciated regions would be likely to perceive the relevance of such apparently marginal phenomena to the problem of the Roads. Agassiz was of course such a person, but Darwin was not.

VI

Darwin's 'Tenacity'

1. *Personal and social factors.* In the narrative sections (II–IV) of this article, I have repeatedly drawn attention to the evidence that Darwin remained deeply committed to his original interpretation of the Roads, even years after his main research interests had moved away from geology. Linked to that commitment, we have seen how reluctant he was, to the last, to abandon his interpretation and concede that the glacial-lake hypothesis gave a more satisfactory explanation.

In seeking to understand Darwin's commitment to his hypothesis, personal and social factors should not be neglected. Darwin first entered English scientific life, after his return from the *Beagle* voyage, as an all-round naturalist, but almost at once he became known as a scientist of promising originality in the field of geology. In the 1830s his principal scientific milieu was the Geological Society, and other men of science thought of him chiefly as a geologist. His paper on Glen Roy was his first full-length scientific paper on any subject; and it was published by, and probably gained him entry to, the most prestigious scientific body in Britain, the Royal Society.

[109] *Op. cit.* note 104, 615–25 ('Addendum to p. 294'.)

It might be suspected, therefore, that he would be naturally reluctant to admit that a paper so important to his reputation had been "one long gigantic blunder". Yet each time his hypothesis was attacked—first by Agassiz, then Milne and finally Jamieson—Darwin's *first* reaction was not to dig his heels in obstinately but to "give up the ghost" and feel "smashed to atoms". Furthermore, the way he tried to improve on Agassiz's hypothesis, after Milne's work had (paradoxically) made him look more favourably on the notion of glacial lakes, does not suggest a man obstinately determined to stick to his own interpretation for reasons of scientific prestige. Finally, if only his reputation had been involved, he could surely have afforded to concede defeat in the late 1840s (when in fact he nearly did so, in his drafted letter to the *Scotsman*). By that time his scientific reputation was perfectly secure on the basis of his substantial books and papers on South America, and he had in addition the personal satisfaction and security of believing that his unpublished work on the species problem was of outstanding importance and originality. Yet as we have seen, he was still making last-ditch attempts to save his hypothesis as late as 1861, after the *Origin of Species* had been published.

Again, in the earlier years Darwin had strong social support for retaining his hypothesis. In particular, its broader implications for a general theory of elevation made Lyell a staunch supporter, even though in his published writing he characteristically hedged his bets between Darwin and Agassiz.

The extravagantly speculative aspects of the early glacial theory acted likewise as a negative factor in Darwin's favour: good middle-of-the-road geologists like Sedgwick were highly sceptical of Agassiz's ideas, and he and many others had a low opinion of the observational thoroughness of both Agassiz and his most ardent English supporter Buckland. Although Darwin and Lyell both adopted a limited part of the glacial theory with enthusiasm, the widespread scepticism about Agassiz's more speculative notions would have given Darwin social support for retaining a hypothesis that gave only minor importance to glacial agencies. This factor was indeed mentioned specifically by Jamieson in his retrospective review of the controversy.[110] Darwin would also have had the strong support of Lyell, and others whom he respected, in his implacable opposition to the *ad hoc* postulates of barriers other than glaciers.

Such negative factors, then, go some way to explain Darwin's adherence to his hypothesis in the earlier years. But once again, this is hardly adequate to account for the long-continued strength of that adherence. Even by the

[110] *Op. cit.* note 87. See also Davies, *op. cit.* note 47.

later 1840s, in the hands of geologists other than Agassiz, the glacial theory was shedding its original trappings of vast ice-sheets, an almost global Ice Age, and mass extinctions of animals and plants. Conversely, by this time Darwin's remaining ally in retaining a marine interpretation was Chambers, whose scientific work he considered almost worthless. Darwin's own exploration of the possibilities of the glacial lake hypothesis at this time shows that he was quite prepared to consider it favourably, and in this he would have had the support of most leading geologists in Britain. Yet even then he drew back, as it were, at the last moment, and reverted to his original hypothesis.

2. *Theoretical factors.* Some further explanation therefore seems to be required, beyond personal pride in his original work and beyond the social support that that work undoubtedly received. I shall suggest that there were reasons of much greater importance why Darwin fought for the validity of his original explanation of the Roads with such persistence over more than twenty years.

We have seen that Darwin's early work, after his return from the *Beagle* voyage, was directed towards the establishment of a general tectonic theory of crustal elevation and subsidence. This theory was Lyellian in derivation. But Darwin's own research, and particularly his interpretation of coral reefs, gave it new and more substantial empirical foundations. Furthermore, his 1838 paper on elevation shows clearly that he was hoping to establish a steady-state theory of elevation and subsidence that would be *causal* in character. He hoped to show that the balanced slow vertical movements of adjacent crustal blocks were due to horizontal migration of fluid magma at depth beneath the crust: phenomena such as earthquakes and volcanic eruptions were to be interpreted as superficial manifestations of this deep-seated process.

The Parallel Roads were exceptionally important within this research programme. I suggest that this accounts for Darwin's decision to interrupt his work in London, not long after reading the paper on elevation, in order to travel to Lochaber to see the Roads for himself. If they could be established as marine in origin, as Darwin clearly believed they could, they would give striking evidence that Scotland, like Chile, had undergone a slow and intermittent elevation from beneath the sea. This in itself would help to demonstrate the global validity of his tectonic theory. Moreover it would demonstrate it, as it were, on the home ground of British geologists, by re-interpreting a phenomenon that was already widely known and discussed. But above all, if Darwin could prove that the Roads were marine beaches, their precise horizontality would become exceptionally important evidence for his hypothesis of the *cause* of elevation. He himself said quite explicitly that this concluding inference was of outstanding importance.[111] The horizontality of the Roads seemed at first sight wholly in favour of the lake hypothesis; but if they were once proved to be marine, that very fact would become evidence that crustal blocks were indeed "floating on molten stone". In the event, as we have seen, he failed to find the unequivocal evidence of marine origin (particularly fossils) that he had hoped for, and his argument therefore had to be more circumstantial than he had anticipated. As a result, the horizontality of the Roads, which he had hoped would lead incontestably to his theoretical conclusion, became something of an Achilles' heel: it could always be cited (and was) by his opponents as important evidence that the Roads were not marine at all, but lake beaches formed while the land was static. Nevertheless, I suggest that this link between the marine hypothesis for the Roads and Darwin's global tectonic theory—a link which was stated quite explicitly by Darwin himself—contributes substantially to our understanding of his adherence to his hypothesis.

This, however, presupposes the continuing importance of this geological theory in Darwin's mind, in the years after the Glen Roy paper was published. It might seem doubtful whether Darwin's commitment to his tectonic theory could have been sustained for so long, since it is well known that his interests shifted increasingly from geology to biology in the late 1830s and early 1840s. But the traditional historiography has tended anachronistically to treat these interests as though they were quite separate disciplines competing for his attention. In fact, however, as several recent writers have pointed out, the context of Darwin's early speculations on the species problem was itself that of biogeography,[112] and that in turn was intimately linked with more purely geological problems. His well-known 'Species' notebooks are full of speculations bearing on the problems of animal and plant distribution. But at the same period (1837–9) he also filled another notebook, less well known to historians, with speculations on elevation and subsidence, and other geological problems.[113] These were not simply parallel but unconnected interests. Many of the biogeo-

[111] *Op. cit.* note 44.
[112] Especially Camille Limoges, *la sélection naturelle* (Paris, 1970); Michael T. Ghiselin, *op. cit.* note 3; M. J. S. Hodge, 'On the Origins of Darwinism in Lyellian historical geography', (paper read to the British Society for the History of Science, July 1971).
[113] See note 101.

graphical puzzles that set Darwin thinking in terms of a mode of speciation had been posed by Lyell himself. Lyell had sought to explain animal and plant distribution in terms of the geological history of the areas in which the organisms lived. Such a history was of course intimately linked to such geological processes as elevation and subsidence.

The same direct connection between tectonic theory and the problems of biogeography was stated explicitly by Darwin in his 1837 paper on coral reefs.[114] Reefs could be used as indicators of rising or subsiding crustal blocks, and they proved that these movements were simultaneous, slow and balanced. But Darwin also outlined the further significance of this tectonic theory for the elucidation of the biogeographical questions that Lyell had posed. Atolls and barrier reefs, by acting as "monuments" of subsiding crustal blocks, could indicate "former centres whence the germs could be disseminated", and hence explain otherwise puzzling patterns of distribution. They would help to show whether distinctive island faunas were "remnants of a former large population, or a new one springing into existence". In other words, if the associated reefs indicated that certain islands were part of a subsiding crustal block, their distinctive terrestial faunas would presumably be relicts, local survivors of a fauna that had been more widespread when the block was a continental area above sea-level. Conversely, if the reef forms indicated elevation the island faunas must somehow be new.

The importance of Darwin's focus on the problems of oceanic island faunas needs no emphasis. But its grounding in his tectonic theory has commonly been overlooked, owing to the tendency of historians to treat him anachronistically as a 'biologist'. Once such a link is conceded, however, Darwin's theoretical commitment to his interpretation of the Parallel Roads takes on greater depth. In his paper on Glen Roy and in his correspondence, there is the same picture of a crustal block rising above sea-level, first emerging as a group of islands and later becoming connected into a larger land area (see Figure 4). His paper gave no occasion to discuss the biogeographical implications of this model. In a formal paper of this kind, even his concluding speculations of the cause of elevation were near the limits of scientific convention, as his referee's comment shows; and speculations still more remote from the matter in hand would not have been acceptable. Nevertheless, Darwin's interpretation of the Roads can be seen, in the light of his other work, not only as a detailed case-study of major importance for his tectonic theory, but also, *through* that theoretical basis,

[114] *Op. cit.* note 105.

as an implicit contribution to an even broader unified research programme. This programme embraced both his geology and his biology, or put more historically, it covered all his natural history. It included both his tectonic theory and his speculations about the species problem; and these areas of interest were linked logically by the problems of biogeography.

Darwin interrupted the compilation of his species notebooks in 1838 to go to Lochaber. Some four years later, while he tackled the doubts raised by Agassiz's new interpretation of the Roads and sought to strengthen his own hypothesis, he was compiling the first sketch of what was to become the *Origin of Species*. In the 1842 *Sketch* and the more polished 1844 *Essay*,[115] the role of Darwin's tectonic theory within the broader framework of his species theory becomes clear. Introducing his hypothesis of natural selection, he wrote in Lyellian style that "Geology proclaims a constant round of change." But when he came to consider the power of his theory to explain the geographical distribution of organisms (Section VI), this geological commonplace was articulated in a specifically Darwinian form, as a model of crustal blocks in ceaseless slow oscillatory movement. Within such a picture of tectonic history, periods of elevation became significant as periods of emergence of islands from beneath the sea. On Darwin's hypothesis, "no point [would be] so favourable for generation of new species" as such isolated oceanic islands;[116] for in such habitats variants could become established as new species, adapted to new environments, without facing competition from other organisms already well adapted to those conditions. Conversely, in the alternating periods of subsidence, although no speciation would be likely to occur, the once-widespread faunas would become isolated on the gradually shrinking island areas, so producing the necessary conditions for speciation as soon as the crustal movement was reversed once more.

Furthermore, Darwin's conception of oscillating crustal blocks gave him an explanatory bonus, in helping to account for the otherwise embarrassing lack of intermediate forms in the fossil record. Lyell had emphasized that the fossil record was extremely imperfect, owing to the haphazard chances of preservation. But in Darwin's view there was in addition a *systematic* imperfection. The chances of preservation were greatest in phases of subsidence and least in phases of elevation (for geological reasons related to the deposition and preservation of subaqueous sediments that could contain the

---

[115] Darwin and Wallace, *Evolution by Natural Selection*, Gavin de Beer (ed.) (Cambridge, 1958), 41–88 and 89–254 respectively.
[116] *Op. cit.* note 115, 69.

fossils). But the speciation would occur in phases of elevation and not in those of subsidence. Hence it would be "wonderful if we should get transitional forms" preserved at all,[117] and the embarrassing lack of such forms was explained.

If Darwin's tectonic theory did indeed play as important a role in his species work as I have suggested, his continuing commitment to that theory is understandable. In fact the same arguments are repeated without change of substance from the 1842 *Sketch* (from which I have quoted) through the 1844 *Essay* and on to the *Origin* itself. It then becomes less surprising that he should have held on to his only *detailed* case-study in tectonic theory— the Parallel Roads—with such tenacity.

3. *The influence of Lyell.* Among methodological factors influencing Darwin's retention of his hypothesis, we may note first the continuing influence of Lyell in the modes of explanation employed by Darwin. Like his 1837 papers, Darwin's paper on Glen Roy is governed by a consistent use of the Lyellian principles of uniformity and continuity. The argument for the marine origin of the Roads themselves, and of other features in Lochaber, was based on actualistic comparisons with the effects of present-day marine processes (erosion, beach formation, tidal action, etc.), particularly those he had studied for himself in South America. As we have seen, this itself was not exclusively Lyellian: both MacCulloch and Lauder had used equally actualistic comparisons, *e.g.*, with the beaches formed around present-day lakes. The difference is more a matter of degree. Darwin was being Lyellian in his steadfast refusal to consider *any* explanation that involved 'catastrophes' (as Lauder's did at one point), or that left certain aspects causally obscure and incomplete (as MacCulloch's did). Above all, Darwin was following Lyell in his vehement rejection of *ad hoc* explanations such as that of barriers across the valleys—at least if they did not serve his own argument.

Darwin's work on Glen Roy was also Lyellian in its emphasis on very gradual change. Thus his discoveries of minor Roads, and his denial of any real distinction between them and the principal Roads, served to 'smooth' the presumptive changes in sea-level from a sequence of abrupt movements into a more nearly continuous process.

Thirdly, the style of Darwin's reconstruction of the geological history of Lochaber is highly Lyellian in its attempt to bridge the gap between past and present with a continuous series of intermediates. Thus he was not

[117] *Op. cit.* note 115, 72.

content merely to establish that the Roads had been marine shore-lines, dating from a period of much higher sea-level. He also felt constrained to show that that period had been linked to the present through an intervening sequence of equally slow and gradual change. This, I suggest, accounts for his continuing emphasis on the importance of the "buttresses" below Road S. They were essential to his whole reconstruction, for they linked the lowest of the Roads to the highest of the incontestably marine raised beaches elsewhere in Scotland, and hence completed the continuity of the series right down to present sea-level (Figure 6).

Finally, a fourth feature of Darwin's style of argument is equally Lyellian. This is his emphasis on the chanciness of preservation through geological time. Darwin took over from Lyell a view of the geological record as "mere pages in chapters, towards the end of a history".[118] In other words, the preservation of evidence of events in the past was always far less likely than their destruction: the record was inevitably highly incomplete (and became still more so in earlier periods). Darwin used this Lyellian emphasis particularly in his 'explaining away' of the localized distribution of the Roads and the absence of marine fossils in their beach material. In both cases, as we have seen, he argued that obliteration (or non-formation) was the norm, and preservation the exception. Hence this apparently negative evidence could not count against his hypothesis, and indeed on the contrary was positively to be expected.

Such 'explaining away' could no doubt be justified in formal terms, since what was negative evidence on the lake hypothesis did indeed become positive within the framework of Darwin's marine explanation.[119] But Darwin's actual use of this manoeuvre shows it to have been as much an *ad hoc* mode of explanation as the barriers he so much abhorred in the rival hypothesis. Thus, for example, the fading away of Roads R1 and R2 near the mouth of the Glen Roy had to be explained away in terms of the non-preservation (or non-formation) of beaches at those levels in the wider (and supposedly more current-swept) channel in Glen Spean; but Darwin failed to account for their non-preservation in other valleys that would have been channels as narrow and sheltered as Glen Roy (particularly Glen Treig). The faint minor Roads he claimed to have seen (K, R1+, R2+) were cited as important evidence for further intermediate sea-levels; but their very faintness (which made others doubt their authenticity) was explained away

[118] *Op. cit.* note 115, 63.
[119] For a summary of the 'expectations' associated with (but not strictly deducible from) the rival hypotheses, see Table 1, p. 138–9.

as further evidence that even minor differences of circumstances could turn the balance between preservation and obliteration.

These were not the only respects in which his hypothesis came close to being unfalsifiable. For example, Darwin stated on the one hand that if the horizontality of the Roads was as precise as had been alleged, it would be evidence in favour of the lake hypothesis; yet almost at the same moment he retracted this potentially decisive test, by asserting that if the Roads *were* perfectly horizontal, it would support his notion of perfectly "equable" elevation and hence support his whole marine hypothesis.[120] This tendency to render his hypothesis unfalsifiable can also, I suggest, be traced to Lyell's example; for he too, in his treatment of the fossil record, used his belief in the chanciness of preservation as a means of making his steady-state (non-progressionist) interpretation of the history of life virtually impregnable.[121]

4. *The influence of Herschel and Whewell.* Lyell's work was not, however, the only philosophical influence on Darwin at the time he wrote his paper on Glen Roy. Darwin's earlier papers of 1837 are straightforwardly Lyellian in their use of actualistic analogy, of continuity of processes between past and present, and of the cumulative use of inductive inference. But the last of these papers, on coral reefs, begins to show a style of deductive inference that is rather rare in Lyell's work. The hypothesis of crustal subsidence, for example, is said to be "almost necessary", and the observed series of reef-forms is said to follow naturally once the hypothesis is conceded.[122] Darwin's paper on elevation (read in March 1838) is also to some extent transitional, particularly in its attempt to infer a single underlying cause for the diverse phenomena of elevation, earthquakes and volcanic activity.[123]

By the time Darwin wrote his paper on Glen Roy later in 1838 there was a much clearer logical structure dominating the whole argument. Here for the first time Darwin constructed a paper around a framework of hypothesis, deductive inference, empirical confirmation, and the consilience of inductions. The argument has a clear *strategy* designed to persuade the reader that the marine hypothesis is the "true theory" of the Roads. Darwin began by stating the problem (Section I) and eliminating the lake

---

[120] *Op. cit.* note 32, Section IX.
[121] See Martin J. S. Rudwick, 'The Strategy of Lyell's *Principles of Geology*', *Isis*, 61 (1970), 4–33.
[122] *Op. cit.* note 105.
[123] *Op. cit.* note 36.

hypothesis (Section II). He then stated his own hypothesis (Section III) and sketched its explanatory effects (Section IV). Objections to it were forestalled (Section V) and its diverse explanatory applications were described (Section VI). After a brief recapitualtion and conclusion, various accessory topics were discussed (Sections VII–IX), and the paper ended with speculations on the possible underlying cause of elevation (Section X).

This tight logical structure gives strong support to a suggestion made by Michael Ruse that during 1838 Darwin was reading and actively reflecting on both Herschel's and Whewell's views on the philosophy of science.[124] This was precisely the time that he was working on Glen Roy. It was in his paper on Glen Roy that Darwin for the first time used the Herschelian term "*vera causa*". A gradually falling relative sea-level was proposed as the "*vera causa*" of the "buttresses" below Road S, and by implication for the Roads themselves. Darwin was clearly contrasting such a causal explanation with the "monstrous" *ad hoc* postulate of barriers across the valleys.

At the same period there is evidence that Darwin was reading Whewell's newly-published *History of the Inductive Sciences* (1837). Whewell did not set out the concept of "consilience" formally until he published the *Philosophy* in 1840, but the notion is implicit in this highly philosophical *History*. Darwin may also have learnt the notion of consilience from personal contact with Whewell. Whewell was president of the Geological Society for two years from February 1837, and would have been involved with Darwin's early papers to the Society; while Darwin himself was Secretary from February 1838, and would have had to work closely with Whewell for at least the following twelve months—the period within which he visited Lochaber and wrote his paper. In these circumstances it would be surprising if Darwin, as a young and relatively inexperienced scientist, had not been influenced through his personal contact with Whewell as well as by the latter's published work.

In his paper on Glen Roy, Darwin did not use the term 'consilience', but the notion itself is clearly deployed in his description of the explanatory applications of his hypothesis (Section VI). Once a falling relative sea-level was accepted as a *vera causa*, and the intermittent elevation of the land accepted as the "true theory" of the Roads, then a number of diverse and apparently unrelated phenomena would receive an intelligible explanation: these included apparently wave-smoothed rock surfaces, "land-

---

[124] M. Ruse, 'Darwin's debt to philosophy', unpublished paper. I am indebted to Professor Ruse for allowing me to consult this paper.

straits", "col-coincidence", and the minor Roads and other terrace-like features. Even the apparent counter-phenomena of the "non-extension" of the Roads and the absence of fossils in the beach material (Section V) could be brought, so Darwin argued, into this consilience of inductions; and the theory also threw light on other less directly related problems such as erratic boulders, the rate of subaerial erosion and the mode of crustal elevation (Sections VII to IX). Even the 'surprise' element of a genuine consilience was present, in Darwin's mind, in the entirely unexpected inferences that the rate of subaerial erosion must be extremely slow and the "erratic block period" extremely remote in time (Section VIII).[125]

The methodological point on which Darwin was later so self-critical was his use of a "principle of exclusion" to move from his rejection of the lake hypothesis to his adoption of the marine hypothesis. But as Hull has pointed out, this eliminative induction was precisely what the most influential English philosophers of science—particularly Herschel and Whewell—recommended (whatever their differences in other matters).[126] Darwin's use of eliminative induction in his paper on Glen Roy cannot therefore be dismissed as a youthful indiscretion or criticized as a logical blunder. He was merely following the most authoritative advice available to him. What is more culpable, and what I suspect really embarrassed him in later years, was the youthfully cocksure manner in which, having dismissed the lake hypothesis, he confidently asserted the self-evident truth of his marine explanation of the Roads. What was at fault was not the "principle of exclusion" itself, but the *degree* of Darwin's trust in it. In other words, had he been more cautious, he would have inserted a proviso: "Since A cannot account for the observed phenomena, while B can, B must be the "true cause"—*unless* there are other alternatives, C . . . etc., which I have not considered." (A would here be the lake hypothesis, B the marine hypothesis, and C the glacial lake hypothesis.)

I have suggested, then, that Darwin's paper on Glen Roy incorporated consciously methodological themes derived from Herschel and Whewell. For the first time Darwin was using a well articulated methodological framework of argument, trying it out on an important and unsolved scientific problem. If this interpretation is correct, we can perhaps read more meaning into his otherwise rather puzzling comment, that the writing of the paper had been "one of the most difficult and *instructive* tasks" he had undertaken (my italics).[127]

---

[125] See also note 40.
[126] Hull, *op. cit.* note 3.   [127] *LLCD*, **1**, 290.

This interpretation would likewise give a further reason for Darwin's later reluctance to concede defeat in the Glen Roy controversy, if his experience with the problem had convinced him that this methodological approach was valid and fruitful. Obviously the best evidence that Darwin was indeed satisfied with his method of reasoning on Glen Roy would be his employment of similar reasoning when tackling other problems.

5. *Glen Roy and the Species Theory.* Darwin's next major project was of course the species theory. It was begun even before the work on Glen Roy but not set out as a connected argument until 1842. I have already suggested links of theoretical content between the two projects. The parallels in structure and style of argument are, I believe, even more striking.

The continuing influence in the species theory of Lyellian modes of reasoning has often been pointed out. Darwin was concerned to show that the undoubted fact of species-origins in the past could be explained actualistically in terms of processes operating in the present (*i.e.*, variation and natural selection). His theory likewise stressed the unbroken continuity of the process of speciation and its insensibly gradual rate of operation. All this can be paralleled, as we have seen, in his explanation of the Parallel Roads of Glen Roy: the blurring of the distinction between principal and minor Roads is analogous to the blurring of that between varieties and species. An even more striking parallel emerges from a comparison of the two works at the point where Darwin had to answer the most formidable objections to his hypothesis. In both works he explained *away* the absence of otherwise expected classes of evidence by arguing that on his hypothesis it was the absence that should be expected. In Lochaber, the non-existence of the Roads and non-existence of marine fossils were no objections to his marine hypothesis, because in both cases their non-preservation was to be expected.[128] In the *Sketch* and *Essay* the absence of intermediate forms of organisms and the apparently sudden changes in the fossil record were no objections to his species theory, because in both cases the non-preservation of intermediates was to be expected and the changes had really been gradual.

There is even the same tendency to make his own position unfalsifiable in the species work that we have already seen in the paper on Glen Roy. For example, he wrote in the *Sketch* that "if as many geologists seem to infer, each separate formation presents even an approach to a consecutive history, my theory must be given up." Yet he immediately went on to

[128] See note 119.

assert that even if the stratigraphical record *were* shown to be relatively complete it would still only record a *local* history of life.[129] In other words, the slow gradual speciation might have occurred elsewhere on earth at any given time, so his theory would *not* in practice have been falsified.

But the most striking parallel of all emerges from a comparison of the overall strategy of the two works (see Table 2). *Glen Roy* began with a statement of the problem, ostensibly in the form of a theory-free description of the Roads but actually containing much anticipatory interpretative comment. There is no parallel in the *Essay* (here and in the following discussion the earlier *Sketch* is taken to be included); but this is surely because the importance of the species problem could be taken for granted, whereas that of the Roads needed more explicit introduction. In *Glen Roy* Darwin went on to consider and reject the alternative solution of the problem, namely the lake hypothesis, again with much of his own interpretative comments interspersed. This section too has no parallel in the *Essay*; but here the reason is presumably that Darwin felt that the only available alternative, namely "creation", was not really an alternative at all—as indeed he argued at the end of the *Essay*[130]—and therefore needed no comparable consideration.

The close parallel commences in the next part of the two works. In *Glen Roy* Darwin approached the problem of the Roads indirectly, by first discussing the "buttresses"—features that were only questionably analogous to the Roads themselves. In the *Essay*, as is well known, he approached the problem of speciation in a similarly indirect manner, by first discussing "artificial selection"—a notion that was only questionably analogous to the "natural selection" he was postulating. In other words, both works opened the main argument with a 'Central Analogy'. With respect to the *Essay* (and of course the much later but closely similar version, the *Origin* itself), there has been much debate recently about the role of the central analogy of artificial selection, whether heuristic, pedagogic or constitutive (or some combination of these).[131] In the case of *Glen Roy*, Darwin maintained that the "buttresses", along with the minor Roads, had been crucially heuristic in his study of Lochaber. Certainly they have pedagogic (or, I would prefer to say, persuasive) value in his argument, for they suggest that the puzzling Roads themselves are only one particular example of a broader class of phenomena that all indicate a formerly higher sea-level. I have

[129] *Op. cit.* note 115, 61.
[130] *Op. cit.* note 115, 248–52.
[131] Limoges, *op. cit.* note 112; Ghiselin, *op. cit.* note 3; Ruse, *op. cit.* note 124.

Table 2. Comparison of the structure of Darwin's paper on Glen Roy and his *Sketch* and *Essay* on the species theory. Roman numerals refer to Darwin's numbering of the sections of each work. Headings quoted (with minor modification) from the originals are shown with inverted commas; headings that are not explicitly stated in the original, or are interpretative, are shown within square brackets.

	GLEN ROY 1839	SKETCH, 1842, and ESSAY, 1844
Statement of Problem	'Description of shelves' (I) [with much anticipatory interpretation]	[importance taken for granted]
Alternative Hypotheses	'Lauder's & MacCulloch's Theories' (II) elimination of alternatives: no lake theory possible	['Creation' unworthy of explicit elimination]
Statement of Hypothesis	'Proofs of Retreat of Sea' (III) 'buttresses' described—CENTRAL ANALOGY [Roads analogous to buttresses] VERA CAUSA'—falling relative sea-level confirmed by minor Roads 'Effects of Elevation in Hypothesis' (IV) 'TRUE THEORY' for origin of Roads: Intermittent elevation of land-mass	[Statement of hypothesis] 'Artificial Selection' (I) : CENTRAL ANALOGY 'Natural Selection' analogous to artificial VERA CAUSA—Natural means of Selection Variation of Instincts and Organs (III) hence: TRUE THEORY for Origin of Species: Selection acting on Variation
Objections Answered	'Objections answered' (v) [by inversion of norms] 'Non-extension of Shelves:' preservation exceptional 'Absence of organic remains:' preservation exceptional	[Objections answered, by inversion of norms] Absence of Intermediate Forms (IV) : preservation exceptional Apparently sudden faunal changes (v) : changes really gradual

Consilience of Inductions	'Applications of Theory' (VI) wave-smoothed rocks 'Land-straits' 'Col-coincidence' minor Roads and terraces	[Applications of hypothesis] 'Geographical distribution' (VI) 'Affinities and Classification' (VII) 'Unity of Type and Embryology' (VIII) 'Abortive and Rudimentary Organs' (IX)
Conclusion	Recapitulation Conclusion: marine origin 'demonstrated'	'Recapitulation' (X) 'Conclusion': species mutable
Accessory Topics	[Further applications and implications] 'Erratic Blocks' (VII) 'Alluvial' action and time-scale (VIII) Horizontality of Roads, and Elevation (IX)	
Underlying Cause	Cause of Elevation (X) Magmatic flotation	[cause of variation: (later) Pangenesis]

already suggested how they are also constitutive in Darwin's argument, in providing the essential connecting link between the Roads and present sea-level and so establishing the "equable" continuity of the causative agencies. Obviously the parallel I am drawing between the "buttresses" and artificial selection has its limitations. In particular, Darwin's interpretation of the "buttresses" was not open to the same criticism as artificial selection, namely that one term of the analogy was not 'natural'. Nevertheless, it was open to a weaker form of the same objection, namely that "buttresses" did not belong to the same *class* of phenomena as Roads, just as critics were later to argue that artificial selection was an invalid analogue to natural selection because it did not belong to the same class of process.

In both works Darwin then moved on, through a statement of the analogy being drawn (in *Glen Roy* this is only implicit at this point), to the assertion of a *vera causa*. In *Glen Roy* as we have seen, the *vera causa* was a gradually falling relative sea-level; in the *Essay* it was 'natural selection' (the Herschelian term was not used explicitly, but the notion is obviously present in the *Essay* and in many of Darwin's subsequent statements about natural selection). The validity of the first was confirmed by the minor Roads; the efficacy of the second was explained by reference to intraspecific variation. In each work, the "true theory" was thereby derived: for the origin of the Roads, the intermittent elevation of the land; for the origin of species, natural selection acting on intra-specific variation.

Darwin then considered the most important objections that might be raised against his 'true theories'. I have already emphasized the parallel in this part of the argument: in both works Darwin bypassed the objections by inverting the norms of expectation. He argued that in both problem-areas the preservation of the evidence of past change was not the rule but the exception, so that its extremely fragmentary preservation was no objection.

In each work the "Applications of the Theory" were then considered. Just as the intermittent elevation of Lochaber explained such diverse phenomena as polished rock surfaces, flattened "land-straits", "col-coincidence" and minor Roads, so the operation of speciation by natural selection explained otherwise puzzling features of biogeographical distribution, classification, embryological development and the morphology of reduced organs. By Whewellian norms, such consilience of inductions served to confirm the validity of Darwin's theories.

Finally, Darwin ended the main part of each work with a Recapitulation and a conclusion: that the marine origin of the Roads had been "demonstrated"; and that species were mutable.

As at the beginning, the end of *Glen Roy* had two final sections without parallel in the *Essay*. The first contained an ill-assorted collection of accessory topics, which, if he had re-written the paper, he might have absorbed into the main argument. Thus his discussion of erratics could well have been added to the other features that confirmed and were illuminated by his theory. His discussion of the low rate of subaerial erosion implied by his interpretation of the Roads could likewise have been absorbed, as an un-expected 'surprise' consilience, as could his suggestion that the horizontality of the Roads confirmed his conception of "equable" elevation. Finally, he tackled the question of the underlying *cause* of elevation, namely magmatic migration beneath the earth's crust. In his work on speciation the analo-gous question of the underlying cause of variation was shelved in the *Essay* and in the *Origin*, as is well known, and Darwin only tackled it later in life in his work on Pangenesis.

I have suggested that there was a very close structural parallel between the *Glen Roy* paper of 1839 and the *Essay* of 1844 (and the *Sketch* of 1842). If the validity of this parallel is accepted, I believe it provides the strongest of all reasons for Darwin's reluctance to admit defeat over *Glen Roy*. Not only had it been his first major scientific paper; not only had it contributed in an important way to his tectonic theory, which in turn played an im-portant role in his theory of speciation. But above all, *Glen Roy* had been as it were a 'trial run' for a particular style of scientific argument, which he followed closely when he wrote out the first draft of his species theory only three or four years later. If *Glen Roy* were shown to have been a "gigantic blunder", might the *Origin of Species* suffer from the same fatal structural weakness?

VII

Conclusion

We can now briefly recapitulate the course of the Glen Roy debate. The mere existence of the Parallel Roads was sufficient to pose an explanatory problem. MacCulloch and Lauder attempted solutions that were based on their perception of an analogy with the lakes still existing in the Highlands. Their hypothesis had substantial explanatory power, but it incorporated one major weakness, namely that there was little or no direct evidence for the former existence of the barriers required to impound the lake or lakes. Darwin proposed a radical alternative, based on his perception of an analogy with unambiguous evidence of raised sea-beaches in South America. He exploited the weakness of the lake hypothesis by emphasizing the *ad hoc* character of the barriers, yet found his own hypothesis required several subsidiary and (in the opinion of later critics) equally *ad hoc* explanations of negative evidence. Agassiz soon afterwards suggested, on the basis of his perception of an analogy with glaciated areas of the Alps, that the barriers required by the lake hypothesis might have been glaciers. Milne then strengthened the original lake hypothesis, particularly with a crucial new observation. This almost persuaded Darwin that the Roads had been formed by lakes of some kind, though he was inclined towards an improved version of Agassiz's *glacial* lake hypothesis. Such an improved version was also proposed by Thomson, and later elaborated still further by Jamieson, who finally convinced Darwin that his marine hypothesis was no longer tenable.

It should be apparent that however critically I have reviewed Darwin's work on Glen Roy, I do not think he was justified in condemning it so extravagantly towards the end of his life. He relied on a "principle of exclusion" because it was recommended by his philosophical mentors Herschel and Whewell. He was incautious in his overconfident use of eliminative induction, but that was simply the result of an understandable personal attachment to a hypothesis that had important connections with the rest of his research programme. Despite his tendency to make his argument unfalsifiable in certain respects, he was not impervious to new empirical evidence. When his major empirical objection to any lake hypothesis was eliminated by Milne's discovery of the missing Col R2, he accepted the implications of this resolved anomaly and tried to improve on existing versions of the hypothesis he had hitherto opposed. The *history* of his opinions on Glen Roy demonstrates what an a-historical analysis of his paper would not: namely that in practice he was open-minded enough to consider abandoning his original interpretation in favour of an improved version of the glacial-lake hypothesis. Yet he still clung to the possibility of retrieving his marine hypothesis until the evidence against it became overwhelming. Clearly he cannot be claimed simply as a 'pragmatist', still less as a 'logical postivist,' nor does a simple 'hypothetico-deductivist' label[132] do justice to the complexity of his part in the debate.

All three hypotheses were based on the perception of analogies considered relevant, and the phenomena were variously 'seen' through interpretative spectacles provided by those hypotheses. Yet the hypotheses were not incommensurable, and individuals (including Darwin) could and did evaluate their respective explanatory advantages and disadvantages, ac-

[132] Ghiselin, *op. cit.* note 3, 4–5.

cording to a set of tacit criteria that were common to all. It should be clear that the history of the Glen Roy debate as a dynamically developing scientific problem fails to fit a Kuhnian description of 'normal' science, yet it was not 'revolutionary'. It spans the alleged establishment of a new (or first?) paradigm in geology, based on Lyell's concept of the science,[133] without any signs of incommensurability. Yet it does not have the 'closed' character of an argument impervious to changes in the wider theoretical framework, since the various interpretations of the Parallel Roads were in fact used in the service of highly diverse global geological theories (compare Darwin's oscillating crustal blocks with Agassiz's drastic Ice Age), which in some senses of the word might be judged to belong to different paradigms of explanation.

Up to a point, the Glen Roy debate matches much better the Feyerabendian notion of an "active interplay of various tenaciously held views".[134] Certainly Darwin's part in the debate is a perfect example of the "principle of tenacity", and I hope that my detailed analysis will be seen to support Feyerabend's view of the *function* of such tenacity. While there may have been psychological and sociological reinforcement for Darwin's tenacity, it was reasonable for him to stick to his own interpretation despite its difficulties, because he knew perfectly well that theories can be improved and that their difficulties can be resolved. In the case of his work on speciation it is conventional wisdom to praise him for his tenacity, so it seems hardly fair to criticize him for the same behaviour in the case of his work on Glen Roy. The difference lies solely—though importantly—in the way his theories stood the test of time. In the former case, Darwin's tenacity was vindicated in the long run by the gradual resolution of many of the original difficulties in his theory. In the latter debate, it was the *other* hypotheses that underwent conspicuous improvement in the course of time, while Darwin's became less and less plausible—as he himself recognized privately.

The "principle of proliferation" can also be seen in operation in the Glen Roy debate, in the introduction of Darwin's novel interpretation closely followed by that of Agassiz's equally radical alternative. The multiplicity of hypotheses clearly fulfilled the Kuhnian function of magnifying anomalies, but this was not simply a passing phase in the revolutionary establish-

[133] Thomas S. Kuhn, *The Structure of Scientific Revolutions* (Chicago: University of Chicago, 1962) 10; Leonard G. Wilson, *Charles Lyell: The years to 1841: The Revolution in Geology* (New Haven: Yale University Press, 1972). For a criticism of this view, see M. J. S. Rudwick, *op. cit.* note 12.
[134] Paul Feyerabend, 'Consolations for the Specialist', *Criticism and the Growth of Knowledge*, Imre Lakatos and Alan Musgrave (eds.) (Cambridge: Cambridge University Press, 1970), 197–230; also earlier papers cited therein.

ment of a new paradigm. The juxtaposition of alternative hypotheses high-lighted the anomalies in them *all*, and led to a search for ways in which those anomalies could be eliminated and the explanations improved. The resolution of the problem of the Parallel Roads was not a part of the con-struction of a new paradigm for geology. It was a problem, limited in itself, which was used to support and illuminate various different but not in-commensurable concepts of the history of the earth.

It is striking that Jamieson's explanation, which Darwin ultimately accepted, was integrated into a reconstruction of recent earth-history that incorporated elements from both Agassiz *and* Darwin himself. A geologi-cally recent period of glacial climate was accepted from Agassiz, but not his concept of a global *Eiszeit;* geologically recent changes in relative sea-level were accepted from Darwin, but not as an explanation of relatively high-altitude features like the Parallel Roads. In both cases the opposing theories were modified in the direction of being less 'powerful' than their original versions, and were combined together in a *synthesis*. It is that synthesis that has 'stood the test of time' for the past hundred years.

I therefore conclude that the history of the Glen Roy debate supports Lakatos's insistence on the co-presence of proliferation and tenacity,[135] and Feyerabend's emphasis on their active interplay in the development of science. But it does not support the subjectivism into which Feyerabend allows his argument to lead him. It would require a bizarre suspension of the critical faculties to believe that the final explanation of the Parallel Roads was not 'better' or more 'rational' in the sense of somehow making better contact with 'reality'. In assessing the history of the debate it is impossible to ignore the continuing 'input' from the world of the phenomena themselves. Certainly the history of the problem illustrates how phenomena may only be seen when hypotheses require them to be found. Darwin's 'blindness' to the valley that might have led him to the missing Col R2, and Milne's later successful discovery of that col, are a case in point. But this must not be allowed to obscure the obvious point that the col *was there* to be found, possessing precisely the characteristics anticipated on the finally successful explanation. Perhaps the history of the Glen Roy prob-lem will thus give some comfort to those who are concerned to defend the place of rational debate and the possibility of objective knowledge in the scientific enterprise.

University of Cambridge

[135] Imre Lakatos, 'Falsification and the Methodology of Scientific Research Programmes', in Lakatos and Musgrave, *op. cit.* note 134, 91–195; also earlier papers cited therein.

APPENDIX A

Darwin's Agenda for Lochaber

This document (University Library, Cambridge, Darwin MS 50, 'Glen Roy notes and scraps') is a single sheet of paper, watermarked 1837. The sides are numbered, in ink, 13 and 14, and the page appears to have been retained from a series on which Darwin had made notes on MacCulloch's and Lauder's papers *before* visiting Lochaber (the final sentence of these notes appears at the top of the sheet). My own notes in explanation of each agendum are appended.

Chief Points to be Attended to

The shelves according to MacCulloch must be due to accumulation. to Dick [Lauder] to corroding.—where cut off (Dick) truncated below level. —but MacCulloch disputes this truncation.

1st Nature of shelves. with respect to foundation

2. Organic remains. *Balani*. Serpula. - calcareous matter

3. Abrupt termination of shelves. - cause - examine hill of Bohuntin - where terminates, is rock corroded as would be from tides. -

4th Is there lip of escape to shelf. 3rd Glen Fintac

5th Are there 2 [nd] and 3 [rd] shelves in Glen G[l]uoy.

6th Relative height of 2nd shelf and Loch Spey

7th Does Alluvium vary above upper shelf

8th Has it much lubricity. -

9th Are there traces of more lines than the three

10th How far are the furrow[s], now occupied by streams, irreconcileable with the idea of lake or sea

11th The relative preservation of the shelves.

12th The great problem. why lines absent in other parts. The Hill of Bohuntine and Glen Turrit must answer this. -

13th Examine Tom-na-fersit and entrance of Loch Treig for Balani. and *smooth* waterworn rocks. also Barnacles on transported blocks. -

14th Shelves correspond to head of plains

15. Form of valleys of Glen Roy and Gluoy, and of Hill of Bohuntine

Notes on Darwin's Agenda (for further explanation, see text of article)

1. *i.e.*, the relation of the Roads to the underlying bedrock, bearing on the question of their formation by erosion of bedrock and/or deposition of débris.

2. Evidence of *marine* organisms: barnacles (*Balani*) and calcareous tube worms (*Serpula*), being adherent organisms, would give unequivocal evidence of marine conditions *in situ*, but *any* fragments of calcareous shelly material would be indicative.

3. Roads R2 and R3 terminate on both sides of the hill of Bohuntine, where Glen Roy and the smaller Glen Collarig open into Glen Spean (Figure 1). On Darwin's hypothesis their termination was due to greater tidal action in the wider valley.

4. Lauder had failed to find any overflow col on the level of Road R2, and Darwin recognized this as a major anomaly. The head of Glen Fintac, which is a small valley tributary to Glen Gluoy, was a plausible point to search for such a col from Glen Roy; in fact the col connecting them is much too high, and the overflow col for R2 was later discovered in a quite different position on the other side of Glen Roy (Glen Glaster).

5. *i.e.*, are the minor Roads reported by MacCulloch in Glen Gluoy (but doubted by Lauder) in fact the fragmentary traces of beaches on the same level as Roads R2 and R3 in Glen Roy?

6. *i.e.*, does Road R1 in Glen Roy pass over into the head of Glen Spey and survive above Loch Spey (as MacCulloch believed) or is it on the same level as the intervening col R1 (as Lauder maintained)? See also note, above, on agendum 4.

7. On both MacCulloch's and Darwin's interpretations, even the highest Roads (G and R1) represented only transient pauses in a fall of relative water-level from even greater heights, whereas on Lauder's explanation the sequence of temporary lakes had begun at the level of Roads G and R1. Hence the uniformity or otherwise of the superficial débris above and below these Roads might provide a critical test between these alternatives.

8. *i.e.*, could the distinction implied in agendum 7 have been obliterated by soil-creep?

9. *i.e.*, intermediate 'minor' Roads, anticipated at many levels and in many places on Darwin's hypothesis.

10. Origin of minor indentations on the sides of the main valleys, followed by the Roads and only deepened to a minor extent by subsequent erosion by present rivulets.

11. On Darwin's hypothesis, the Roads represented successive pauses in an extremely lengthy process of crustal elevation, so that the lower and more recent Roads might be expected to be better preserved than the higher and older ones.

12. Darwin's hypothesis required that the localized distribution of the Roads should be attributed to their localized formation and/or preservation, since the sea-level had stood at the same levels on many hillsides on which no Roads were found. The Hill of Bohuntine is at the mouth of Glen Roy (Figure 1), and Roads R1 and R2 terminate on its flanks (see also Note to agendum 3). Glen Turrit is a tributary valley near the head of Glen Roy, and has a broad terrace (interpreted by MacCulloch as a delta) on the same level as Road R3.

13. Tom-na-Fersit is a rocky knoll at the mouth of Glen Treig; when the water-level was at Road R3/S it would have been a small island, on the shores of which there might be exceptionally clear evidence of marine erosion and/or marine organisms, since it would not have been obscured by any subsequent soil-creep from higher hillsides (see also Note to agendum 2).

14. *i.e.,* Lauder's assertion that at least some Roads are coincident in height with particular cols of flattened form (see also Notes on agenda 4 and 6).

15. The form of these valleys, containing one or more Roads at high levels, might suggest why such Roads occurred there and not elsewhere (see also Notes on agenda 3 and 12).

APPENDIX B

Sedgwick's Report on Darwin's Paper

This referee's report (Royal Society Library, RR.1.46, folios 88–90) is transcribed here in full. The extent to which Darwin revised his paper in the light of this report is not clear, except that he did provide a map and divide the paper into sections. The page numbers referred to by Sedgwick are of course those of Darwin's manuscript, not the published paper.

Report on Mr. Darwin's paper on the parallel roads of Glenroy &c &c.

I have read the paper of Mr. Darwin and think that it contains much original research, much ingenious speculation, & some new and very important conclusions. I think therefore that it would be for the advantage of the Royal Society that it should appear in their transactions, in some form or other. But I do not think that it should appear in its present form. The descriptions of the phenomena are good but require condensation, & they are too much scattered. They ought to appear all together: & to make the description understood, there ought to be a small map of the river courses

X

in addition to the wood cuts. It would be well to divide the paper into sections, each with its proper heading. This might be easily done, as Mr. Darwin has himself indicated such natural divisions at p. 60. of this memoir. Thus the phenomena of the parallel roads & the physical geography would appear in the first section; and after what has been written by others ought to be in the most condensed form—The discussion of the [theories *deleted*] labours [*inserted*] of Dr M'Culloch & Sir Thomas Dick Lauder ought to be as short as possible, and only in way of reference—& in invalidating their conclusions (which I think Mr. Darwin has completely done) he might state the additional facts & unexplained phenomena on which he rests his own opinion - In short, all the paper as far as p. 36 might be compressed, perhaps, into half the room, with advantage & without hurting the force of the reasoning.

Again, pp. 41, 44, 46, where the author discusses the evidence for a gradual rise of the land, are far more diffuse than necessary. In all cases, except when the observation is original, it is enough to state the fact & quote the authority; & the pages above mentioned contain no facts which might not be well expressed in one half the space. —

The hypothetical view commencing p. 44 is again, in my opinion far too diffuse. It is only a mode of representing such [facts *deleted*] phenomena, as are, or ought to be, recorded in the first or descriptive parts, & should be as brief as possible. A part of it, however, is very ingenious.

Again p. 49 &c. In considering objections to his theory the author runs unnecessarily into details; which if essential to his conclusions ought to have been stated before, in the descriptive part of his memoir. Again his descriptions can hardly be followed without maps and plans.

In several subsequent pages Mr. D. speculates with great ingenuity on the conditions essential to the formation of "parallel roads"; on their frequent obliteration, on the absence of sea shells &c. This part of the paper again might be compressed and the quotations ought to be left out: as in each case a reference & statement of the fact might be made almost in a single line. From this remark I would except a quotation from a M.S. of Mr. Lyell as I understand it has never been printed. —

The discussion (p. 62 . . . p. 69) of the erratic blocks found in the district is very good, & has a direct bearing on Mr Darwin's argument; & his conclusions derived from his own observations in South America, have on this subject tended to remove one of the greatest preliminary difficulties in geology.

In subsequent pages (p. 77 . . . 78 &c) he quotes Sir T. Lauder to prove the horizontality of the "parallel roads". The passage is far too long. It is quite enough to assert the fact & quote the authority. Again the discussion of the "theory of expansion" (p. 81.82.83) appears almost unnecessary. At least it is far too long. A portion of it is almost an abstract of another paper, instead of a mere allusion to it. There are some speculations at the end which are of doubtful value; but they are ingenious and short, and therefore ought I think to be printed.

I fear I have hardly made myself understood. The author appears to me to have upset the hypotheses of those who have preceded him and to have given very weighty reasons for his own; and in spite of the objections I have stated, I think the memoir has much original & important matter that ought to appear in the transactions of the Royal Society.

Signed A Sedgwick

London March 26
    1839
To the President & Council of the Royal Society.

APPENDIX C

Naming of Roads

The Table on page 184 shows the terminology of the principal Roads, in the major papers discussed in this article. The variety of these schemes is my reason for adopting a distinct and uniform one. Darwin uses the 'shelf' terminology in his text and the letters A–D on his map. Agassiz's numbers 1–3 are used on his map. Darwin correctly suspected, from his own barometric measurements, that MacCulloch's heights were systematically too high: note the modern estimates in the first column (the discrepancy does not affect the argument about the origin of the Roads).

APPENDIX D

Place Names in Lochaber

A full appreciation of the reasoning employed by Darwin and others cannot be gained without studying the same features at first hand. To help those who may wish to examine Darwin's paper (or others) in the light of their own perception of the phenomena, I give here a list, in alphabetical order, of some of the place-names that are used in the papers I have discussed (excluding the larger features), with their modern, generally

Term used	THIS ARTICLE Roads	MACCULLOCH 1819 lines	LAUDER 1821 shelves	DARWIN 1839 lines or shelves	AGASSIZ 1842 roads	MILNE 1847 shelves	JAMIESON 1863 roads or lines
Glen Gluoy Road (1165 feet)	G	line 1 (1266 feet)	1st shelf	1st shelf (D)	–	shelf 1	Glen Gluoy
highest Glen Roy Road (1149 feet)	R1		2nd shelf	2nd shelf (C)	3	shelf 2	upper-most
middle Glen Roy Road (1068 feet)	R2	line 2 (1184 feet)	3rd shelf	3rd shelf (B)	2	shelf 3	middle
lowest Glen Roy, and Glen Spean Road (857 feet)	R3/S	line 3 (972 feet)	4th shelf	4th shelf (A)	1	shelf 4	lowest

Gaelic, equivalents, and grid-references to the kilometre squares in which *names* are printed on the current edition of the Ordnance Survey One-Inch map. The whole area is on Sheet 36 (Fort Augustus), except the Pass of Muckul which is on Sheet 37 (Kingussie). Bartholomew's 'Half-Inch' map, Sheet 51 (Grampians) covers the whole area on a smaller scale, but has coloured contouring. Another valuable source, later in date than the work I have discussed in this article, in the first detailed official survey of the area, which was also published separately with illustrations and a short memoir: Maj-Gen. Sir Henry James, *Notes on the Parallel Roads of Lochaber . . . with illustrative maps and sketches from the Ordnance Survey of Scotland* (Southampton, 1874).

19th-century name	*Modern name on map*	*Grid Reference*
Ben Erin	Beinn Iaruinn	2990
Bohuntine	Bohuntine	2883
Bohuntine, Hill of	Bohuntine Hill/Beinn Mhonicag	2884
Collarig, Glen	Caol Lairig	2783
[Col G]	Col 1172 [ft]	3193
[Col R1]	Col 1151 [ft]	4094
[Col R2]	Col 1075 [ft]	3383
Fintac, Glen	Glen Fintaig	2688
Glaster, Glen	Glean Glas Dhoire	3084
Highbridge	Highbridge	1981
Insh	Insh	2680
Kilfinnin	Kilfinnan	2795
Lowbridge	near Glenfintaig Lodge	2286
Meal Derry	Meall Doire	2881
Muckul, Pass of [Col S]	Unnamed channel near Feagour	5790
Roy Bridge	Roybridge	2781
Spean Bridge	Spean Bridge	2282
Spey, Loch	Loch Spey	4193
Tombhran	unnamed hillside N.N.E. of Turret Bridge	3493
Tom-na-Fersit	unnamed hillock N. of Fersit	3578
Turrit, Glen	Glen Turret	3393

INDEX

Actual causes, Actualism: I 212, 221–5; II 3, 6–8, 10, 12; III 206, 220, 222, 228–31, 233, 241–2; IV 298–9; V 89–90, 103–4; VII 544–5, 550; VIII 256–7

Adanson, Michel: IV 303

Agassiz, Louis: I 225; IX 192; X 100, 101n, 127, 130–37, 140, 142–151, 153, 158–61, 164, 176–8, 183

Alps: X 130

Altertumswissenschaft: V 92

Andes: I 219; X 115

Anning, Mary: VIII 242

Antiquity of Man: II 19

Ardêche: III 215

Arkaig, Glen: X 151

Arnold, Thomas: V 92, 94

Ascribed competence, gradient of: IX 190–91, 206

Athenaeum (London club): II 2

Austen, Robert: IX 192

Auvergne (*see also* Massif Central): II 6, 7; III 207–13, 216, 220, 223, 225–7, 230–40; IV 289; V 93, 95; VI 234; VIII 250, 252, 257

Babbage, Charles: VIII 239, 244, 246, 249, 251–2

Baltic shorelines: I 225

Basterot, Barthélemy de: VI 229–32, 234–5

Beagle Channel: X 123

Beagle voyage: IV 289; IX 186–9, 193–4, 196, 198, 200–202, 205; X 113, 118, 130, 154, 159, 161

Beaumont, Élie de: *see* Élie de Beaumont

Beche, De la: *see* De la Beche

Benett, Etheldred: VIII 242

Ben Nevis: X 133–5, 151

Blake, William: VII 553

Blomfield, Charles: VIII 234, 240, 244

Bonelli, Franco: VI 231

Bove, Val del: IV 292, 294, 302–4

British Association: VIII 247; X 131, 137

British Critic: VIII 233

Brocchi, Giovanni Battista: II 7, 9; V 98; VI 228–35, 237; VIII 246

Broderip, William: VII 554

Brongniart, Adolphe: I 216; IX 192

Brongniart, Alexandre: II 5; VI 226–8; VIII 246

Bronn, Heinrich: VI 234–6

Buch, Leopold von: I 219, 223; IV 292, 303; VIII 258; IX 190

Buchdahl, Gerd: X 99n

Buckland, Mary: VIII 242

Buckland, William: I 218, 223; II 2, 4–5, 8–9; III 207, 210, 233; IV 288, 290–92; V 93; VII 537, 540, 550, 557; VIII 235, 237, 239, 246, 258; X 131, 160

Buffon, Georges Louis Leclerc, (Comte) de: I 213; II 4, 6

Burnet, Thomas: III 229

Byron, George Gordon, (Lord): VII 539, 559

Caesar, Julius: III 223

Caledonian Canal: X 110

Cambridge Philosophical Society: IX 193

Cambro-Silurian controversy: II 17

Candolle, Auguste-Pyrame de: VI 229, 231–2, 234

Cannon, Susan [Walter]: V 90

Cardwell, Donald: VIII 231

Catastrophes, geological; Catastrophism: I 210, 213, 217, 220–24; II 15; III 221–2, 229; V 93; VI 229; VII 536; VIII 234, 246–7, 250, 256, 258

Censuses: *see* Demography

Central heat: I 214

Chambers, Robert (*see also Vestiges*): X 138, 142, 145–7, 149, 161

Champollion, Jean: V 95

Chamonix: X 131, 134, 153
Charpentier, Jean de: X 127, 130,158
Characteristic fossils: I 216; VI 226,
    240
Chile: X 114–15, 123, 153, 156, 161
Chronometers, natural: II 13–14;
    III 235–6; IV 300; VI 230,
    233–4, 236, 238, 241–2
Consilience of inductions: X 167–9,
    174–5
Conybeare, William: VII 553, 555;
    VIII 237, 239, 255
Copleston, Edward: VIII 235–7, 239,
    251
Coquimbo: VIII 257; X 114–15, 117,
    153, 155
Cordier, Louis: I 214
Corsica: VI 232–3
Costa, Oronzio: IV 298
Crag formations: VI 234
Cuvier, Georges: I 215–18, 221;
    II 4–7, 13; III 233; V 92–4,
    96–7; VI 226–9, 231; VIII 245,
    247–9, 253

Daniell, John Frederic: VIII 231
Darwin, Charles; Darwinism (see also
    Essay, Origin, Sketch): I 209;
    II 1, 8, 19; III 206, 238, 242;
    IV 288–9, 298; VI 240–41;
    VII 547; VIII 251, 257;
    IX 186–206; X 97–102,
    104–5, 108, 110, 113–16,
    118–34, 136–8, 140–49,
    151–72, 174–83
Daubeny, Charles: I 214, 223;
    III 206–212, 215–17, 220,
    223–4, 227, 231, 240;
    IV 290–91, 293, 295–7
Davy, Humphry: VII 534
De la Beche, Henry: II 15–16;
    VII 535–8, 540, 542, 544–5,
    547–8, 550, 553–7, 559–60;
    VIII 238; IX 190
Deluge, biblical: see Flood
Deluge, geological: see Diluvial
    theory
Demography: III 239; V 97–9, 102;
    VI 238
Deshayes, Paul: II 13; V 95;
    VI 234–5, 237–41; VIII 250
Desmarest, Nicolas: III 207–13, 217
Desnoyers, Jules-Pierre: VI 234–5,
    237
Devonian controversy: II 17
Diluvial theory; Diluvium: I 217–19,

221; II 4; III 210–11,
    216–17, 220–26, 233–4, 240;
    VI 227; VII 550–53; VIII 237,
    246, 251, 258
Directionalist theory: I 212–16,
    222–5; II 9, 19; III 222, 232,
    242; VII 547, 557
Dumont, André-Hubert: IZ 192

Edinburgh Review: II 7
Eiszeit: see Glacial theory
Elements of Geology: II 2, 17–19;
    VI 225; VIII 238, 241
Elevation, craters of, epochs of:
    I 219; VIII 258
Élie de Beaumont, Léonce:
    I 219–20, 223; IX 190
Eocene: see Tertiary
Erratic blocks: I 218; X 127, 130–31,
    134, 146, 157–9, 169, 175
Essay of 1844 (Darwin's):
    IX 187–8, 201, 203;
    X 170–72, 174–5
Etna: II 14; IV 288, 291, 293–7,
    301–4; VI 240; VIII 250–51,
    258
Evolution: see Transformist theory

Faujas de St Fond, Barthélemy:
    III 207
Feyerabend, Paul: X 177
Fitton, William: VIII 239, 244–5
Fleming, John: VIII 253
Flood, biblical: I 217; II 5; III 210;
    VI 227; VII 550; VIII 235–6
Forbes, James: X 147–8
Fourier, Joseph: I 214, 217, 221

Garnett, Thomas: VII 534
Gemmellaro, Carlo: IV 293, 295
Genesis: II 9; III 210; VIII 237
Geoffroy St Hilaire, Étienne: I 217,
    220, 223
Geological Society (London): II 2,
    6–7, 9, 20; III 206, 226–8;
    V 95, 103; VI 226; VII 535,
    540, 545, 547, 553, 557, 560;
    VIII 234, 241, 259; IX 189–90,
    192–3, 195–6, 198, 200–201,
    205; X 102, 104–5, 129, 137,
    140, 155–6, 159, 168
Geological Survey of Great Britain:
    VII 535
Geothermal gradient: I 214
Gillray, James: VII 534, 545, 560
Glacial theory: II 19; X 100, 127,

130–31, 133–5, 157–9, 161, 177–8
Glaster, Glen: X 111, 140–41, 143, 148, 150–51, 180, 185
Gluoy, Glen and Loch: X 103, 105, 107–10, 113, 117, 126–7, 135–6, 141–2, 148, 151, 179–80
Gneiss, origin of: I 215
Göttingen: I 222; II 6
Gradualism: I 212, 224; III 222
Granite, origin of: I 215
Great Glen: X 103, 107, 110, 113, 118, 120, 123, 136, 148, 151
Greenland: X 152
Greenough, George: X 102, 104
Grimm, Jacob: V 97
Guettard, Jean Étienne: III 207–8

Hall, Basil: X 114
Hall, James: II 4; X 106, 147
Henslow, John: III 206; IX 193, 197; X 114
Herbert, Sandra: IX 194–5, 203
Herculaneum: V 96
Herschel, John: I 224; IV 290; VIII 249; IX 192, 197–8; X 97, 154, 157, 167–9, 176
Historiography: V 91, 102
Hoff, Karl von: I 222; II 6, 10; V 90, 101–2
Hooker, Joseph: IX 186; X 152
Hooykaas, R.: I 211; V 89–90, 103
Horner, Leonard: II 7; VIII 233; X 137, 140
Horner, Mary: see Lyell, Mary
Howley, William: VIII 234, 239
Hull, David: X 169
Humboldt, Alexander von: II 9; V 101; IX 189
Huskisson, William: VII 538
Hutton, James; Huttonian theory: I 215; II 3, 9–10, 14–16; III 232–3; IV 288, 304; V 100–101; VI 227; VIII 237, 243; X 106, 147
Hypertropical climates: I 215

Ice Age: see Glacial theory
Ichthyosaur: II 16; VII 539, 553–9
Iguanodon: VII 558

James, Henry: X 185
Jameson, Robert: III 207, 210; X 145n
Jamieson Thomas: X 138, 148–53,

160, 176, 178
Jones, William: V 97

Kelvin, William Thomson, (Lord): II 19
King's College London: II 16; VII 538, 557; VIII 231–5, 237–42, 245, 247, 249, 251–5, 258–60
Kirkdale cave: II 4
Kuhn, Thomas: I 210; V 103; X 177

Lakatos, Imre: X 178
Lamarck, Jean-Baptiste de: I 217, 221, 225; II 12; III 230, 233, 235, 242; VI 232; VII 559; VIII 235
Lartet, Edouard: I 225
Lauder, Thomas Dick: X 109–21, 124–6, 130, 132, 134, 136, 140–42, 147, 149, 153–4, 156–7, 165, 175, 179–82
Lisbon earthquake: II 8
Lindley, John: VIII 233
Linguistics: see Philology
Lochaber: X 102, 106, 109–10, 113, 115–18, 121, 123–5, 127–9, 131, 134, 136–7, 140, 142, 144, 146, 148–51, 153–4, 157–9, 161, 164–5, 168, 170, 174, 179, 183
Lockhart, John Gibson: III 205; VIII 235
Lomonosov, Mikhail: V 90
Lubbock, John: VIII 245
Lyell, Charles (see also Antiquity, Elements, Principles): I 209, 211, 214–15, 217, 219–26; II 1–20; III 205–6, 212, 220–21, 226–35, 239; IV 288–92, 295–304; V 90–104; VI 225–42; VII 535–8, 540, 542–5, 547–50, 553–6, 558–60; VIII 231–60; IX 189–90, 195, 197, 202; X 97, 114–15, 117, 123, 125, 127–30, 133, 137, 140, 142, 144, 146, 148–9, 152–8, 160–61, 163–7, 170, 177, 182
Lyell, Mary: VIII 252

MacCulloch, John: X 97, 103–10, 112–15, 117–18, 120–21, 124–6, 132, 134, 138, 140, 149, 153–4, 156–7, 165, 175,

MacCulloch, John (cont.): X 179–83
Mackenzie, George: X 147, 149
Magellan, Straits of: X 127
Malthusianism: III 238, 242; V 99, 100; IX 202
Man, origin, history, status of: I 216; III 230, 233, 242; IX 201, 204
Mantell, Gideon: VII 547, 558; VIII 234
Marsupials, fossil: I 216
Massif Central: III 206–7, 214, 221, 226, 232–4, 236; IV 289–90; V 93, 98; VI 230–32; VIII 250
Metamorphic rocks: II 14, 18; VIII 243
Milne, David: X 111–12, 140–45, 147, 149, 156–7, 160, 176, 178
Miocene: see Tertiary
Mont Blanc: X 134
Montlosier, François: III 207–8, 211–12, 215, 217, 220, 237n
Mosaic geology: see Scriptural geology
Moseley, Henry: VIII 231
Muckul, Pass of: X 110, 117, 124, 136, 140, 151, 185
Murchison, Charlotte: VIII 242
Murchison, Roderick: I 225; II 8–9; III 212, 232; IV 290–91, 297–8; V 95; VI 230–32, 237; VII 554; VIII 239, 250; IX 190; X 125, 133, 137, 140
Murray, John: II 14–17; VIII 238

Napoléon Bonaparte: VII 545
Neptunism: III 208
Newton, Isaac; Newtonianism: II 2–3, 10; III 237
Niebuhr, Barthold: V 91–2, 94
Noetics: VIII 235
Noto, Val di: IV 294–7, 302

Ordnance Survey: X 137, 185
Origin of species: see Transformist theory
Origin of Species: II 1, 19; IX 187, 201, 205; X 160, 175
Otter, William: VIII 242, 244–6, 254

Palaetiological sciences: I 211; IV 89–90
Pantin, Carl: X 101n
Pennant, Thomas: X 102
Patagonia: X 131
Phillips, John: VIII 254; IX 192

Philology: V 94–7, 102
Philosophical Transactions: X 129, 181
Pitt, William: VII 545
Playfair, John: II 3, 10; IV 288; V 100; VI 227; VII 553; VIII 243; X 128
Pleistocene: II 13; VI 238
Plesiosaur: VII 553–5
Pliocene: see Tertiary
Political economy: III 237–42; V 99–100, 102; VIII 235
Pope, Alexander: VII 548
Prestwich, Joseph: X 152
Prévost, Constant: II 5–6, 8, 14; V 90; VI 227
Primary rocks: VI 226; VIII 242, 254
Primates, fossil: I 225
Principles of Geology: I 209–10, 213, 223, 225; II 1–2, 7–9, 11, 14–17, 19–20; III 206, 226, 228, 232–3, 236, 239, 242; IV 288, 292, 294, 298–9, 302, 304; V 91, 96, 98–9; VI 225, 228–9, 233, 235–7, 240; VII 535, 537, 540, 543, 547, 550, 553, 555–8; VIII 232–5, 237–9, 241–7, 250, 252–3, 256, 258–60
Progressionist theory: I 213, 215–17; IV 299; VII 557, 559; VIII 253
Providentialism: I 223–4

Quarterly Review: II 2, 7, 14–15; III 205, 220, 226, 228; V 92; VIII 233, 235, 237
Quetelet, Adolphe: VI 238

Relative privacy, scale of: IX 187, 200–201, 205–6
Religious Tract Society: VII 538
Rennie, James: VIII 231
Revolutions in earth-history: I 217, 221
Risso, Giovanni: VI 231
Roy, Glen: VIII 257; IX 194–5, 197, 199, 201, 204; X 97–105, 107, 109–10, 112–18, 120–21, 124–7, 129, 132, 134, 136–7, 140–48, 150–53, 155, 157, 159, 162–3, 165–72, 174–7, 179–81
Royal Institution: VII 534; VIII 252, 254–5, 258–9
Royal Society (London): II 15; III 227; VII 535; VIII 245;

IX 190, 194, 197; X 129–30,
159, 181, 183
Royal Society of Edinburgh: X 140
Ruse, Michael: X 168

Saltatory changes: I 212, 217, 224
Scandinavia: I 218
Scharf, George: VIII 230
Scott, Walter: VIII 249
Scriptural geology: II 9; III 226;
VIII 237, 244, 258; IX 191,
193
Scrope, George Poulett: I 214, 221–3;
II 6–8, 15; III 206–42;
IV 289, 291; V 99–101;
VI 227, 230, 234; VII 559;
VIII 237, 239, 250
Secondary formations, periods,
faunas: I 216; II 14; VI 226,
228, 241; VII 555; VIII 254;
IX 196
Sedgwick, Adam: I 219; III 206;
VII 540, 554; VIII 233–4, 237,
239, 244; IX 190, 193–4, 196;
X 113, 129–30, 160, 181, 183
Selection, natural and artificial:
X 171, 174
Senior, Nassau William: VIII 235
Serapis, Temple of: II 10–11; IV 291;
VIII 258
Sicily: II 8, 14; III 206; IV 290–98,
301–4; VI 232–3; VIII 246,
250–51
Sidonius Apollinaris: III 223
Silurian formations, period: I 225
Sketch of 1842 (Darwin's):
IX 187–8, 201, 203; X 170–72,
175
Smith, Adam: V 100
Smith, William: I 216; VI 226;
VIII 254
Société Géologique de France: IX 192
Society for the Diffusion of Useful
Knowledge: VII 538
Somerville, William: VIII 238
Spean, Glen and Loch: X 103, 105,
107–8, 110, 112, 117,
121–2, 124–5, 131–4, 136,
142–5, 148, 150–51, 157, 166,
180
Speciation: see Transformist theory
Spey, Glen and Loch: X 103, 106–7,
110, 112, 179–80, 185
Spitzbergen: X 152
Statistics: VI 229–31, 236–7, 241
Steady-state theories: I 212, 224–5;

III 230, 232; VII 559–60;
VIII 243
Stonesfield: II 10
Subapennine formations: II 7; IV 296;
VI 228–9, 232, 234
Subiaco, Lake: X 109
Sweden: X 128

Taphonomy: II 12
Tertiary formations, periods, faunas:
I 216; II 7, 10, 13–15, 17;
III 216, 223, 231–6; IV 291,
295–6, 300–304; V 98–9;
VI 225–41; VII 550, 555;
VIII 246–7, 250, 253–4;
X 151, 154
Theology, natural: I 223
Thomson, James: X 147–8, 151, 176
Tierra del Fuego: X 131, 143
Timescale, geological: I 220; III 223,
229–32, 240–42; IV 288, 304;
VI 240, 242; VII 549–50;
VIII 232, 234, 239, 250, 252
Toulmin, George: VIII 237
Touraine: VI 234
Tractarians: VIII 235
Transition formations: I 215–16;
II 14; VI 226; IX 196
Transformist theory: I 217, 220–21,
223, 225; II 12, 17, 19;
III 230, 233, 235; IV 299, 302;
V 98; VI 241; VII 559; X 137,
163–5, 170–72, 174–5, 177
Transmutation of species: see
Transformist theory
Treig, Glen and Loch: X 103, 105,
117–18, 125, 131, 134–6, 143,
148, 151, 166, 179, 181
Turrit, Glen: X 105
Tsunamis: I 219; II 4, 8; X 106
Turner, Edward: VII 554; VIII 233

Uniformitarianism (see also Steady-
state theories): I 210, 225;
II 15; IV 298, 300–301; V 93;
VI 225; VII 559–60; VIII 234,
247
Uniformity, meanings of: I 211–12;
IV 291; V 89; VIII 257
University College London: VIII 231,
233

Velay (see also Massif Central):
III 217
Venetz, Ignace: X 127, 130, 158
Verae causae: II 3, 8, 12; X 122, 157,

Verae causae (cont.):X 159, 168, 174
Vercingetorix: III 223
Vestiges: X 146
Vesuvius: III 207; IV 291, 301; V 96; VIII 258
Vicentin: IV 291
Visual language: X 102
Vivarais (*see also* Massif Central): III 207, 215; IV 289, 301
Viviani, Domenico: VI 232, 234
Von Buch: *see* Buch, von
Von Hoff: *see* Hoff, von

Vulcanism: III 208
Wallace, Alfred: III 237; IX 186–7
Webster, Thomas: VI 226–7; VIII 246
Werner, Abraham: III 207; VI 226
Westminster Review: III 228
Whewell, William: I 210–11, 217–18; II 15; V 89; VI 238; VIII 233–4, 237–8, 244; IX 192, 195, 197–8; X 97, 167–9, 174, 176
Whiston, William: III 229
Wilson, Leonard: I 209; V 90

T - #0489 - 101024 - C0 - 224/150/18 - PB - 9781138375666 - Gloss Lamination